T0327370

INTRODUCTION TO CATALYSIS AND INDUSTRIAL CATALYTIC PROCESSES

INTRODUCTION TO CATALYSIS AND INDUSTRIAL CATALYTIC PROCESSES

ROBERT J. FARRAUTO

Earth and Environmental Engineering Department
Columbia University
New York, New York

LUCAS DORAZIO

BASF Corporation
Iselin, New Jersey

C.H. BARTHOLOMEW

Department of Chemical Engineering
Brigham Young University
Provo, Utah

WILEY

A joint publication of the American Institute of Chemical Engineers, Inc. and John Wiley & Sons, Inc.

Published by John Wiley & Sons, Inc., Hoboken, New Jersey.
Published simultaneously in Canada.

For general information on our other products and services or for technical support, please contact our Customer Care Department within the United States at (800) 762-2974, outside the United States at (317) 572-3993 or fax (317) 572-4002.

Wiley also publishes its books in a variety of electronic formats. Some content that appears in print may not be available in electronic formats. For more information about Wiley products, visit our web site at www.wiley.com.

Library of Congress Cataloging-in-Publication Data:

Names: Farrauto, Robert J., 1941- author. | Dorazio, Lucas, author. |
    Bartholomew, C. H., author.
Title: Introduction to catalysis and industrial catalytic processes / Robert
    J. Farrauto, Lucas Dorazio, C. H. Bartholomew.
Description: Hoboken : John Wiley & Sons, Inc., 2016. | Includes index.
Identifiers: LCCN 2015042209 (print) | LCCN 2015044585 (ebook) | ISBN
    9781118454602 (cloth) | ISBN 9781119089155 (pdf) | ISBN 9781119101673 (epub)
Subjects: LCSH: Catalysis.
Classification: LCC QD505 .F37 2016 (print) | LCC QD505 (ebook) | DDC
    660/.2995–dc23
LC record available at http://lccn.loc.gov/2015042209

10 9 8 7 6 5 4 3 2 1

To my wife Olga (Olechka) who has been a partner, friend, and critic over the precious years we have been together. She has provided love, understanding, focus, and a new vision to life. I thank my loving daughters, Jill Marie and Maryellen, and their husbands Glenn and Tom. I am fortunate to have inspiring grandchildren Nicky, Matt, Kevin, Jillian, Owen, and Brendan and stepdaughters Elena and Marina. I want to acknowledge my brother John (wife Noella) and sister Marianna (husband Ron) who have supported me emotionally through all of our years together. I am forever grateful to my parents who raised me as a proud Italian-American with a desire to help others.

Robert J. Farrauto

To my wife Cara, whose encouragement and support helped complete this project, and to my young children Lauren and Zach for their support and genuine interest in my career.

Lucas Dorazio

To my loving wife, friend, and critic, Karen, of over 49 years, who has supported me in all good things and forgiven my faults and mistakes; my 5 children and 10 grandchildren who have brought me mostly joy, been a constant source of fun and entertainment, and have given me understanding, support, love, excitement, inspiration, and challenges that have led to my growth; my loving, supportive brothers and sisters (all 7); and my dad and mom who taught me to love learning, life, and the Christian way. I especially dedicate this work (an offspring of our earlier book) to my son Charles who died unexpectedly on September 25, 2014 and who greatly touched and brightened the lives of his family, friends, and coworkers.

Calvin H. Bartholomew

CONTENTS

PREFACE   XV

ACKNOWLEDGMENTS   XVII

LIST OF FIGURES   XIX

NOMENCLATURE   XXVII

CHAPTER 1   *CATALYST FUNDAMENTALS OF INDUSTRIAL CATALYSIS*   1

1.1   Introduction   1
1.2   Catalyzed versus Noncatalyzed Reactions   1
    1.2.1   Example Reaction: Liquid-Phase Redox Reaction   2
    1.2.2   Example Reaction: Gas-Phase Oxidation Reaction   4
1.3   Physical Structure of a Heterogeneous Catalyst   6
    1.3.1   Active Catalytic Species   7
    1.3.2   Chemical and Textural Promoters   7
    1.3.3   Carrier Materials   8
    1.3.4   Structure of the Catalyst and Catalytic Reactor   8
1.4   Adsorption and Kinetically Controlled Models for Heterogeneous Catalysis   10
    1.4.1   Langmuir Isotherm   11
    1.4.2   Reaction Kinetic Models   13
        1.4.2.1   Langmuir–Hinshelwood Kinetics for CO Oxidation on Pt   14
        1.4.2.2   Mars–van Krevelen Kinetic Mechanism   17
        1.4.2.3   Eley–Rideal (E–R) Kinetic Mechanism   18
        1.4.2.4   Kinetic versus Empirical Rate Models   18
1.5   Supported Catalysts: Dispersed Model   19
    1.5.1   Chemical and Physical Steps Occurring during Heterogeneous Catalysis   19
    1.5.2   Reactant Concentration Gradients within the Catalyzed Material   22
    1.5.3   The Rate-Limiting Step   22
1.6   Selectivity   24
    1.6.1   Examples of Selectivity Calculations for Reactions with Multiple Products   25
    1.6.2   Carbon Balance   26
    1.6.3   Experimental Methods for Measuring Carbon Balance   27
Questions   27
Bibliography   29

CHAPTER 2   *THE PREPARATION OF CATALYTIC MATERIALS*   31

2.1   Introduction   31
2.2   Carrier Materials   32

2.2.1   Al$_2$O$_3$ **32**
2.2.2   SiO$_2$ **34**
2.2.3   TiO$_2$ **34**
2.2.4   Zeolites   **35**
2.2.5   Carbons   **37**
2.3   Incorporating the Active Material into the Carrier   **37**
   2.3.1   Impregnation   **37**
   2.3.2   Incipient Wetness or Capillary Impregnation   **38**
   2.3.3   Electrostatic Adsorption   **38**
   2.3.4   Ion Exchange   **38**
   2.3.5   Fixing the Catalytic Species   **39**
   2.3.6   Drying and Calcination   **39**
2.4   Forming the Final Shape of the Catalyst   **40**
   2.4.1   Powders   **40**
      2.4.1.1   Milling and Sieving   **41**
      2.4.1.2   Spray Drying   **42**
   2.4.2   Pellets, Pills, and Rings   **43**
   2.4.3   Extrudates   **43**
   2.4.4   Granules   **44**
   2.4.5   Monoliths   **44**
2.5   Catalyst Physical Structure and Its Relationship to Performance   **45**
2.6   Nomenclature for Dispersed Catalysts   **45**
Questions   **46**
Bibliography   **46**

CHAPTER 3   *CATALYST CHARACTERIZATION*   **48**

3.1   Introduction   **48**
3.2   Physical Properties of Catalysts   **49**
   3.2.1   Surface Area and Pore Size   **49**
      3.2.1.1   Nitrogen Porosimetry   **49**
      3.2.1.2   Pore Size by Mercury Intrusion   **51**
   3.2.2   Particle Size Distribution of Particulate Catalyst   **51**
      3.2.2.1   Particle Size Distribution   **51**
      3.2.2.2   Mechanical Strength   **53**
   3.2.3   Physical Properties of Environmental Washcoated Monolith Catalysts   **54**
      3.2.3.1   Washcoat Thickness   **54**
      3.2.3.2   Washcoat Adhesion   **54**
3.3   Chemical and Physical Morphology Structures of Catalytic Materials   **54**
   3.3.1   Elemental Analysis   **54**
   3.3.2   Thermal Gravimetric Analysis and Differential Thermal Analysis   **55**
   3.3.3   The Morphology of Catalytic Materials by Scanning Electron Microscopy   **56**
   3.3.4   Structural Analysis by X-Ray Diffraction   **57**
   3.3.5   Structure and Morphology of Al$_2$O$_3$ Carriers   **58**
   3.3.6   Dispersion or Crystallite Size of Catalytic Species   **58**
      3.3.6.1   Chemisorption   **58**
      3.3.6.2   Transmission Electron Microscopy   **61**
   3.3.7   X-Ray Diffraction   **62**

3.3.8   Surface Composition of Catalysts by X-Ray Photoelectron Spectroscopy **62**
3.3.9   The Bonding Environment of Metal Oxides by Nuclear Magnetic
Resonance **64**
3.4   Spectroscopy **65**
Questions **66**
Bibliography **67**

CHAPTER 4   *REACTION RATE IN CATALYTIC REACTORS*   **69**

4.1   Introduction **69**
4.2   Space Velocity, Space Time, and Residence Time **69**
4.3   Definition of Reaction Rate **71**
4.4   Rate of Surface Kinetics **72**
    4.4.1   Empirical Power Rate Expressions **72**
    4.4.2   Experimental Measurement of Empirical Kinetic Parameters **73**
    4.4.3   Accounting for Chemical Equilibrium in Empirical Rate Expression **77**
    4.4.4   Special Case for First-Order Isothermal Reaction **77**
4.5   Rate of Bulk Mass Transfer **78**
    4.5.1   Overview of Bulk Mass Transfer Rate **78**
    4.5.2   Origin of Bulk Mass Transfer Rate Expression **79**
4.6   Rate of Pore Diffusion **80**
    4.6.1   Overview of Pore Diffusion **80**
    4.6.2   Pore Diffusion Theory **81**
4.7   Apparent Activation Energy and the Rate-Limiting Process **82**
4.8   Reactor Bed Pressure Drop **83**
4.9   Summary **84**
Questions **84**
Bibliography **87**

CHAPTER 5   *CATALYST DEACTIVATION*   **88**

5.1   Introduction **88**
5.2   Thermally Induced Deactivation **88**
    5.2.1   Sintering of the Catalytic Species **89**
    5.2.2   Sintering of Carrier **92**
    5.2.3   Catalytic Species–Carrier Interactions **95**
5.3   Poisoning **96**
    5.3.1   Selective Poisoning **96**
    5.3.2   Nonselective Poisoning or Masking **97**
5.4   Coke Formation and Catalyst Regeneration **99**
Questions **101**
Bibliography **103**

CHAPTER 6   *GENERATING HYDROGEN AND SYNTHESIS GAS BY CATALYTIC
HYDROCARBON STEAM REFORMING*   **104**

6.1   Introduction **104**
    6.1.1   Why Steam Reforming with Hydrocarbons? **104**
6.2   Large-Scale Industrial Process for Hydrogen Generation **105**
    6.2.1   General Overview **105**

6.2.2   Hydrodesulfurization **106**
6.2.3   Hydrogen via Steam Reforming and Partial Oxidation **106**
 6.2.3.1   Steam Reforming **106**
 6.2.3.2   Deactivation of Steam Reforming Catalyst **110**
 6.2.3.3   Pre-reforming **111**
 6.2.3.4   Partial Oxidation and Autothermal Reforming **111**
6.2.4   Water Gas Shift **112**
 6.2.4.1   Deactivation of Water Gas Shift Catalyst **116**
6.2.5   Safety Considerations During Catalyst Removal **116**
6.2.6   Other CO Removal Methods **116**
 6.2.6.1   Pressure Swing Absorption **116**
 6.2.6.2   Methanation **117**
 6.2.6.3   Preferential Oxidation of CO **117**
6.2.7   Hydrogen Generation for Ammonia Synthesis **119**
6.2.8   Hydrogen Generation for Methanol Synthesis **120**
6.2.9   Synthesis Gas for Fischer–Tropsch Synthesis **120**
6.3   Hydrogen Generation for Fuel Cells **121**
 6.3.1   New Catalyst and Reactor Designs for the Hydrogen Economy **122**
 6.3.2   Steam Reforming **123**
 6.3.3   Water Gas Shift **124**
 6.3.4   Preferential Oxidation **125**
 6.3.5   Combustion **125**
 6.3.6   Autothermal Reforming for Complicated Fuels **126**
 6.3.7   Steam Reforming of Methanol: Portable Power Applications **126**
6.4   Summary **126**
Questions **127**
Bibliography **128**

CHAPTER 7   *AMMONIA, METHANOL, FISCHER–TROPSCH PRODUCTION*   **129**

7.1   Ammonia Synthesis **129**
 7.1.1   Thermodynamics **129**
 7.1.2   Reaction Chemistry and Catalyst Design **130**
 7.1.3   Process Design **132**
 7.1.4   Catalyst Deactivation **134**
7.2   Methanol Synthesis **134**
 7.2.1   Process Design **136**
  7.2.1.1   Quench Reactor **136**
  7.2.1.2   Staged Cooling Reactor **137**
  7.2.1.3   Tube-Cooled Reactor **137**
  7.2.1.4   Shell-Cooled Reactor **138**
 7.2.2   Catalyst Deactivation **139**
7.3   Fischer–Tropsch Synthesis **140**
 7.3.1   Process Design **142**
  7.3.1.1   Bubble/Slurry-Phase Process **142**
  7.3.1.2   Packed Bed Process **143**
  7.3.1.3   Slurry/Loop Reactor (Synthol Process) **143**
 7.3.2   Catalyst Deactivation **143**
Questions **144**
Bibliography **145**

CHAPTER 8   *SELECTIVE OXIDATIONS*   146

8.1   Nitric Acid   146
    8.1.1   Reaction Chemistry and Catalyst Design   146
        8.1.1.1   The Importance of Catalyst Selectivity   147
        8.1.1.2   The PtRh Alloy Catalyst   147
    8.1.2   Nitric Acid Production Process   148
    8.1.3   Catalyst Deactivation   150
8.2   Hydrogen Cyanide   151
    8.2.1   HCN Production Process   152
    8.2.2   Deactivation   152
8.3   The Claus Process: Oxidation of H_2S 154
    8.3.1   Clause Process Description   154
    8.3.2   Catalyst Deactivation   155
8.4   Sulfuric Acid   155
    8.4.1   Sulfuric Acid Production Process   155
    8.4.2   Catalyst Deactivation   158
8.5   Ethylene Oxide   159
    8.5.1   Catalyst   159
    8.5.2   Catalyst Deactivation   160
    8.5.3   Ethylene Oxide Production Process   160
8.6   Formaldehyde   160
    8.6.1   Low-Methanol Production Process   162
        8.6.1.1   Fe + Mo Catalyst   162
    8.6.2   High-Methanol Production Process   163
        8.6.2.1   Ag Catalyst   164
8.7   Acrylic Acid   164
    8.7.1   Acrylic Acid Production Process   164
    8.7.2   Acrylic Acid Catalyst   165
    8.7.3   Catalyst Deactivation   166
8.8   Maleic Anhydride   166
    8.8.1   Catalyst Deactivation   166
8.9   Acrylonitrile   166
    8.9.1   Acrylonitrile Production Process   167
    8.9.2   Catalyst   168
    8.9.3   Deactivation   168
Questions   168
Bibliography   169

CHAPTER 9   *HYDROGENATION, DEHYDROGENATION, AND ALKYLATION*   171

9.1   Introduction   171
9.2   Hydrogenation   171
    9.2.1   Hydrogenation in Stirred Tank Reactors   171
    9.2.2   Kinetics of a Slurry-Phase Hydrogenation Reaction   174
    9.2.3   Design Equation for the Continuous Stirred Tank Reactor   176
9.3   Hydrogenation Reactions and Catalysts   177
    9.3.1   Hydrogenation of Vegetable Oils for Edible Food Products   177
    9.3.2   Hydrogenation of Functional Groups   180
    9.3.3   Biomass (Corn Husks) to a Polymer   183

9.3.4   Comparing Base Metal and Precious Metal Catalysts **183**
9.4   Dehydrogenation **185**
9.5   Alkylation **187**
Questions **188**
Bibliography **189**

CHAPTER 10   *PETROLEUM PROCESSING*   **190**

10.1   Crude Oil **190**
10.2   Distillation **191**
10.3   Hydrodemetalization and Hydrodesulfurization **193**
10.4   Hydrocarbon Cracking **197**
      10.4.1   Fluid Catalytic Cracking **197**
      10.4.2   Hydrocracking **200**
10.5   Naphtha Reforming **200**
Questions **202**
Bibliography **203**

CHAPTER 11   *HOMOGENEOUS CATALYSIS AND POLYMERIZATION CATALYSTS*   **205**

11.1   Introduction to Homogeneous Catalysis **205**
11.2   Hydroformylation: Aldehydes from Olefins **206**
11.3   Carboxylation: Acetic Acid Production **208**
11.4   Enzymatic Catalysis **209**
11.5   Polyolefins **210**
      11.5.1   Polyethylene **210**
      11.5.2   Polypropylene **212**
Questions **213**
Bibliography **213**

CHAPTER 12   *CATALYTIC TREATMENT FROM STATIONARY SOURCES: HC, CO, NO_X, AND O_3*   **215**

12.1   Introduction **215**
12.2   Catalytic Incineration of Hydrocarbons and Carbon Monoxide **216**
      12.2.1   Monolith (Honeycomb) Reactors **218**
      12.2.2   Catalyzed Monolith (Honeycomb) Structures **219**
      12.2.3   Reactor Sizing **220**
      12.2.4   Catalyst Deactivation **222**
      12.2.5   Regeneration of Deactivated Catalysts **224**
12.3   Food Processing **225**
      12.3.1   Catalyst Deactivation **226**
12.4   Nitrogen Oxide (NO_x) Reduction from Stationary Sources **226**
      12.4.1   SCR Technology **227**
      12.4.2   Ozone Abatement in Aircraft Cabin Air **229**
      12.4.3   Deactivation **229**
12.5 CO_2 Reduction **230**
Questions **231**
Bibliography **233**

CHAPTER 13 *CATALYTIC ABATEMENT OF GASOLINE ENGINE EMISSIONS* 235

13.1 Emissions and Regulations 235
    13.1.1 Origins of Emissions 235
    13.1.2 Regulations in the United States 236
    13.1.3 The Federal Test Procedure for the United States 238
13.2 Catalytic Reactions Occurring During Catalytic Abatement 238
13.3 First-Generation Converters: Oxidation Catalyst 239
13.4 The Failure of Nonprecious Metals: A Summary of Catalyst History 240
    13.4.1 Deactivation and Stabilization of Precious Metal Oxidation Catalysts 241
13.5 Supporting the Catalyst in the Exhaust 242
    13.5.1 Ceramic Monoliths 242
    13.5.2 Metal Monoliths 245
13.6 Preparing the Monolith Catalyst 246
13.7 Rate Control Regimes in Automotive Catalysts 247
13.8 Catalyzed Monolith Nomenclature 248
13.9 Precious Metal Recovery from Catalytic Converters 248
13.10 Monitoring Catalytic Activity in a Monolith 248
13.11 The Failure of the Traditional Beaded (Particulate) Catalysts for Automotive Applications 250
13.12 NO_x, CO and HC Reduction: The Three-Way Catalyst 251
13.13 Simulated Aging Methods 255
13.14 Close-Coupled Catalyst 256
13.15 Final Comments 258
Questions 259
Bibliography 261

CHAPTER 14 *DIESEL ENGINE EMISSION ABATEMENT* 262

14.1 Introduction 262
    14.1.1 Emissions from Diesel Engines 262
    14.1.2 Analytical Procedures for Particulates 264
14.2 Catalytic Technology for Reducing Emissions from Diesel Engines 265
    14.2.1 Diesel Oxidation Catalyst 265
    14.2.2 Diesel Soot Abatement 266
    14.2.3 Controlling NO_x in Diesel Engine Exhaust 267
Questions 272
Bibliography 273

CHAPTER 15 *ALTERNATIVE ENERGY SOURCES USING CATALYSIS: BIOETHANOL BY FERMENTATION, BIODIESEL BY TRANSESTERIFICATION, AND H_2-BASED FUEL CELLS* 274

15.1 Introduction: Sources of Non-Fossil Fuel Energy 274
15.2 Sources of Non-Fossil Fuels 276
    15.2.1 Biodiesel 276
        15.2.1.1 Production Process 276
    15.2.2 Bioethanol 277
        15.2.2.1 Process for Bioethanol from Corn 278
    15.2.3 Lignocellulose Biomass 278

15.2.4   New Sources of Natural Gas and Oil Sands **279**

15.3   Fuel Cells **279**

    15.3.1   Markets for Fuel Cells **281**

        15.3.1.1   Transportation Applications **281**

        15.3.1.2   Stationary Applications **282**

        15.3.1.3   Portable Power Applications **282**

15.4   Types of Fuel Cells **283**

    15.4.1   Low-Temperature PEM Fuel Cell **284**

        15.4.1.1   Electrochemical Reactions for H_2-Fueled Systems **284**

        15.4.1.2   Mechanistic Principles of the PEM Fuel Cell **286**

        15.4.1.3   Membrane Electrode Assembly **287**

    15.4.2   Solid Polymer Membrane **288**

    15.4.3   PEM Fuel Cells Based on Direct Methanol **289**

    15.4.4   Alkaline Fuel Cell **290**

    15.4.5   Phosphoric Acid Fuel Cell **290**

    15.4.6   Molten Carbonate Fuel Cell **291**

    15.4.7   Solid Oxide Fuel Cell **293**

15.5   The Ideal Hydrogen Economy **293**

Questions **294**

Bibliography **295**

INDEX   **297**

PREFACE

"Simplicity is the ultimate sophistication."

THESE WORDS of Leonardo da Vinci were recently quoted by Steve Jobs of Apple in the book by Walter Isaacson. *Simplicity* was the first guiding principle in the preparation of this introductory book. The second guiding principle was to *share our considerable industrial and academic experience* in working with and teaching about catalysis fundamentals and industrial catalytic processes.

All of us authors have worked in industry and academia, two of us as technical consultants. Dr. Farrauto was affiliated with BASF (formerly Engelhard), Iselin, New Jersey for 37 years having worked in environmental, chemical, petroleum, and alternative energy fields and is now Professor of Practice in the Earth and Environmental Engineering Department at Columbia University in the City of New York. Dr. Dorazio, a research engineer at BASF (New Jersey), has worked in catalysis research and in scale-up of catalysts for the chemical, petroleum, and environmental fields. He is also Adjunct Professor in the Chemical Engineering Department at New Jersey Institute of Technology (NJIT). Dr. Bartholomew, Professor Emeritus in the Chemical Engineering Department at Brigham Young University, Provo, Utah, worked for a year at Corning Glass (with Dr. Farrauto) in auto emissions control after which he taught and conducted research and consulting for 41 years in catalyst design/deactivation and reactor/process design for environmental cleanup and synthetic fuel production. He continues to be active in writing, teaching short courses, and consulting. All of us have been widely engaged in various degrees of teaching industrial catalysis at the undergraduate and graduate levels. Bartholomew and Farrauto have coauthored the widely used text and reference book entitled "Fundamentals of Industrial Catalytic Processes," a more advanced, in-depth version of the topics in the current book and a likely sequel to this book.

Industrial catalytic applications are seldom taught in undergraduate chemistry and chemical engineering programs, a surprising fact, given the large number of commercial processes that utilize catalysis. Thus, we accepted the challenge of writing a book that would introduce senior level undergraduates and new graduate students to this exciting field of catalytic processes, which is fundamental to chemical engineering and chemistry as practiced in industry. The need for a thorough understanding of fundamental principles of chemistry and catalysis is given. The transition of this knowledge to their commercial applications is our objective, especially for the many chemistry and chemical engineering students who spend much of their careers working in industry with catalytic processes. We also include the many professionals of varying disciplines who suddenly find themselves with a new assignment of working on a catalytic process without previous training in the basics of catalysis and catalytic processes.

Our goal is to explain the fundamental principles of catalysis and their applications of catalysis in a simple, introductory textbook that excites those contemplating an industrial career in chemical, petroleum, alternative energy, and environmental fields in which catalytic processes play a dominant role. The book focuses on non-proprietary, basic chemistries and descriptions of important, currently used catalysts and catalytic processes. Considerable practical examples, recommendations, and cautions located throughout the book are based on authors' experience gleaned from teaching, research, commercial development, and consulting, including feedback from many students and associates. Suggested readings (reviews, books, and journal articles) are included at the end of each chapter to encourage interested readers to deepen their knowledge of these topics. Process diagrams have been simplified to provide an overview of principal process units (e.g., reactors and separation units) and important process steps, including reactant and product streams. Nevertheless, it should be recognized that commercial engineering process flow sheets include many other details and specifications, for example, piping, pumps, valves, heat exchangers, and other process equipment needed to operate and control the plant, including special equipment for plant start-up, catalyst pretreatment, purges, safety, regeneration, and so on.

Chapters 1–5 introduce the reader to basic principles of catalysis, including reaction kinetics, simple reactor design concepts, and catalyst preparation, characterization, and deactivation. Accompanying each chapter are questions and suggested readings. Chapters 6–15 describe by category applications and practice in the industry, including process chemistry, conditions, catalyst design, process design, and catalyst deactivation problems for each catalytic process. Chapter 6 describes hydrogen and syngas generation processes for different end applications. Processes for the synthesis of ammonia, methanol, and hydrocarbon liquids (Fischer–Tropsch process) are presented in Chapter 7. Processes for selective catalytic oxidation to (a) commodity chemicals, including nitric, cyanic, and sulfuric acids, formaldehyde, and ethylene oxide, and (b) specialized products such as acrylic acid, maleic anhydride, and acrylonitrile are presented in Chapter 8. Catalytic processes for hydrogenation of vegetable oils, olefins, and functional groups for highly specialized products are presented in Chapter 9. Catalytic processes in refining of petroleum to fuels are presented in Chapter 10. Selected commercial processes utilizing (a) homogeneous catalysts, (b) commercial enzymes, and (c) polymerization catalysts are described in Chapter 11. Chapters 12, 13, and 14 summarize features of important processes for catalysts used in environmental control of gaseous emissions from (a) stationary sources (e.g., power plants) and mobile sources, including (b) gasoline- and (c) diesel-fired vehicles. The final chapter 15 gives a brief summary of (1) catalytic processes for production of bio diesel and ethanol fuels from edible biomass which will ultimately find application to production of similar fuels from non-edible cellulosic biomass and (2) catalyst technology for the emerging hydrogen economy with emphasis on fuel cell technology.

New York, New York   Robert J. Farrauto
Iselin, New Jersey   Lucas Dorazio
Provo, Utah   Calvin H. Bartholomew
22 November 2015

ACKNOWLEDGMENTS

Drs. Farrauto and Dorazio acknowledge BASF (and Engelhard) for their strong leadership in the field of catalysis. We also acknowledge our students at Columbia University and NJIT, respectively, who have provided course and teaching evaluations that have been invaluable in showing us the need for a simple approach to catalysis and industrial processes.

Dr. Bartholomew is grateful for the financial support of his research, teaching, and writing endeavors by Brigham Young University and of his research by DOE, NSF, GRI, and many companies. He wishes to acknowledge the opportunity to work with distinguished colleagues and friends on the Faculty (especially in the Chemical Engineering Department and Catalysis Laboratory) and some 200+ bright, creative, hardworking graduate and undergraduate students and postdoctoral fellows who worked with him under his direction at BYU. He has also enjoyed the stimulation of teaching more than 750 company professionals during dozens of short courses on catalysis, deactivation, and Fischer–Tropsch synthesis. He wishes to acknowledge the collaboration with and friendship of Dr. Robert Farrauto over the past 42 years, first at Corning Glass, then on a landmark paper, and now two books addressing industrial catalytic processes; he is especially grateful for Bob's patience with him during the long process of preparing the first and second editions of *Fundamentals of Industrial Catalytic Processes*.

LIST OF FIGURES

Chapter 1

Figure 1.1   Catalyzed and uncatalyzed reaction energy paths illustrating the lower energy barrier (activation energy) associated with the catalytic reaction compared with the noncatalytic reaction   2

Figure 1.2   Illustration of catalyzed versus noncatalyzed reactions   2

Figure 1.3   Catalytic Fe–Ce redox reaction catalyzed by Mn   3

Figure 1.4   Activation energy diagram for (a) noncatalytic thermal reaction of CO and O_2 and (b) the same reaction in the presence of Pt. Activation energy for the noncatalyzed reaction is E_{nc}. The Pt-catalyzed reaction activation energy is designated E_c. Note that heat of reaction ΔH is the same for both reactions   4

Figure 1.5   Conversion of CO versus temperature for a noncatalyzed (homogeneous) and catalyzed reaction   5

Figure 1.6   Particulate catalysts for fixed bed reactors: spheres, extrudates, and tablets. Powdered catalysts for batch slurry phase processors. A cartoon of a fixed bed reactor loaded with catalyst tablets   9

Figure 1.7   Adsorption isotherm (θ_{CO}) for CO on Pt for large, moderate, and low partial pressures of CO. The slope at low partial pressures of CO equals the adsorption equilibrium constant K_{CO} 12

Figure 1.8   Illustration of Langmuir–Hinshelwood reaction mechanism   13

Figure 1.9   Illustration of Mars–van Krevelen reaction mechanism   14

Figure 1.10   Illustration of Eley–Rideal reaction mechanism   14

Figure 1.11   L–H kinetics applied to increasing P_{CO} at constant P_{O2}. Maximum rate was achieved when an equal number of CO molecules and O atoms are adsorbed ($\theta_O = \theta_{CO}$) on adjacent Pt sites   16

Figure 1.12   Ideal dispersion of Pt atoms on a high surface area Al_2O_3 carrier   17

Figure 1.13   Illustration of the sequence of chemical and physical steps occurring in heterogeneous catalysis   20

Figure 1.14   Conversion versus temperature profile illustrating regions for chemical kinetics, pore diffusion, and bulk mass transfer control   21

Figure 1.15   Relative rates of bulk mass transfer, pore diffusion, and chemical kinetics as a function of temperature. Chemical kinetics controls the rate between temperatures A and B. Pore diffusion controls from B to C temperatures, while bulk mass transfer controls at temperatures greater than C 22

Figure 1.16   Reactant concentration gradients within a spherical structured catalyst for three regimes controlling the rate of reaction   23

Chapter 2

Figure 2.1	(a) SEM of γ-Al$_2$O$_3$ (80,000× magnification) and (b) SEM of α-Al$_2$O$_3$ (80,000× magnification) **33**
Figure 2.2	Three zeolites: (a) mordenite, (b) ZSM-5, and (c) Beta **36**
Figure 2.3	Ceramic and metallic (center image) monoliths of different shapes and cell geometries **41**
Figure 2.4	Ceramic washcoated monoliths **44**

Chapter 3

Figure 3.1	(a) Adsorption isotherm for nitrogen for BET surface area measurement. (b) Linear plot of the BET equation for surface area measurement. (c) Nitrogen adsorption/desorption isotherm for pore size measurement **50**
Figure 3.2	Mercury penetration as a function of pore size of catalyst **52**
Figure 3.3	Differential porosimetry for a porous catalyst **52**
Figure 3.4	Particle size measurement using laser light scattering analysis **53**
Figure 3.5	Thermal gravimetric analysis and differential thermal analysis of the decomposition of barium acetate on ceria **55**
Figure 3.6	Electron microprobe showing a two-washcoat-layer monolith catalyst. The top layer is Rh on Al$_2$O$_3$ and the bottom layer is Pt on Al$_2$O$_3$ **57**
Figure 3.7	SEM of γ-Al$_2$O$_3$ with its highly porous network **58**
Figure 3.8	X-ray diffraction patterns of γ- and α-Al$_2$O$_3$ **59**
Figure 3.9	(a) Chemisorption isotherm for determining surface area of the catalytic component. (b) Pulse chemisorption profiles for the dynamic chemisorption method **60**
Figure 3.10	Transmission electron micrograph of Pt on TiO$_2$ **61**
Figure 3.11	Transmission electron micrograph of Pt on CeO$_2$ **62**
Figure 3.12	X-ray diffraction profile for different crystallite sizes of CeO$_2$ **63**
Figure 3.13	An XPS spectrum of various oxidation states of palladium on Al$_2$O$_3$ **64**
Figure 3.14	NMR profile of a Y faujasite zeolite **65**
Figure 3.15	DRIFT spectra of CO chemisorbed on different precious metal particles of catalysts prepared in different ways. The CO chemisorption followed by FT-IR measurements was performed at room temperature after the catalysts were treated at 400 °C for 1 h with 7% H$_2$ in Ar gas **66**

Chapter 4

Figure 4.1	Illustration of the three processes that can limit the reaction rate during heterogeneous catalysis **70**
Figure 4.2	Illustration showing how experimental rate measurements can be plotted in order to determine the concentration dependence used in the power rate law **74**
Figure 4.3	Illustration showing how experimental rate measurements can be plotted in order to determine the activation energy and pre-exponential factor used in the Arrhenius expression **75**

Figure 4.4   Conversion versus temperature at different space velocities. Experiment is performed to determine the rate constant at various temperatures   76

Figure 4.5   Arrhenius plot for determining activation energies   83

Chapter 5

Figure 5.1   Idealized cartoon of perfectly dispersed Pt on a high-surface γ-Al_2O_3 89

Figure 5.2   Conceptual diagram of sintering of the catalytic component on a carrier   90

Figure 5.3   TEM of fresh and sintered Pt on Al_2O_3 in an automobile catalytic converter application. "Black dots" are platinum crystallites. The size difference in crystallites between the two pictures is the result of sintering   90

Figure 5.4   Idealized conversion versus temperature for various aging phenomena   91

Figure 5.5   Illustration of the sintering of the catalyst carrier occluding the catalytic component   92

Figure 5.6   Microscopy images of low surface area rutile (a) and high surface area anatase (b). Each set of four photos show the structure at increasing magnification   93

Figure 5.7   (a) NMR profile of a thermally aged zeolite showing the loss of the Si–O–Al bridges. Si(3Al), Si(2Al), and Si(Al) are seen to decrease in intensity with the progressively more severe thermal aging. (b) Growth of penta- and octahedral coordination sites in a thermally deactivated zeolite   94

Figure 5.8   Conceptual cartoon showing selective poisoning of the catalytic sites   96

Figure 5.9   Conceptual cartoon showing masking or fouling of a catalyst washcoat   97

Figure 5.10   XPS spectrum of the surface of a contaminated Pt on Al_2O_3 catalyst   98

Figure 5.11   Electron microprobe showing the deposition location of the poisons within the washcoat of a monolith catalyst used in an automobile catalytic converter. The X-ray beam is scanned perpendicular to the axial direction through thickness of the washcoat   98

Figure 5.12   TGA/DTA in air of coke burn-off from a catalyst   100

Figure 5.13   TGA/DTA profile for desulfation of Pd on Al_2O_3 catalyst   100

Chapter 6

Figure 6.1   Illustration of industrial hydrogen generation process   105

Figure 6.2   A series of metallic tubes filled with particulate catalysts bathed in a furnace of burning natural gas providing the required heat of reaction. The rate of reaction and temperature are highest near the heat source   108

Figure 6.3   Reduction or activation of Ni SR catalyst: H_2O (steam)/H_2 as a function of temperature for redox of NiO/Ni   109

Figure 6.4 H_2O/C versus temperature: a high H_2O/CH_4 ratio allows higher temperatures for coke-free operation. To the right of the line is the coke forming regime   110

Figure 6.5   WGS equilibrium: free energy and equilibrium constant for WGS as a function of temperature   114

Figure 6.6   Typical performance of a HTS WGS catalyst with respect to exit CO   115

Figure 6.7   Reformer schematic for pure H_2 118

Figure 6.8   Overall process flow diagram for preformed natural gas to H_2 and N_2 for NH_3 production   **119**

Figure 6.9   Monolith catalysts for H_2 generation using PSA or PROX   **123**

Figure 6.10   Illustration of a highly simplified catalyzed double pipe heat exchanger where a combustion catalyst is applied to the inside surface and a steam reforming catalyst is applied to the outside surface of the inner tube   **124**

Figure 6.11   Preferential oxidation of 0.5% CO using a (Pt, Fe, Cu)/Al_2O_3 monolith catalyst   **125**

Figure 6.12   Various catalytic processes for generating H_2 and synthesis gas from desulfurized natural gas (methane) and methanol   **127**

Chapter 7

Figure 7.1   Simplified flow sheet for NH_3 synthesis illustrating a "quench"-type ammonia converter and two-stage feed gas compression   **130**

Figure 7.2   Simplified illustration of a single-stage radial flow ammonia converter   **134**

Figure 7.3   Illustration of methanol quench reactor design   **136**

Figure 7.4   Illustration of staged cooling design   **137**

Figure 7.5   Illustration of cooled tube reactor design   **138**

Figure 7.6   Illustration of shell-cooled reactor design   **138**

Figure 7.7   Flow sheet for methanol synthesis   **139**

Figure 7.8   Bubble slurry reactor for Fischer–Tropsch   **142**

Figure 7.9   Loop reactor for Fischer–Tropsch   **144**

Chapter 8

Figure 8.1   Surface roughening (sprouting of PtRh gauze)   **148**

Figure 8.2   High-pressure NH_3 oxidation/HNO_3 plant with Pd getter gauze   **149**

Figure 8.3   An expanded view of the reactor containing the stacks of oxidation and getter gauzes   **150**

Figure 8.4   HCN process flow diagram   **153**

Figure 8.5   The Claus process with staged reaction and liquid sulfur removal   **154**

Figure 8.6   Elemental sulfur is reacted with dry air at 900 °C producing SO_2. Staged air injection into the second and third stages for cooling is shown in Figure 8.8   **156**

Figure 8.7 SO_2/SO_3 equilibrium as a function of temperature   **157**

Figure 8.8   Quench reactor for SO_3 production with staged air injection for cooling for stages 2 and 3   **158**

Figure 8.9   The O_2 process for ethylene oxide production   **161**

Figure 8.10   Process for low methanol concentration process to formaldehyde over a (Fe, Mo)/SiO_2 catalyst   **162**

Figure 8.11   Process using Ag catalyst   **163**

Figure 8.12   Propylene to acrolein to acrylic acid process flow diagram. Tubular reactor with a diameter of about 2.5 cm and a length of about 4 m cooled by a molten carbonate   **165**

Figure 8.13   Process for converting propylene to acrylonitrile   **167**

Chapter 9

Figure 9.1   Illustration comparing difference between a semibatch stirred tank reactor and a continuous stirred tank reactor   172

Figure 9.2   Illustration of a semibatch stirred tank reactor (STR). The sparger (or also called dip tube) is used for continuous addition of a reactant, which is hydrogen for hydrogenation reactions. Not shown is the removal of unreacted hydrogen from the headspace, which must occur to maintain the desired reactor pressure   173

Figure 9.3   Illustration showing hydrogen consumption versus time during typical hydrogenation reaction   173

Figure 9.4   Illustration of the mass transfer path taken by hydrogen as it diffuses from the gas bubble, through the bulk liquid, and ultimately to the catalyst particle. In most hydrogenation reactions, the rate of this diffusion process limits the overall rate of reaction   174

Figure 9.5   Kinetic rate for a catalytic slurry-phase batch reaction   176

Figure 9.6   Linolenic oil shown as an example of an unsaturated fat molecule   178

Figure 9.7   Sequential reactions at 140 and 200 °C   179

Figure 9.8   CATOFIN propane dehydrogenation to propylene using Cr_2O_3/Al_2O_3 catalyst   185

Figure 9.9   Flow diagram for dehydrogenation of ethyl benzene to styrene   187

Chapter 10

Figure 10.1   Simplified illustration of the crude oil refining process. The desalting process (removal of inorganic components in the crude using a water wash) is not shown   192

Figure 10.2   Examples of metal-containing (nickel porphyrin) and sulfur-containing (thiophene) species typically found in crude oil   193

Figure 10.3   The HDM/HDS process flow diagram. Inset shows presulfided catalyst and its positive effect on decreasing excessive gas make and hydrogen consumption   195

Figure 10.4   Catalyst deactivation by HDM metal deposition (masking) and coking   196

Figure 10.5   Controlled O_2 addition in coked catalyst regeneration   196

Figure 10.6   Faujasite zeolite   198

Figure 10.7   Schematic of FCC reactor with catalyst regenerator   199

Figure 10.8   Process flow diagram for naphtha reforming   201

Figure 10.9   Regeneration and rejuvenation of $(Pt, Re)/\gamma\text{-}Al_2O_3 + Cl^-$ reforming catalyst   202

Chapter 11

Figure 11.1   Hydroformylation process using a cobalt homogeneous catalyst   207

Figure 11.2   Dow (Davy McKee) LP Oxo Selector process using the Rh catalyst   207

Figure 11.3   Monsanto acetic acid process   209

Figure 11.4   Phillips loop reactor   211

Figure 11.5 $TiCl_3/MgCl_2$ process for polyethylene   212

Chapter 12

Figure 12.1   (a) VOC abatement process with heat integration. (b) VOC abatement with supplemental heating   **223**

Figure 12.2   Slipstream reactor concept used for VOC abatement design   **224**

Figure 12.3   Catalyst abatement of food processing fumes   **225**

Figure 12.4   SCR with V_2O_5 and a metal-exchanged zeolite: 1.1 NH_3/NO and zeolite   **228**

Figure 12.5   SCR reactor schematic. It would be worth mentioning that the widening, that is, lower velocity, increases contact time   **229**

Figure 12.6   Ozone abatement reactor design   **230**

Chapter 13

Figure 13.1   Gasoline-relative engine emissions and temperature as a function of air/fuel ratio   **236**

Figure 13.2   Monolith catalyst housed in a metal canister secured in the exhaust   **239**

Figure 13.3   Optical micrographs of double-layered washcoated ceramic monoliths   **243**

Figure 13.4   Conversion proceeding axially down the channel of a monolith with poisoning. Units of time are arbitrary units   **249**

Figure 13.5   Temperature profiles ($\Delta T/\Delta L$) for an exothermic reaction down the axial length of a catalyzed monolith channel caused by sintering   **250**

Figure 13.6   Simultaneous conversion of HC, CO, and NO_x for TWC as a function of air/fuel ratio   **252**

Figure 13.7   Oxygen sensor response output as a function of air/fuel ratio   **253**

Figure 13.8   Electron microprobe scan of an automotive catalyst contaminated with P and S from lubricating oil   **255**

Figure 13.9   Close-coupled TWC catalyst, under-floor TWC, and oxygen sensors connected to electronic feedback to control air/fuel ratio close to stoichiometric ($\lambda = 1$)   **258**

Chapter 14

Figure 14.1 NO_x–particulate trade-off with emission regulations   **263**

Figure 14.2   Electron microprobe scans of the washcoat of an aged diesel oxidation catalyst. (a) Zn and Ca. (b) P and S   **266**

Figure 14.3   Wall flow filter. Soot particulates deposit on the porous wall, while the gaseous components (CO_2, H_2O, NO, and NO_2) and air pass through. The soot is combusted periodically by raising the inlet temperature to >500 °C when a small amount of diesel fuel is injected into the DOC   **267**

Figure 14.4   SCR with Cu and Fe zeolites   **268**

Figure 14.5   Schematic of simplified diesel exhaust aftertreatment system. A diesel oxidation catalyst and wall flow filter (or diesel particulate filter) are contained in one canister, a dosing system for injecting urea to the SCR catalyst. An ammonia decomposition catalyst (Pt/γ-Al_2O_3//ceramic monolith) is installed at the outlet of SCR. The DOC catalyzes the oxidation of CO, HC, and some of the NO to NO_2 and generates sufficient heat (~500 °C) by oxidizing injected diesel fuel to initiate combustion of the soot collected on the wall flow filter. The NO and NO_2 exiting the filter are mixed with inject urea, which hydrolyzes to NH_3 and enters the SCR catalyst. EGR may be used to further reduce the engine out NO. Not shown is a turbocharger than compresses the air as it enters the combustion chamber. This further enhances the power of the engine to move heavy loads   **269**

Figure 14.6   Chemistry of NO_x reduction in using BaO to capture NO_2 during lean operation   **270**

Figure 14.7   Deactivation of the NO_x trap by sulfur oxide poisoning   **271**

Figure 14.8   Driving profile for a LNT. During lean mode (fuel economy), NO is converted to NO_2 over a Pt catalyst, which is adsorbed in the alkaline trap. Regeneration (solid area) occurs when the engine is commanded to a stoichiometric mode where the NO_x is desorbed from the BaO and the Rh in the TWC reduces it to N_2 **271**

Chapter 15

Figure 15.1   Biodiesel production process   **277**

Figure 15.2   A comparison of power generation for a coal-fired power plant, gasoline/ diesel internal combustion engine, and a H_2–O_2 low-temperature fuel cell   **280**

Figure 15.3   A single cell of the PEM fuel cell   **285**

Figure 15.4   Voltage–current profile for the PEM fuel cell. The curve with the maxima represents the power profile   **286**

Figure 15.5   A PEM single cell and arranged in a "stack." Each cell is separated by an electrically conductive impermeable bipolar plate that serves as a gas manifold for the cells connected in series   **288**

Figure 15.6   Ideal H_2 economy with the sun providing energy for a photovoltaic device generating sufficient voltage to electrolyze water to H_2 and O_2 **294**

NOMENCLATURE

SYMBOLS

A	Frontal area (m^2)
a_s	Geometric surface area = ratio of surface area to volume (m^{-1})
a_{Sh}	Fitted parameter used for mass transfer coefficient
B	Integral breadth (Scherrer's formula) (°)
C	Concentration (mol/m^3)
C_{BET}	BET constant
C_{WP}	Weisz–Prater criterion
D	Diffusivity (m^2/s)
D_e	Effective diffusivity (m^2/s)
D_k	Knudsen diffusivity (m^2/s)
d_p	Particle diameter (d_h) = monolith channel diameter (m)
d_{pore}	Pore diameter (m)
E_A	Activation energy (J/mol)
E_{cell}	Net voltage of fuel cell (V)
E_{cell}°	Standard state cell voltage (V)
E_o	Standard state voltage in the Nernst equation (V)
f	Friction factor
F	Molar flow (mol/s)
F_c	Faraday constant (C/V)
g_c	Gravitational constant (m/s^2)
GHSV	Gas hourly space velocity (h^{-1})
ΔG_{rxn}	Free energy change during reaction (J/mol)
H	Henry's law constant
ΔH_{rxn}	Enthalpy change during reaction (J/mol)
J	Molar flux ($mol/(m^2\ s)$)
k	Reaction rate constant (mol/(s l))
K_{eq}	Equilibrium constant
k_f, k_d	Rate constants for absorption and desorption (mol/(s l))
k_{MT}	Mass transfer coefficient (m/s)
k_o	Pre-exponential factor (g)
M_{flow}	Mass flow rate (kg/h)
MW	Molecular weight (mol/g)
n	Number of electrons transferred
N_A	Avogadro's number (mol^{-1})
P	Pressure (atm)
P_i	Partial pressure of species i (atm)

P_o	Saturation pressure (atm)
r	Reaction rate per unit volume (mol/(s l))
r'	Reaction rate per unit mass (mol/(s g))
R	Ideal gas constant (J/(mol K))
Re	Reynolds number
r_k	Rate of reaction on catalyst surface (mol/(s l))
r_{MT}	Rate of bulk diffusion (mol/(s l))
$r_{MT,g}$	Rate of bulk diffusion through gas phase (mol/(s l))
$r_{MT,g-l}$	Rate of diffusion through gas–liquid interface (mol/(s l))
$r_{MT,l-s}$	Rate of diffusion through liquid–solid interface (mol/(s l))
r_p	Pore radius (m)
R_{pellet}	Radius of catalyst pellet (m)
r_{pore}	Rate of pore diffusion (mol/(s g))
Sc	Schmidt number
Sh	Sherwood number
ΔS_{rxn}	Entropy change during reaction (J/K)
SV	Space velocity (h^{-1})
t	Residence time (min or h)
T	Temperature (°C or K)
t_{crys}	Thickness of crystallite (Scherrer's formula) (nm)
u	Bulk stream velocity (m/s)
V	Reactor volume (m^3)
V_{ad}	Volume adsorbed at P (m^3)
V_{flow}	Volumetric flow (m^3/s)
V_m	Molar volume (mol/m^3)
V_{mono}	Adsorbed volume at monolayer coverage (m^3)
v_o	Volumetric flow rate (m^3/h)
V_{rxtr}	Reactor volume (m^3)
W_{cat}	Catalyst mass (g)
WHSV	Weight hourly space velocity (h^{-1})
X	Fractional conversion
x_{Sh}	Fitted parameter used for mass transfer coefficient
Z	Reactor length (m)

GREEK SYMBOLS

β	Approach to equilibrium
γ	Surface tension (N/m)
δ_M	Thickness of mass transfer boundary layer (m)
ε	Binary interaction energy (K)
ε	Void fraction in the bed
ε_p	Particle porosity
θ	Contact angle (°)
θ_d	Bragg angle (°)

θ_i Surface coverage of species i

λ Molecule mean free path (m)

λ X-ray wavelength (Å)

ρ Catalyst bulk density (kg/m^3)

ρ_g Gas density (kg/m^3)

σ Interaction length (m)

σ Surface tension of liquid nitrogen (N/m)

σ_p Pore constriction

τ Space time (h)

τ_p Pore tortuosity

υ Kinematic viscosity (m^2/s)

Ω Diffusion collision integral

SUBSCRIPTS

A   Generic species "A"

B   Generic species "B"

i Species

i   Inlet position

o   Outlet position

o   Denotes saturation pressure

CHAPTER *1*

CATALYST FUNDAMENTALS OF INDUSTRIAL CATALYSIS

1.1 INTRODUCTION

Chemical reactions occur by breaking the bonds of reactants and forming new bonds and new compounds. Breaking stable bonds requires the absorption of energy, while making new bonds results in the liberation of energy. The combination of these energies results in either an exothermic reaction in which the conversion of reactants to products liberates energy or an endothermic process in which the conversion process requires energy. In the former case, the energy of the product is lower than that of the reactants with the difference being the heat liberated. In the latter case, the product energy is greater by the amount that must be added to conserve the total energy of the system. Under the same reaction conditions, the heat of reaction (ΔH) being a thermodynamic function does not depend on the path or rate by which reactants are converted to products. Similarly, ΔG of the reaction is not dependent on the reaction path since it too is a thermodynamic state function. This will be emphasized once we discuss catalytic reactions. The rate of reaction is determined by the slowest step in a conversion process independent of the energy content of the reactants or products.

1.2 CATALYZED VERSUS NONCATALYZED REACTIONS

In the most basic sense, the purpose of the catalyst is to provide a reaction pathway or mechanism that has a lower activation barrier compared to the noncatalyzed (E_{nc}) pathway, as illustrated in Figure 1.1. Also shown is the catalyzed barrier (E_{Mn}). In any reaction, catalyzed or noncatalyzed, the reaction sequence occurs through a series of elementary steps. In a noncatalyzed reaction, the species that participate in the reaction sequence are derived solely from the reactants. In a catalyzed reaction, the catalyst is simply an additional species that participates in the reaction sequence by lowering the activation energy and hence enhances the kinetics of the reaction. Finally, during the catalyzed reaction sequence, the catalyst species returns to its original state. It is the regeneration of the catalyst species to its original state that makes a catalyst a "catalyst"

Introduction to Catalysis and Industrial Catalytic Processes, First Edition. Robert J. Farrauto, Lucas Dorazio, and C.H. Bartholomew.
© 2016 John Wiley & Sons, Inc. Published 2016 by John Wiley & Sons, Inc.

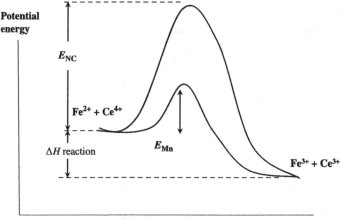

Figure 1.1   Catalyzed and uncatalyzed reaction energy paths illustrating the lower energy barrier (activation energy) associated with the catalytic reaction compared with the noncatalytic reaction. (Reproduced from Chapter 1 of Heck, R.M., Farrauto, R.J., and Gulati, S.T. (2009) *Catalytic Air Pollution Control: Commercial Technology*, 3rd edn, John Wiley & Sons, Inc., New York.)

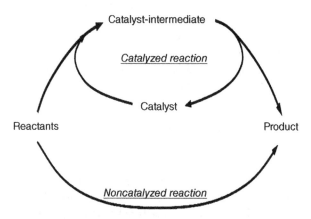

Figure 1.2   Illustration of catalyzed versus noncatalyzed reactions.

and not a "reactant." Thus, a catalyst is a species that participates in the reaction sequence—it interacts with the "reactants" to form an intermediate species that undergoes further reaction to form the "product" with the catalyst returning to its original state. This basic sequence of events is illustrated in Figure 1.2.

1.2.1   Example Reaction: Liquid-Phase Redox Reaction

Let us consider the simple redox reaction between Fe^{2+} and Ce^{4+} in aqueous solution. The reaction below excludes the H_2O present in the coordination sphere for each

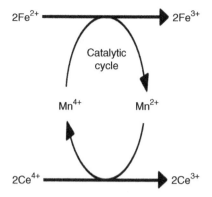

Figure 1.3 Catalytic Fe–Ce redox reaction catalyzed by Mn.

species since it does not directly participate in the reaction.

$$Fe^{2+} + Ce^{4+} \rightarrow Fe^{3+} + Ce^{3+} \tag{1.1a}$$

$$K_e = \frac{[Fe^{3+}][Ce^{3+}]}{[Fe^{2+}][Ce^{4+}]} \tag{1.1b}$$

This reaction involves a direct electron transfer from the Fe^{2+} to the Ce^{4+} and by itself occurs very slowly because the electron transfer process occurs slowly. However, in the presence of Mn^{4+} species, the rate dramatically increases because the electron transfer is now facilitated through the Mn^{4+}/Mn^{2+} couple. The Mn^{4+} species is a catalyst, not a reactant. While it does directly participate in the reaction, the reaction pathway results in no overall change in the chemical state of the Mn ion (Figure 1.3).

The reaction profile of both the catalyzed and noncatalyzed reactions can be described kinetically by the Arrhenius profile in which reactants convert to products by surmounting the energy barrier called the activation energy. According to the Arrhenius expression (Equation 1.2), the rate of reaction is proportional to the exponential of absolute temperature (T) and inversely proportional to the exponential of the activation energy (E). The remaining terms in Equation 1.2 include the universal gas constant (R), pre-exponential factor (k_0), and the rate constant (k). Thus, the rate of reaction (i.e., rate constant) will increase as the temperature increases or the activation energy decreases:

$$k = k_0 \cdot \exp\left(-\frac{E}{R \cdot T}\right) \tag{1.2}$$

Referring back to our redox example, the catalyzed pathway has the lower activation energy and, therefore, will have a higher reaction rate at a given temperature. The energy barrier was lowered by the Mn catalyst providing a chemical shortcut to products. Although it is greater, the reactants and products are the same as the noncatalyzed reaction. Thus, thermodynamic properties remain unchanged and both reaction pathways will have the same reaction enthalpy (ΔH) and reaction free energy (ΔG) and equilibrium constant. The catalyst can only influence the rate at which reactants are converted to products as the equilibrium constant is approached and cannot make thermodynamically unfavorable reactions occur. In industrial practice reaction conditions, such as temperature, pressure and reactant compositions are varied to maximize product formation consistent with the equilibrium constant.

1.2.2   Example Reaction: Gas-Phase Oxidation Reaction

Consider the conversion of carbon monoxide (CO), a known human poison, to CO_2, a reaction of great importance to the quality of air we breathe. The overall rate of the noncatalytic reaction is controlled by the dissociation of the O_2 molecule to O atoms (rate-limiting step) that rapidly react with CO forming CO_2. The temperature required to initiate the dissociation of O_2 is greater than 700 °C and once provided, the reaction rapidly goes to completion with a net liberation of energy (the heat of reaction is exothermic). The requirement to bring about the O_2 dissociation and ultimately the conversion of CO to CO_2 has activation energy (E_{nc}) in Figure 1.4. Reaction occurs when a sufficient number of molecules (O_2) possess the energy necessary (as determined by the Boltzmann distribution) to surmount the activation energy barrier. The rate of reaction is expressed in accordance with the Arrhenius equation (Equation 1.2).

Let us now discuss the effect of passing the same gaseous reactants, CO and O_2, through a reactor containing a solid catalyst. Since the process is carried out in two

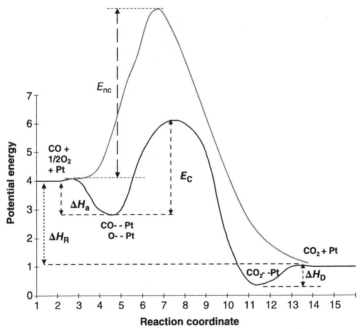

Figure 1.4   Activation energy diagram for the non-catalytic thermal reaction of CO and O_2 (E_{nc}) and the same reaction in the presence of Pt (E_c). Activation energy for the non-catalyzed reaction is E_{nc}. Activation energy for the non-catalyzed reaction is E_{nc}. The Pt catalyzed reaction activation energy is designated Ec. Note that heat of reaction ΔH is the same for both reactions. *Catalytic Air Pollution Control: Commercial Technology*, Chapter 1, Wiley and Sons 2009

separate phases, the term heterogeneous catalytic reaction is used. In the presence of a solid catalyst such as Pt, gaseous O_2 and CO adsorb on separate Pt sites in a process called chemisorption in which a chemical partial bond is formed between reactants and the catalyst surface. Dissociation of chemisorbed O_2 to chemisorbed O atoms rapidly occurs at room temperature. Adsorbed O atoms react with chemisorbed CO on adjacent Pt sites producing CO_2 that desorbs from the Pt completing the reaction and freeing the catalytic site for another cycle. The presence of the Pt catalyst greatly facilitates the dissociation of oxygen, which was the slow energy-intensive step associated with the noncatalytic gas-phase oxidation of CO. Thus, the activation energy for the Pt-catalyzed reaction (E_c), shown in Figure 1.4, is considerably smaller than that for the noncatalyzed reaction, enhancing the conversion kinetics. This difference in activation energy can be easily observed when comparing the light-off temperature of the catalyzed versus the gas-phase reactions, which is illustrated in Figure 1.5. The noncatalyzed reaction has a considerably higher light-off temperature (around 700 °C) due to its higher activation energy. More input energy is necessary to provide the molecules the necessary energy to surmount the activation barrier so that light-off occurs at higher temperatures. It should be noted, however, that the non-catalyzed reaction has a greater sensitivity to temperature (slope of plot). Thus, the reaction with the higher activation energy has greater sensitivity to temperature, making it increase to a greater extent with temperature than that with a lower activation energy. This is a serious problem for highly exothermic reactions, such as hydrocarbon oxidations, where noncatalytic free radical reactions, with large activation energies, can lead to very undesirable products.

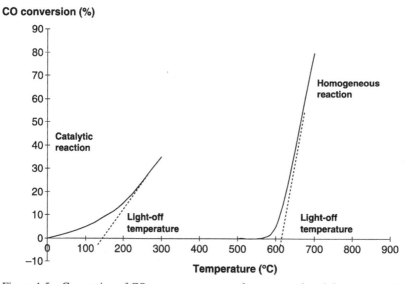

Figure 1.5   Conversion of CO versus temperature for a noncatalyzed (homogeneous) and catalyzed reaction. (Reproduced from Chapter 1 of Heck, R.M., Farrauto, R.J., and Gulati, S.T. (2009) *Catalytic Air Pollution Control: Commercial Technology*, 3rd edn, John Wiley & Sons, Inc., New York.)

Equations relating reaction rates to activation energies will be discussed in considerable detail in Chapter 4, but for now it is sufficient to understand that an inverse relationship exists between the activation energy and the reaction rate.

Relative to one that is not catalyzed, kinetic rate studies indicate that the rate-limiting step is the reaction of chemisorbed CO with chemisorbed O atoms on adjacent Pt sites. The reaction occurs around $100\,°C$ far below the $700\,°C$ required for the noncatalytic process described above. Thus, the catalyst provides a new reaction pathway in which the rate-limiting step is altered at a significantly lower temperature. This shows the great importance of catalysis in enhancing rates of reaction allowing them to occur at moderate temperatures. Lower operating temperature translates to energy savings, less expensive reactor materials of construction, and preferred product distributions with greater rates of production with smaller size reactors. For this reason, catalysts are commonly used in many industrial applications: petroleum processing, chemical and energy synthesis, and environmental emission control.

Inspection of Figure 1.4 indicates that an energy decrease associated with the adsorption of CO and O_2 on the Pt surface (ΔH_a) is due to its exothermic nature. This is a consequence of the decreased entropy (ΔS) when molecules are confined in an adsorbed state with the commensurate loss in a degree of freedom. Since ΔG_a must be negative and $-T\Delta S_a$ is positive, ΔH_a must be negative in accordance with $\Delta G_a = \Delta H_a - T\Delta S_a$. Desorption is always endothermic.

1.3 PHYSICAL STRUCTURE OF A HETEROGENEOUS CATALYST

All heterogeneous catalysts are solid substances that can be classified into two general categories: unsupported or supported. Most heterogeneous catalysts fall under the "supported" category, where the active catalytic component is dispersed into an inorganic porous high surface area carrier (often called a support) physically similar to a household porous sponge. The primary function of the carrier is to maintain the dispersion of the active phase; however, it can also participate in the reaction. As an example of supported and unsupported catalysts, consider the process for ammonia synthesis (Equation 1.3) where both supported and unsupported catalysts are used. One catalyst used for this reaction is *unsupported* iron derived from iron oxide. Another catalyst used for this reaction is ruthenium *supported* on a high surface area carbon carrier. In the preparation of this catalyst, a small amount of ruthenium is dissolved into water and impregnated into the carbon carrier. Thus, the bulk of the catalyst is comprised of carbon with small "islands" of well-dispersed ruthenium. Both types of catalyst structures are widely used and the optimal catalyst structure is often a function of several variables, including the nature of the active species, catalyst manufacturing cost, and the desired catalyst form to be used in the industrial process.

$$3H_2 + N_2 \leftrightarrow 2NH_3 \tag{1.3}$$

Whether supported or unsupported, the catalyst will be comprised of two basic components: *active species* and *promoters* in addition to the carrier. The species that adsorbs the gas-phase reactant and on which the surface reaction occurs is called the *active species.* Within the active species, the specific atomic location where adsorption and reaction occur is called the *active site*. It is on the active site where the short-lived reaction intermediates form is called the *active center*. Beyond the active species, there are often other species incorporated into the catalyst structure, called *promoters*, that either improve the activity of the active species by positively participating in the surface reaction sequence (*chemical promoter*) or help maintain the catalyst activity over time by stabilizing the catalyst structure (*textural promoter*).

1.3.1   Active Catalytic Species

Group VIIIB metals and their oxides such as Fe, Co, and Ni are catalytic as are Cu and Ag (Group 1b), V (Group Vb), and Cr and Mo (Group V1b) and in specific combinations are mainly used in the chemical and petroleum industries. These are referred to as base metals (oxides). Precious metals such as Pt, Pd, Ru, and Rh are also in Group VIII that are very commonly used broadly in all industries. Ironically, the precious metals are also referred to as noble metals for their resistance to oxidation, various poisons, and high temperatures, yet they are some of the most catalytically active elements in nature due to their ability to chemisorb and convert adsorbed species with high rates. They are rare and very expensive and thus when no longer performing satisfactorily, they are recycled, purified, and reused. They are primarily mined in South Africa and Russia with small deposits in Canada and the United States. In most cases, both base and precious metals (or their oxides) are deposited on high surface area carriers in order to maximize their accessible catalytic sites.

It should be understood that the active catalytic component is often not present in its native elemental state, but may be present as an oxide. For the oxidation of many hydrocarbons, Pd is catalytically active as PdO; while for the hydrogenation reactions, Ni, Cu, and Pd metals are most active. Vanadium pentoxide (V_2O_5) is an active catalyst for oxidizing SO_2 to SO_3 in the manufacture of sulfuric acid.

Throughout this book you will see the breadth of applications for all of these metals and their oxides as catalysts that enhance activity and selectivity.

1.3.2   Chemical and Textural Promoters

While catalyst activity and selectivity are dominated by the main catalytic component, other metals or metal oxides may also contribute to the overall activity or selectivity. When this is the case, these materials are called *promoters*. Promoters are generally classified as two types: chemical promoters and textural (structural) promoters. In the case of the chemical promoter, the material facilitates the surface reaction. In the case of the textural promoter, the material stabilizes the metal dispersion or the structure of the carrier, which prevents or slows catalyst deactivation over time. There are many examples for a wide variety of petroleum, chemical, environmental, and alternative energy processes that we will discuss in subsequent chapters, but just to give a preview here are some examples. The addition of CeO_2 to a precious metal catalyst such as Pt or Pd in the automotive catalyst promotes the oxidation of hydrocarbons and carbon monoxide decreasing the minimum temperature needed to initiate catalytic oxidation during the cold start portion of the automobile cycle requirement.

Adding Cl^- to a naphtha reforming catalyst $PtRe/\gamma-Al_2O_3$ used to make high octane gasoline enhances the acidity of the catalyst and catalyzes isomerization reactions generating branched hydrocarbons with high octane. The addition of sulfur to a hydro-desulfurization catalyst ($Co, Mo/Al_2O_3$) diminishes excessive activity leading to undesired gaseous products in crude oils. The addition of alkali (K_2O) to a Ni/Al_2O_3 steam-reforming catalyst decreases acidity and promotes steam gasification minimizing carbon formation. In the generation of H_2, the gas stream is enriched in H_2 by the oxidation of CO by water in the water gas shift reaction using a Cu-containing catalyst. The addition of 5–10% ZnO to the catalyst suppresses the methanation reaction and avoids the consumption of H_2 and the large exotherm associated with the reaction:

$$CO + H_2O \rightarrow H_2 + CO_2 \tag{1.4}$$

$$CO + 3H_2 \rightarrow CH_4 + H_2O \tag{1.5}$$

1.3.3 Carrier Materials

In most industrial reactions, the number of reactant molecules converted to products in a given time is directly related to the number of catalytic sites available to the reactants. It is necessary to maximize the number of active sites by dispersing the catalytic components onto a porous material. Maximizing the surface area of the catalytic components, such as Pt, Fe, Ni, Rh, Pd, CuO, PdO, CoO, and so forth increases the number of sites upon which chemisorption and catalytic reaction can occur. The catalytic components are introduced into the carrier by impregnation from aqueous solutions. Maximizing dispersion of the catalytic component requires carriers with a high surface area, such as Al_2O_3, SiO_2, TiO_2, C, $SiO_2-Al_2O_3$, zeolites, and CeO_2. In most cases, carriers themselves are not catalytically active for the specific reaction in question, but can play a major role in promoting the activity, selectivity, and maintaining the overall stability and durability of the finished catalyst. Such is the case in the hydrogenation of organic functional groups where high surface area carbons are used as carriers for precious metals and Ni. In these cases, the carrier adsorbs the organic compound prior to hydrogenation by the H_2 dissociated on the metallic catalytic component. For now when we speak of the catalyst, it is understood to be composed of catalytic components dispersed on high surface area porous carriers. Exceptions to this broad definition will be discussed in the application section.

1.3.4 Structure of the Catalyst and Catalytic Reactor

In chemical and petroleum applications, many different catalyst structures can be used. The structure depends on a number of factors such as volume of product to be produced, addition or withdrawal of heat, thermodynamic equilibrium, need for frequent regenerations, and life. The rationale for each reactor type will be discussed in the specific application section. For now, it is sufficient to say the catalyst materials are loaded and supported into a tubular reactor through which the reactants pass, interact with the catalyst, and convert to products. Figure 1.6a shows particulate structures, while a cartoon of a fixed bed reactor is shown in Figure 1.6b.

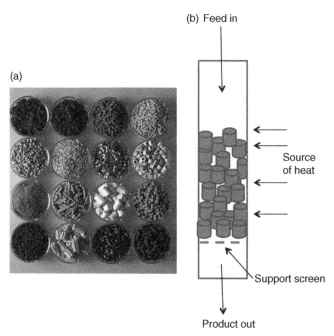

Figure 1.6   Particulate catalysts for fixed bed reactors: spheres, extrudates, and tablets. Powdered catalysts for batch slurry phase processors. A cartoon of a fixed bed reactor loaded with catalyst tablets. (Picture of particulates printed with permission from BASF.)

The catalyst is present in a preshaped particulate containing the catalytic component dispersed on the carrier. The size and shape of the catalyst depend on the nature of flow, permitted pressure drop, requirement for heat management (exothermic or endothermic reactions), mechanical strength, and so on. Spherical and tablet shapes 3–10 mm are commonly used. The desired reaction gas mixture flows through the bed of packed catalyst particulates and reaction occurs. The effluent gas is often monitored with some type of analytical device, such as a gas chromatograph, to monitor conversion. The temperature inside the bed is often measured at several points to monitor the reaction using thermocouples.

The preparation and properties of these materials and catalyzed monolith structures and their influence on catalytic reactions will be discussed in Chapter 2, but, for now, γ-Al$_2$O$_3$, most commonly used in catalysis, will be used to develop a model of a heterogeneous catalyst.

The pores of γ-Al$_2$O$_3$ are typically 2–10 nm in diameter, but irregularly shaped. The simplistic drawing shown in Figure 1.6 has circular catalytic components dispersed on the walls similar to raisins in a cake. The physical surface area of the carrier is the sum of all internal areas from all the walls of each and every pore. It is upon these internal walls that the catalytic components are bound or dispersed. The catalytic surface area is the sum of all the areas of the active catalytic components in this example. The smaller the individual size of the active catalytic material (higher catalytic surface area), the more the sites available for the reactants to interact. As a rough approximation, one assumes the higher the catalytic surface area, the higher the rate of reaction for a process controlled by kinetics. This is often the case, but there are exceptions in which a particular reaction is said to be structurally sensitive and the rate is maximum when interacting with a catalytic crystal size of a specific size range.

This procedure maximizes the catalytic surface area but also introduces other physical processes such as mass transfer of the reactants to the catalytic sites. Each of these processes has a rate influenced by the hydrodynamics of the fluid flow, the pore size and structure of the carrier, and the molecular dimensions of the diffusing molecule.

1.4 ADSORPTION AND KINETICALLY CONTROLLED MODELS FOR HETEROGENEOUS CATALYSIS

In heterogeneous catalysis, the initiating and terminating steps for the surface reaction involve the adsorption of the reactant onto and the desorption of the product from the active surface. In chemical terms, adsorption is the formation of chemical bonds between the adsorbing species (adsorbate) and the adsorbing surface (adsorbent). In general, there are two kinds of adsorptions, chemical and physical, typically referred to as chemisorption and physisorption, respectively. Physisorption is weak, nonselective adsorption of gaseous molecules on a solid at relatively low temperatures. In contrast, chemisorption is the relatively strong, selective adsorption of chemically reactive species on available sites of metal or metal oxide surfaces. In chemisorption, the adsorbent–adsorbate interaction involves the formation of chemical bonds and heats of reaction on the order of 40–300 kJ/mol. It is the formation of these strong bonds that alters the chemical nature of the adsorbate, making the newly formed structure more reactive and more easily transformed than the "free" molecule.

The maximum number of reactant molecules adsorbed onto a unit area of active catalyst surface will be the number of available catalytic sites per unit area. In the case where all catalytic sites are covered with an adsorbate molecule, the surface is said to be saturated and the fractional site coverage (θ) is unity. Adsorption and desorption continuously occur even when the net number of adsorbed species reaches a steady state. Thus, a dynamic equilibrium is established, where the fractional coverage at equilibrium is a function of temperature, adsorbate partial pressure, and the chemical nature of the adsorbate.

In heterogeneous catalysis, it is essential to know the amount of reactants that adsorb onto a given area of active catalyst surface. The quantity of adsorbed species is generally characterized by an isotherm, which is a mathematical relationship between fractional site coverage (θ) and temperature. The rate of reaction on the catalyst surface is going to be a function of the surface concentration of reactants, not necessarily the gas-phase reactant concentrations. Thus, the isotherm provides us with the link between the gas-phase concentration (measurable) and the surface concentration (immeasurable directly).

1.4.1 Langmuir Isotherm

There are many different isotherm forms, but the Langmuir isotherm is the simplest and most widely used and is applicable to many reactions. It is based on the key assumption that all sites on the adsorbent surface are of equal energies. Consider the example reaction presented earlier where CO is oxidized on a platinum catalyst. In this reaction, CO and O_2 must adsorb onto the platinum surface for reaction to occur. Further, the O_2-Pt must dissociate to form two Pt-O species. We can use the Langmuir isotherm to predict the surface concentration of each reactant and how the surface concentrations will vary with reaction conditions. We first consider the strong adsorption of CO in equilibrium with the surface of Pt:

$$CO + Pt \leftrightarrow CO\text{-}\text{-}Pt \tag{1.6}$$

The forward rate of Equation 1.6 (CO adsorption) is given by

$$r_{fCO} = k_{fCO}P_{CO}(1 - \theta_{CO}) \tag{1.7}$$

k_{fCO} is the forward rate constant of CO adsorption on Pt, P_{CO} is the partial pressure of CO, and θ_{CO} is the fraction of the surface of Pt covered by CO. The term $(1 - \theta_{CO})$ is the fractional number of sites available for additional CO adsorption. The isotherm also assumes each site is occupied by only one adsorbate molecule and full coverage is a monolayer.

The rate or reverse of Equation 1.6 (CO desorption) is given by

$$r_{dCO} = k_{dCO}\theta_{CO} \tag{1.8}$$

At equilibrium, the forward and desorption rates are equal and the ratio of the forward rate to reverse rate is defined as the adsorption equilibrium constant (Equation 1.9):

$$\frac{k_{fCO}}{k_{dCO}} = K_{CO} \tag{1.9}$$

$$k_{fCO}P_{CO}(1 - \theta_{CO}) = k_{dCO}\theta_{CO} \tag{1.10}$$

$$\theta_{CO} = \frac{K_{CO}P_{CO}}{(1 + K_{CO}P_{CO})} \tag{1.11}$$

Thus, Equation 1.11 provides the surface concentration of adsorbed CO as a function of CO partial pressure and the adsorption equilibrium constant. The adsorption equilibrium constant captures the influence of temperature and chemical nature of the adsorbate since the forward and reverse rate constants are a function of these parameters. Plotting Equation 1.11 generates Figure 1.7.

We can now consider two limiting cases: one where the partial pressure of CO is very low (P_{CO} is small) and one where the partial pressure of CO is very large (P_{CO} is large).

When P_{CO} is large, the denominator in Equation 1.11 $(1 + K_{CO}P_{CO})$ is approximately equal to $(K_{CO}P_{CO})$ since $K_{CO}P_{CO} \gg 1$. Thus, Equation 1.11 reduces to $\theta_{CO} = 1$ that signifies all the catalytic sites are covered with CO molecules and we say the surface is saturated with CO.

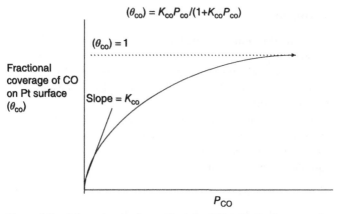

Figure 1.7   Adsorption isotherm (θ_{CO}) for CO on Pt for large, moderate, and low partial pressures of CO. The slope at low partial pressures of CO equals the adsorption equilibrium constant K_{CO}. (Reproduced from Chapter 1 of Heck, R.M., Farrauto, R.J., and Gulati, S.T. (2009) *Catalytic Air Pollution Control: Commercial Technology*, 3rd edn, John Wiley & Sons, Inc., New York.)

When P_{CO} is small, the denominator in Equation 1.11 $(1 + K_{CO}P_{CO})$ is approximately equal to 1 since $K_{CO}P_{CO} \ll 1$ and Equation 1.11 reduces to $\theta_{CO} = K_{CO}P_{CO}$. Under these conditions, the surface concentration responds linearly with increasing gas-phase concentration. Recall the equilibrium constant is defined as the ratio of the forward to reverse adsorption rates (Equation 1.9). An adsorbate that strongly adsorbs onto the surface will have a relatively large equilibrium constant and surface coverage will increase rapidly with increasing gas partial pressure. This is the case for CO on Pt, which is illustrated in Figure 1.4 as the steeply sloped trace at low P_{CO}.

At moderate values of P_{CO}, Equation 1.11 applies as written with the curve becoming more shallow as P_{CO} increases. $\theta_{CO} = 1$ at large CO.

The isotherm for dissociative chemisorption of O_2 on Pt is similarly generated where θ_O refers to the fractional coverage by O atoms consistent with the stoichiometry for CO oxidation of one O atom for each CO:

$$O_2 + Pt \leftrightarrow 2O\text{--}Pt \tag{1.12}$$

The rate of the forward (adsorption of O_2 on Pt)

$$r_{fO_2} = k_{fO_2} P_{O_2} (1 - \theta_O)^2 \tag{1.13}$$

The rate of the reverse reaction (O desorption from Pt) is

$$r_{dO_2} = k_{dO_2} \theta_{O_2}^2 \tag{1.14}$$

The square term for both the forward and reverse rates is due to the lower probability that two adjacent Pt sites will be available to accommodate two oxygen atoms from the dissociative chemisorption of O_2. Similarly, two adsorbed O atoms on Pt must be adjacent for desorption and recombination to diatomic O_2 to occur. Also, the adsorption equilibrium constant (K_{O_2}) for O_2 on Pt is k_{fO_2}/k_{dO_2}. Equating forward and reverse rates,

$$k_{fO_2}P_{O_2}(1-\theta_O)^2 = k_{dO_2}\theta_O^2 \tag{1.15}$$

$$\theta_O = \frac{K_{O_2}^{1/2}P_{O_2}^{1/2}}{\left(1 + K_{O_2}^{1/2}P_{O_2}^{1/2}\right)} \tag{1.16}$$

Plotting the fractional coverage of oxygen atoms versus $P_{O_2}^{1/2}$ generates a similar plot as Figure 1.7. The slope at low O_2 pressures is equal to $KO_2^{1/2}$, while at high pressures $\theta = 1$.

1.4.2 Reaction Kinetic Models

The isotherm provides us with the relationship for reactant surface concentration as a function of reaction conditions (i.e., temperature and reactant partial pressure). The next step is to use this information to estimate the rate of the chemical reaction occurring on the catalyst surface. To develop an expression for reaction rate, we need to know or assume a mechanism for the surface reaction. Specifically, the mechanism considers which reactants adsorb onto the surface and on which active species do they adsorb? While anything is possible, three general mechanisms are used to describe adsorption and reaction in heterogeneous catalysis:

> *Langmuir–Hinshelwood Mechanism:* Reactants adsorb onto the same active species and thus compete for available active sites (Figure 1.8).

> *Mars–van Krevelen Mechanism:* Reactants adsorb onto different sites, and the two surface intermediates interact and reaction occurs. Since each reactant adsorbs onto different surface species, there is no site competition between reactants (Figure 1.9).

> *Eley–Rideal Mechanism:* One reactant adsorbs onto the catalyst surface forming an intermediate surface species, which then reacts with a gas-phase reactant. Thus, there is no competition for catalytic sites (Figure 1.10).

Each of these three mechanisms is described in more detail in the following sections. The critical question is determining which mechanism a particular reaction follows. For each mechanism, a rate equation can be derived in terms of the appropriate surface and gas-phase reactant concentrations. Once derived, experiments can be conducted to see which rate equation best describes the experimental data. One possibility will be that neither of these three mechanisms describes the

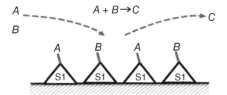

Figure 1.8 Illustration of Langmuir–Hinshelwood reaction mechanism.

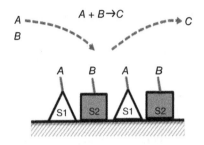

Figure 1.9 Illustration of Mars–van Krevelen reaction mechanism.

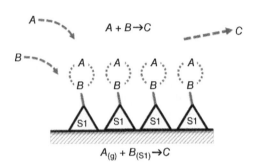

Figure 1.10 Illustration of Eley–Rideal reaction mechanism.

experimental data, which suggests the existence of a complex combination of mechanisms occurring simultaneously on the surface or the existence of an entirely different mechanism.

1.4.2.1 Langmuir–Hinshelwood Kinetics for CO Oxidation on Pt

The widely accepted kinetic model for the CO oxidation reaction on Pt is based on the Langmuir isotherm from which is derived Langmuir–Hinshelwood (LH) kinetics. The adsorption isotherms for CO and O_2 were considered separately in Section 1.4.1, but for the oxidation of CO by O_2 it is necessary to consider both gases each competing for the same sites on Pt. We will use k as the rate constant for the oxidation of CO where the net rate of reaction for CO oxidation will be

$$r_{CO} = k\theta_{CO}\theta_O \qquad (1.17)$$

The surface coverage of CO and O, θ_{CO} and θ_O, respectively, must be modified to account for the competitive adsorption that occurs on the active Pt sites. Specifically, the fraction of open sites used to determine the rate of adsorption must include terms for both O and CO, as written in Equation 1.18:

$$r_{fCO} = k_{fCO}P_{CO}(1 - \theta_{CO} - \theta_O) \qquad (1.18)$$

However, the desorption rate for CO is still dependent only on the sites occupied by CO and is the same as that given in Equation 1.8:

$$r_{dCO} = k_{dCO}\theta_{CO} \qquad (1.19)$$

Equating adsorption (1.18) and desorption (1.19) rates at equilibrium and recognizing that $k_{fCO}/k_{dCO} = K_{CO}$ yields an expression for the surface concentration of CO,

$$K_{CO}P_{CO} = \frac{\theta_{CO}}{(1 - \theta_{CO} - \theta_O)} \tag{1.20}$$

However, Equation 1.20 contains two unknowns, θ_{CO} and θ_O. Before we can solve it, we need an expression for θ_O. Following a similar approach to that used previously, for the rate of adsorption and desorption of O_2, we obtain

$$K_{O_2}^{1/2}P_{O_2}^{1/2} = \frac{\theta_O}{(1 - \theta_{CO} - \theta_O)} \tag{1.21}$$

Simplification is achieved by dividing (1.20) by (1.21):

$$\theta_O = \theta_{CO} \frac{K_{O_2}^{1/2}P_{O_2}^{1/2}}{K_{CO}P_{CO}} \tag{1.22}$$

We now substitute (1.22) into (1.20) to yield an expression for θ_{CO} in terms of known quantities.

$$\theta_{CO} = \frac{K_{CO}P_{CO}}{\left(1 + K_{CO}P_{CO} + K_{O_2}^{1/2}P_{O_2}^{1/2}\right)} \tag{1.23}$$

Similarly, we substitute (1.22) into (1.20) to yield an expression for θ_O in terms of known quantities:

$$\theta_O = \frac{K_{O_2}^{1/2}P_{O_2}^{1/2}}{\left(1 + K_{CO}P_{CO} + K_{O_2}^{1/2}P_{O_2}^{1/2}\right)} \tag{1.24}$$

Now we have an expression for θ_{CO} and θ_O that can be substituted into Equation 1.17 to develop a rate expression for CO oxidation in terms of the surface concentration of CO and O:

$$r_{CO} = k\theta_{CO}\theta_O \tag{1.17}$$

$$r_{CO} = k\frac{K_{O_2}^{1/2}P_{O_2}^{1/2}K_{CO}P_{CO}}{\left(1 + K_{CO}P_{CO} + K_{O_2}^{1/2}P_{O_2}^{1/2}\right)^2} \tag{1.25}$$

Let us now consider the concentration extremes when the concentration of CO is very low and very high. For low P_{CO}, Equation 1.25 reduces to Equation 1.26:

$$r_{CO} = k\frac{K_{O_2}^{1/2}P_{O_2}^{1/2}K_{CO}P_{CO}}{\left(1 + K_{O_2}^{1/2}P_{O_2}^{1/2}\right)^2} \tag{1.26}$$

This shows a direct relationship between the rate and P_{CO} when P_{O_2} is constant. For large P_{CO}, Equation 1.25 reduces to Equation 1.27:

$$r_{CO} = k\frac{K_{O_2}^{1/2}P_{O_2}^{1/2}}{K_{CO}P_{CO}} \tag{1.27}$$

$$\text{Rate} = k \, [K_{CO} P_{CO} K_{O2}^{1/2} P_{O2}^{1/2}]/[1 + K_{CO} P_{CO} + K_{O2}^{1/2} P_{O2}^{1/2}]^2$$

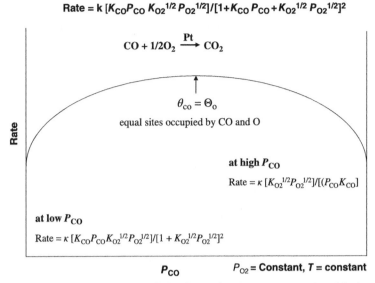

Figure 1.11   L–H kinetics applied to increasing P_{CO} at constant P_{O2}. Maximum rate was achieved when an equal number of CO molecules and O atoms are adsorbed ($\theta_O = \theta_{CO}$) on adjacent Pt sites. (Reproduced from Chapter 1 of Heck, R.M., Farrauto, R.J., and Gulati, S.T. (2009) *Catalytic Air Pollution Control: Commercial Technology*, 3rd edn, John Wiley & Sons, Inc., New York.)

Equation 1.28 indicates that the rate of reaction will decrease as the gas-phase concentration of CO becomes high. This means high CO concentrations inhibit the rate of reaction, which is due to CO saturating the catalyst surface and displacing adsorbed oxygen. Thus, the reaction rate approaches zero at high and low CO concentrations, which indicates a maximum of rate exists when $\theta_O = \theta_{CO}$, which is shown graphically in Figure 1.11.

Applying this model was very useful in designing an optimum system for the first gasoline oxidation catalyst for the automobile converter. When the CO was high during the cold start portion of the driving cycle, the addition of extra O_2 (from air) decreased the P_{CO} more than P_{O_2} and the rate of the reaction for CO oxidation increased. Thus, understanding kinetics and the rate expressions helped design a workable system to meet regulations.

From a fundamental point of view, it should be noted that the assumption of uniform energy sites on the catalyst in the Langmuir isotherm is not correct. A heterogeneous catalytic surface consists of a distribution of strong, moderate, and weak sites upon which the reactant molecules adsorb. Increasing temperature allows only the stronger sites to be retained. This results in a change in the overall energy of the adsorbed states on the activation energy profile of Figure 1.2. Fundamentally, with increasing temperature, this causes a small change in the activation energy, but for all intensive purposes it can be ignored when making activation energy measurements. Measurements of activation energies will be discussed later.

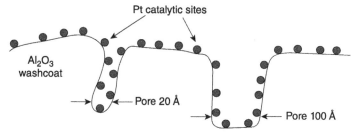

Figure 1.12 Ideal dispersion of Pt atoms on a high surface area Al_2O_3 carrier. (Reproduced from Chapter 1 of Heck, R.M., Farrauto, R.J., and Gulati, S.T. (2009) *Catalytic Air Pollution Control: Commercial Technology*, 3rd edn, John Wiley & Sons, Inc., New York.)

1.4.2.2 Mars–van Krevelen Kinetic Mechanism There are a number of reactions that proceed by a surface redox mechanism where oxygen is provided by one catalytic component to an adsorbed reactant on another catalyst site leading to oxidation. Examples for this mechanism apply when V_2O_5 is the catalyst for the oxidation of sulfur dioxide to sulfur trioxide. It also applies to oxidation of benzene to benzoquinone. We will use an example of the oxidation of CO over copper oxide and iron oxide to demonstrate the mechanism and rate equation. These mechanisms are known as Mars–van Krevelen (MvK). In this case, gas-phase CO reduces Cu^{2+} to Cu^{1+} (Equation 1.29). The FeO contributes its O to reoxidize Cu to Cu^{2+}, with reduced Fe forming gas-phase O_2 that then adsorbs and oxidizes Fe to FeO (Equation 1.30).

$$CO + 2CuO \rightarrow Cu_2O + CO_2 \tag{1.28}$$

$$Cu_2O + FeO \rightarrow 2CuO + Fe \tag{1.29}$$

$$\frac{1}{2}O_2 + Fe \rightarrow FeO \tag{1.30}$$

$$\text{Net reaction}: \quad CO + \frac{1}{2}O_2 \rightarrow CO_2$$

The overall rate of reaction can be written recognizing that Θ equals the fractional coverage of O atoms on the Fe species.

Overall oxidation of CO:

$$r_{CO} = k_1 P_{CO}\theta \tag{1.31}$$

Rate of reoxidation of the Fe:

$$r_o = k_2 P_{O_2}^{1/2}(1 - \theta) \tag{1.32}$$

The overall oxidation rate of the CO can be no faster that the rate at which Fe is reoxidized at steady state, so we can equate the two equations:

$$k_1 P_{CO}\theta = k_2 P_{O_2}^{1/2}(1 - \theta) \tag{1.33}$$

$$\theta = \frac{k_2 P_{O_2}^{1/2}}{k_1 P_{CO} + k_2 P_{O_2}^{1/2}} \tag{1.34}$$

The MvK rate equation for CO oxidation can be written as

$$r_{CO} = \frac{k_1 k_2 P_{CO} P_{O_2}^{1/2}}{k_1 P_{CO} + k_2 P_{O_2}^{1/2}} \tag{1.35}$$

This expression recognizes that the rate is dependent on the gas-phase partial pressure of both CO and O_2:

At low CO/O_2, the rate reduces to $k_1 P_{CO}$.

At high CO/O_2, the rate reduces to $k_2 PO_2^{1/2}$.

1.4.2.3 Eley–Rideal (E–R) Kinetic Mechanism

This model represents a case where a gas-phase reactant adsorbs onto another reactant adsorbed onto the catalyst site:

$$A_{(s)} + B_{(g)} \rightarrow C_{(g)} + D_{(g)} \tag{1.36}$$

Assuming C and D do not adsorb and A adsorbs with an equilibrium constant K_A and gas species B does not compete with A for catalytic sites and only adsorbs on adsorbed A, the expression reduces to

$$r = k\theta_A P_B \tag{1.37}$$

$$r = kP_B \left(\frac{K_A P_A}{(1 + K_A P_A)} \right) \tag{1.38}$$

where $\Theta_A = [K_A P_A/(1 + K_A P_A)]$ is derived by equating the forward and reverse rates of adsorption and desorption at equilibrium. The slope of rate versus P_B is the rate constant k when K_A is very large. Rate versus P_A at low concentrations of P_A relative to P_B is linear with the slope equal to $K_A P_B k$.

The hydrogenation of CO_2 to methane at 300 °C over a supported Ru catalyst obeys the Eley–Rideal mechanism. The empirical rate equation is essentially first order in H_2 and almost zero order in CO_2. The CO_2 is strongly adsorbed onto the Ru sites, while H_2 reacts with adsorbed CO_2 producing methane. In this case, $P_B = P_{H_2}$, $P_A = P_{CO_2}$ and $K_A = K_{CO_2}$.

1.4.2.4 Kinetic versus Empirical Rate Models

Kinetic models are useful in giving us insight into the precise mechanism by which a reaction occurs. It is a goal of catalytic chemistry to fit the experimental data to a model such as the three described above in Sections 1.4.2.1–1.4.2.3. However, often catalytic reactions are much more complicated and no single model can be applied due to many competing rate-limiting steps. For these cases, it is necessary to use what is referred to as an empirical model that permits the calculation of kinetic parameters such as reaction orders and activation energies that allow the process to be designed for the optimum reactant and product concentrations and temperatures for maximizing selectivity. Empirical rate models will be described in Chapter 4.

1.5 SUPPORTED CATALYSTS: DISPERSED MODEL

Most heterogeneous catalyst materials are supported on a high surface area (and highly porous) carrier such as Al_2O_3, carbon, SiO_2, TiO_2, and zeolites composed of SiO_2–Al_2O_3. This disperses the catalytic components into nanosized clusters that generate a high catalytic surface area, which maximizes their accessibility to reactants. Figure 1.12 shows Pt atoms ideally dispersed on high surface area Al_2O_3.

The catalytic sites (Pt in Figure 1.12) are bound moderately to surface functional groups (i.e., OH— not shown) of the carrier. One can envision chocolate chips well dispersed within the cookie. Commonly, the surface of the carriers has functional groups that provide some "anchoring" of the catalytic components maintaining them as nanosized clusters as they function in the reaction. Carriers and catalyst preparation methods will be discussed in Chapter 2. Chapter 3 will cover various characterization methods to provide understanding of their physical and chemical properties and changes that may occur during their operation. Chapter 5 will show how catalysts can undergo changes due to the environmental factors that occur during reaction.

Although supporting a catalytic component offers the benefit of maximizing the catalytic surface area, it does introduce transport issues whereby reactants must diffuse to the dispersed catalytic sites within the porous network of the carrier. These issues are discussed in the next section.

1.5.1 Chemical and Physical Steps Occurring during Heterogeneous Catalysis

To maximize reaction rates, it is essential to ensure accessibility of all reactants to the active catalytic sites dispersed within the internal pore network of the carrier. Once again, let us consider the physical and chemical steps occurring during heterogeneous CO oxidation in a packed bed catalytic reactor.

CO and O_2 molecules are flowing through a bed of a solid particulate catalyst. To be converted to CO_2, the following physical and chemical steps must occur:

1. CO and O_2 must make contact with the outer surface of the carrier. To do so, they must diffuse through a stagnant thin layer of gas or boundary layer forming around the outer surface of the supported catalyst particles. Bulk molecular diffusion rates vary approximately with $T^{3/2}$ and typically have an "apparent" activation energies, $E_1 = 8$–$16\,kJ/mol$ (2–4 kcal/mol).

   The term "apparent" activation energy is used here to distinguish the physical phenomena of diffusion from the truly activated chemical processes that occur at the catalytic site. Diffusion reactions are a physical phenomenon and thus are not activated processes. Thus, the term "apparent" activation energy is as convenient as a figure of merit for reaction sensitivity to temperature.

2. Since the bulk of the catalytic components are internally dispersed within the carrier, the CO and O_2 molecules must diffuse through the porous network toward the active catalytic sites. The "apparent" activation energy for pore diffusion, E_2, is approximately 1/2 that of a chemical reaction or about 25–35 kJ/mol (6–9 kcal/mol).

3. Once CO and O_2 arrive at the catalytic site, chemisorption of both O_2 and CO occurs on adjacent catalytic sites. The kinetics generally follow exponential dependence on temperature, that is, $\exp(-E_3/RT)$, where E_3 is the activation energy, which for chemisorption is typically greater than 40 kJ/mol (>10 kcal/mol).

4. An activated complex forms between adsorbed CO and adsorbed O with an energy equal to that at the peak of the activation energy profile since this is the rate-limiting step. At this point, the activated complex has sufficient energy to convert to CO_2 that remains adsorbed on the catalytic site. Kinetics also follow exponential dependence on temperature, that is, $(-E_4/RT)$ with activation energies typically greater than 40 kJ/mol (10 kcal/mol).

5. CO_2 desorbs from the site following exponential kinetics, that is, $\exp(-E_5/RT)$ with activation energies typically greater than 40 kJ/mol (10 kcal/mol).

6. The desorbed CO_2 diffuses through the porous network toward the outer surface with an "apparent" activation energy and kinetics similar to step 2.

7. CO_2 must diffuse through the stagnant layer and, finally, into the bulk gas. Reaction rates follow $T^{3/2}$ dependence. "Apparent" activation energies are also similar to step 1: less than 8–16 kJ/mol (2–4 kcal/mol).

Steps 1 and 7 represent bulk mass transfer, which is a function of the specific molecules, the dynamics of the flow conditions, and the geometric surface area (outside or external area of the catalyst/carrier). Pore diffusion, illustrated in steps 2 and 6, depends primarily on the size and shape of both the pore and the diffusing reactants and products. Steps 3–5 are related to the chemical interactions of reactants and products (i.e., CO, O_2, and CO_2) at the catalytic site(s). The sequence of these three processes are illustrated in Figure 1.13.

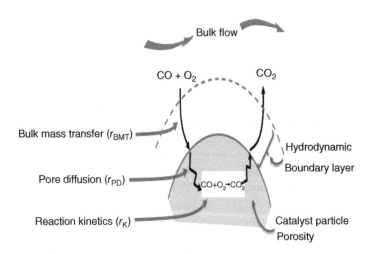

Figure 1.13   Illustration of the sequence of chemical and physical steps occurring in heterogeneous catalysis.

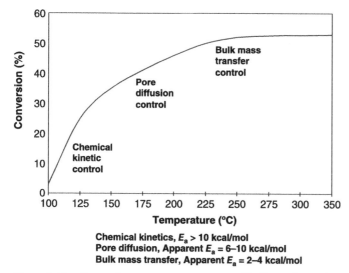

Chemical kinetics, $E_a > 10$ kcal/mol
Pore diffusion, Apparent $E_a = 6$–10 kcal/mol
Bulk mass transfer, Apparent $E_a = 2$–4 kcal/mol

Figure 1.14   Conversion versus temperature profile illustrating regions for chemical kinetics, pore diffusion, and bulk mass transfer control. (Reproduced from Chapter 1 of Heck, R.M., Farrauto, R.J., and Gulati, S.T. (2009) *Catalytic Air Pollution Control: Commercial Technology*, 3rd edn, John Wiley & Sons, Inc., New York.)

Any of the seven steps listed above can be rate limiting and control the overall rate of reaction. The rate-controlling process then dictates the overall rate of reaction observed experimentally. In catalytic processes, it is essential to understand what process limits the overall rate of reaction. The rate of each of these processes is a function of reaction conditions, the design of the catalyst bed, and properties of the catalyst. As reaction conditions change, so will the rate of each of these processes and the rate-limiting step may change. For example, consider the effect of temperature on the rate of a hypothetical reaction described in the temperature–conversion plot in Figure 1.14.

At a low temperature, the rate of reaction is chemically controlled. However, the chemical reaction rate is very sensitive to temperature (high activation energy), so its rate increases more with temperature than the diffusion processes. Pore diffusion then becomes rate limiting as the temperature increases, but eventually the least temperature-sensitive bulk mass transfer process becomes rate limiting. This is also demonstrated in Figure 1.15 that depicts the three relative rates of reaction. It shows that of the three rate-limiting phenomena, bulk mass transfer (BMT) is the fastest process at low temperatures but has a shallow dependence on temperature due to its low "apparent" activation energy. Pore diffusion has a lower rate than BMT due to its higher "apparent" activation energy but greater temperature dependence. The highest temperature dependence occurs for a reaction controlled by chemical kinetics, but because of its higher activation energy, the rate is small at low temperatures relative to those controlled by diffusion and thus is the rate-limiting step.

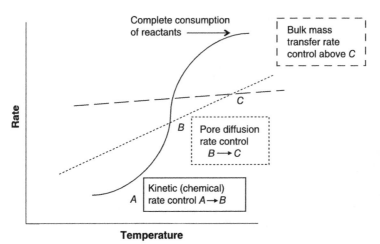

Figure 1.15 Relative rates of bulk mass transfer, pore diffusion, and chemical kinetics as a function of temperature. Chemical kinetics controls the rate between temperatures *A* and *B*. Pore diffusion controls from *B* to *C* temperatures, while bulk mass transfer controls at temperatures greater than *C*. (Reproduced from Chapter 1 of Heck, R.M., Farrauto, R.J., and Gulati, S.T. (2009) *Catalytic Air Pollution Control: Commercial Technology*, 3rd edn, John Wiley & Sons, Inc., New York.)

1.5.2 Reactant Concentration Gradients within the Catalyzed Material

In the chemical kinetic control region, the reaction of chemisorbed CO with chemisorbed O is slow relative to diffusion and thus is rate limiting. As the temperature is further increased, control of the overall rate will shift to pore diffusion. Here, the surface reaction between CO and O is faster than the rate at which gaseous CO and O_2 can be supplied to the sites and a concentration gradient exists within the carrier. This is referred to as intraparticle diffusion in which the catalytic components deep within the carrier are not being completely utilized or have an effectiveness factor less than 1. The effectiveness factor is the ratio of the actual rate versus the theoretical maximum rate if all catalytic sites are functioning. It can be thought of as a measure of the utilization of the catalytic component(s). At higher temperatures, the rate of diffusion of the CO and O_2 from the bulk gas to the external surface of the catalyst is slow relative to the other processes and the rate becomes controlled by bulk mass transfer. In this regime, the CO and O_2 are converted to CO_2 as soon as they arrive at the external surface of the carrier. The concentration of reactant and product is essentially zero at the external interface of the catalyst and the bulk fluid. The effectiveness factor is close to zero. Figure 1.16 graphically shows the relative gradients in concentration for reactants for the three rate-controlling processes in the catalyzed carrier.

1.5.3 The Rate-Limiting Step

The efficiency with which a catalyst functions in a process depends on what controls the overall reaction rate. If the kinetics of a process are measured and found

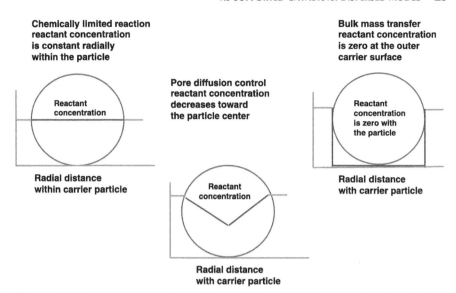

Chemically limited reaction reactant concentration is constant radially within the particle

Pore diffusion control reactant concentration decreases toward the particle center

Bulk mass transfer reactant concentration is zero at the outer carrier surface

Reactant concentration

Reactant concentration is zero with the particle

Reactant concentration

Radial distance within carrier particle

Radial distance with carrier particle

Radial distance with carrier particle

Figure 1.16   Reactant concentration gradients within a spherical structured catalyst for three regimes controlling the rate of reaction.

to be in a regime where chemical kinetics are rate controlling, the catalyst should be made with as high a catalytic surface area as possible. This is accomplished by increasing the catalytic component loading and/or dispersion so that every catalytic site is available to the reactants with an effectiveness factor approaching 1. Process parameters, such as an increase in temperature, promote a reaction controlled by chemical kinetics.

When it is known that a process will have significant pore diffusion limitations, the carrier should be selected with large pores and locate the catalytic components as close to the surface as possible to improve the effectiveness factor. Alternatively, a special geometry such as a carrier with holes (donut shape) can be used that also decreases the diffusion path. To enhance the transport rate, one can decrease the diameter of the carrier to decrease the diffusion path of reactants and products. A temperature increase will have some effect of enhancing the rate, but to a lesser extent than for those reactions controlled by chemical kinetics.

The rate of mass transfer is enhanced by increasing turbulence in the bulk gas and by increasing the geometric surface area (i.e., external area) by decreasing the particle size of the catalyst. This can be accomplished by selecting a catalyst with a high geometric surface area (small particle size or density of channels per unit area as is possible for monolith catalysts). Clearly, increasing the catalytic components surface area, the loading of the catalytic components or the size of the pores will have no effect on enhancing the rate of mass transfer since these catalyst properties do not participate in the rate-limiting step. Also, temperature will have virtually no impact on the BMT rate. This kinetics of all these rate-limiting steps will be more thoroughly discussed Chapter 4.

1.6   SELECTIVITY

The selectivity (S) of a reaction is defined as the ratio of the amount of desired product formed compared to the amount of reactant converted.

$\%S$ = moles of a specific product/moles of reactant converted
$\times$ (reactant/product stoichiometry) $\times$ 100

A general equation for selectivity of reactant A converting to product C is

$$xA + zB \rightarrow yC \tag{1.39}$$

$$S_C = \frac{(\text{moles } C \text{ produced})(x/y)}{(\text{moles } A \text{ consumed})} \tag{1.40}$$

The catalyst has the potential to greatly influence selectivity by preferentially lowering the activation energy for a particular step in the reaction sequence and increasing the rate at which this step proceeds. For the same reactants at the same conditions, different catalysts will influence the product distribution (selectivity) differently. By choosing the proper catalyst, the selectivity toward the desired product can be enhanced even if other products are more thermodynamically favored. For example, consider the oxidation of ammonia for the production of nitric acid (Chapter 8). Two ammonia oxidation reactions are possible, which are given in Equations 1.41 and 1.42. For the production of nitric acid, it is the formation of NO that is the desired reaction. From a thermodynamic perspective, the equilibrium constant for the N_2-forming reaction is four orders of magnitude more favorable than the desired NO reaction. However, when this reaction occurs in the presence of a Pt–Rh catalyst, the formation of NO greatly exceeds that of N_2. Thus, the Pt–Rh catalyst greatly favors the reaction sequence to form NO and selectively enhances the rate of this reaction over the formation of N_2.

$$4NH_3 + 5O_2 \rightarrow 4NO + 6H_2O \tag{1.41}$$

$$4NH_3 + 3O_2 \rightarrow 2N_2 + 6H_2O \tag{1.42}$$

Another example of a catalyst influencing product selectivity is the oxidation of ethylene. Consider the comparison between the products produced when Pt is used as a catalyst as opposed to oxides of vanadium. In the case of Pt, the carbon in the ethylene is completely oxidized to CO_2 with a selectivity of 100%:

$$C_2H_4 + 3O_2 \xrightarrow{\text{Pt}} 2CO_2 + 2H_2O \tag{1.43}$$

However, the V_2O_5 catalyst favors the selective partial oxidation of only one carbon in the ethylene yielding an aldehyde because this reaction pathway has the lowest activation energy compared to complete combustion to CO_2 and H_2O:

$$C_2H_4 + 3O_2 \xrightarrow{V_2O_5} CH_3CH = O \tag{1.44}$$

For Pt, the reaction products are exclusively CO_2 and H_2O; thus, selectivity is essentially 100%, making it a good catalyst for pollution abatement. For V_2O_5, the selectivity is about 80–90% toward the aldehyde with the balance 20–10% being

CO_2 and H_2O. Clearly, V_2O_5 would not be desirable for conversion of hydrocarbons such as ethylene to harmless CO_2 and H_2O, but is used commercially for selective partial oxidation reactions to desirable chemicals. So it is the function of the catalyst and reaction conditions to provide the right path that will yield the most desirable product. The ability to (i) enhance reaction rates and (ii) direct reactants to specific products makes catalysis extremely important in the environmental, petroleum, chemical, and alternative energy industry.

An important reaction in the automotive catalytic converter is the reduction of NO by H_2 during a specific driving mode. Two parallel reaction pathways are possible: one desirable leading to N_2 formation and the other undesirable producing toxic NH_3:

$$2H_2 + 2NO \xrightarrow{\text{Rh}} N_2 + 2H_2O \tag{1.45}$$

$$5H_2 + 2NO \xrightarrow{\text{Pt}} 2NH_3 + 2H_2O \tag{1.46}$$

Clearly, Rh is more selective and dominates the NO to N_2 pathway with a rate considerably higher than that with undesired pathway leading to NH_3 formation when Pt is used.

The reaction conditions also have a pronounced effect on product distribution depending on activation energies for all possible reactions. For example, NO (component of acid rain and a contributor to ozone formation) emitted from automobile engines and power plant exhausts can be reduced using a V_2O_5-containing catalyst with high selectivity provided the temperature is maintained between 250–300 °C:

$$2NH_3 + NO + O_2 \xrightarrow{250-300\,°C} 3/2N_2 + 3H_2O \tag{1.47}$$

$$2NH_3 + NO + 5/2O_2 \xrightarrow{>300\,°C} 3NO + 3H_2O \tag{1.48}$$

Ammonia also decomposes to N_2 above 300 °C and thus is not available to reduce the NO:

$$2NH_3 + 3/2O_2 \xrightarrow{>300\,°C} N_2 + 3H_2O \tag{1.49}$$

The desired reaction is favored below 300 °C since it has the lowest activation energy of the other two reactions. Once the temperature exceeds 300 °C, the reactions with the higher activation energy (greater temperature sensitivity) become favored and mixed products form.

1.6.1 Examples of Selectivity Calculations for Reactions with Multiple Products

General equation for selectivity:

$$xA + x/2B \rightarrow yC \tag{1.50}$$

$$S_C = \frac{(\text{moles } C \text{ produced})(x/y)}{(\text{moles } A \text{ consumed})} \tag{1.40}$$

To convert to a percent, multiply by 100.

In the example below, we assume 1 mol of NO is converted producing 0.45 mole of N_2:

$$2NO + 2H_2 \rightarrow N_2 + 3H_2O + N_2O + NH_3 \qquad (1.51)$$
$$\underset{(1\ mol)}{} \qquad \qquad \underset{(0.45\ mol)}{}$$

$$S_{N_2} = \frac{(\text{mole } N_2 \text{ formed})(2)}{(\text{mole NO converted})}$$

$$S_{N_2} = \frac{(0.45\ mol)(2)}{(1\ mol)} = 0.90$$

The balance of the NO is converted to a mixture of N_2O and NH_3.

1.6.2 Carbon Balance

In many organic processes, there are several thermodynamically favorable pathways to which reactants can be converted. It is the function of the catalyst and process conditions to direct the reactants to the most desired products. Seldom is selectivity 100% and, therefore, it is necessary to know what other products are being formed and to account for all of the carbon. This is especially important when the feedstock is expensive. The desired product must have a certain purity for commercial use and thus separation from the other products must be conducted, which can be costly. For these cases, it is necessary to know the other compounds containing carbon.

Take for example the catalytic partial oxidation of ethylene to ethylene oxide where the latter is used for the production of terephthalate polyesters and ethylene glycol (antifreeze). An undesired side product is CO_2.

$$H_2C{=}CH_2 + 1/2O_2 \longrightarrow \overset{O}{\underset{H_2C-CH_2}{\diagup\diagdown}} \text{ (desired)} \qquad (1.52)$$
$$\longrightarrow CO \text{ (undesired)}$$

The general carbon balance equation can be written as shown below (1.53) and should approach 100%. Although not shown, there may be carbon-containing compounds present on the catalyst in the form of "coke" and should be included in the products. However, often its value is small and is not directly considered in the balance, but is certainly important for the life of the catalyst:

$$\text{Carbon balance} = \frac{(C \text{ atoms in ethylene oxide} + C \text{ atoms in } CO_2)}{(C \text{ atoms in ehtylene})} \qquad (1.53)$$

To convert to a percent, multiply by 100.

In the example above, the carbon in the product is distributed between the desired ethylene oxide and undesired CO_2. Detailed product analysis permits determination of carbon in each, which can then be used to account for all the carbon introduced in the process from ethylene.

1.6.3 Experimental Methods for Measuring Carbon Balance

Many gas-phase reactions lead to either an increase or a decrease in product volume. A conventional way to perform a material balance is to use an internal standard that undergoes no reaction. Commonly N_2 is used. The ratio of grams/time of N_2 to grams/time of total C distributed among many different products will not change. Thus, if the N_2 concentration signal in the product decreases (as measured in a gas chromatograph (GC) by its calibrated peak area), it means there has been a volume expansion that will dilute all the gas components. The ratio of the N_2 concentration in the feed $N_{2\,(in)}$ to the $N_{2\,(prod)}$ concentration in the product gives a measure of the extent of volume increase. The total product flow rate is determined as shown below:

$$\text{(Total flow rate in)}\frac{N_{2,in}}{N_{2,prod}} = \text{Total flow rate produced} \qquad (1.54)$$

Equation 1.55 is used for the carbon material balance in which grams/time of carbon in is equal to grams/time of carbon in the product where [] is concentrations:

Flow rate [liter/time] $\times$ Concentration [grams/liter] = grams/time
$$\text{(Flow rate)}_{in} \times \text{[carbon in reactant]}_{in} = \text{(Flow rate)}_{prod} \times \text{[carbon in products]}_{prod}$$
$$(1.55)$$

The material balance is achieved by comparing all of the carbon distributed in the products with the carbon introduced from the reactants. Ideally, it should be 100%. If less, there is some carbon that has not been measured, which may be a small amount of coke (i.e., carbon) deposited on the catalyst.

QUESTIONS

1. Distinguish a homogeneous catalyst and process from a heterogeneous catalyst and process.

2. Why is it important to maximize the number of active sites in a heterogeneous catalyst?

3. In your own words, what is activation energy and how is it measured experimentally?

4. What is the practical value in preparing a catalyst or adjusting the process conditions by knowing the activation energy of a reaction?

5. How would you change the process conditions by knowing that in a hydrocarbon oxidation reaction, the hydrocarbon has a very large inhibition effect as in the following equation:
$$\text{Rate} \sim k[(\text{Kads}_{HC}\, PP_{HC})(\text{Kads}_{O_2}\, PP_{O_2})]/(\text{Kads}\, PP_{HC})$$

6. **a.** How does the presence of a catalyst change the thermodynamic equilibrium constant of a reaction?

   b. Two reactions are thermodynamically feasible (both have negative free energies), but one is much more negative than the other. Can a catalyst direct the reactants to the least favorable?

7. What are the benefits of using a catalyst for abating emissions?

8. Derive the L–H rate expression for the decomposition of PH_3 on a W metallic catalyst where no products adsorb:

$$PH_3 \rightarrow P + 3/2H_2$$

9. Write the L–H for N_2O decomposition on Pt where both the N_2O and the product O_2 strongly adsorb onto the Pt. Make a plot of rate versus conversion of N_2O.

10. What catalyst properties (assume the catalyst is Pt dispersed on porous Al_2O_3 present as a thin coating (washcoat) on a ceramic monolith) and process conditions will enhance the rate of a gas-phase reaction if the reaction is controlled by

   a. chemical kinetics?

   b. pore diffusion?

   c. bulk mass transfer?

11. What temperature is required for a catalyst with an activation energy of 15,000 cal/mol to operate with the same volume as one with an activation energy of $E = 12,000$ cal/mol (operation temperature is $227\,^{\circ}C$) with the same conversion? Assume k_0 is the same for both catalysts. $R = 2\,cal/^{\circ}K$.

12. How will the rate constant (and hence the reaction rate) change when the temperature is increased from 100 to $110\,^{\circ}C$? Assume an activation energy of 20 kcal/mol with the universal gas constant $= 2\,cal/(deg\,mol)$. Feel free to use joules also for activation energy with $(R = 8.3\,J/(mol\,^{\circ}K))$.

13. **a.** Consider a reaction with two possible product distributions: one desired and the other not. Why is it important to know their respective activation energies?

   b. What conditions temperature and concentrations would you use in the process to favor the desired reaction? Plot their respective relative rates versus temperature.

14. What properties of the particulate catalyst and process conditions can be changed to enhance the rate-limiting step for each of the four separate cases of rate control?

   1. Bulk mass transfer

   2. Pore diffusion

   3. Chemical

   4. Heat transfer

15. Consider a scenario where one reactant could undergo two different reactions leading to two products. Why is it important to know the respective activation energy of each possible reaction?

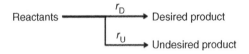

16. You are hired to troubleshoot a fixed bed catalyst, oxidizing a pollutant (CO) with air (O_2) to CO_2. The operator tells you that conversion does not increase significantly as the temperature is increased by $30\,^{\circ}C$.

   a. What could cause such an observation?

   b. How would you verify this by doing one experiment?

17. Sketch out the conversion–temperature profile and Arrhenius plot for the above catalyst if the reaction operates up to 200 °C in kinetic control, from 200–250 °C in pore diffusion control, and above 250 °C in bulk mass transfer control. A maximum conversion of 99% is achieved at 300 °C.

BIBLIOGRAPHY

General References

Bartholomew, C. and Farrauto, R. (2006) *Fundamentals of Industrial Catalytic Processes*, 2nd edn, John Wiley & Sons, Inc., Hoboken, NJ.

Bond, G., Lous, C., and Thompson, D. (2006) *Catalysis by Gold*, Imperial College Press, London.

Farrauto, R.J. (2012) Industrial catalysis: a practical guide, in *Handbook of Industrial Chemistry and Biochemistry*, Vol. 1, 12th edn (ed. J.A. Kent), Springer, New York, pp. 201–230.

Hagen, J. (2006) *Industrial Catalysis*, 2nd edn, Wiley-VCH Verlag GmbH, Weinheim, Germany.

Rase, H. (2000) *Handbook of Commercial Catalysts: Heterogeneous Catalysts*, CRC Press, Boca Raton, FL.

Selectivity

Avgouropoulos, G., Ioannides, T., Papadopoulos, C., Hocevar, S., and Matralis, H. (2002) A comparative study of Pt/Al_2O_3, $Au/\alpha\text{-}Fe_2O_3$ and $CuO\text{-}CeO_2$ catalysts for the selective oxidation of carbon monoxide in excess hydrogen. *Catalysis Today* 75, 157.

Korotkikh, O. and Farrauto, R. (2000) Selective catalytic oxidation of CO in H_2: fuel cell applications. *Catalysis Today* 62, 2.

Korotkikh, O., Ruettinger, W., and Farrauto, R. (2003) *Suppression of methanation activity by water gas shift reaction catalyst*. U.S. Patent 6,562,315.

Liu, X., Korotkikh, O., and Farrauto, R. (2002) Selective catalytic oxidation of CO in H_2: a structural of Fe oxide promoted Pt/alumina catalyst. *Applied Catalysis B: Environmental* 226, 293.

Catalyst Design

Bartholomew, C. and Farrauto, R. (2006) *Fundamentals of Industrial Catalytic Processes*, 2nd edn, John Wiley & Sons, Inc., Hoboken, NJ.

Bond, G., Lous, C., and Thompson, D. (2006) *Catalysis by Gold*, Imperial College Press, London.

Grisel, R. and Nieuwenhuys, B. (2001) A comparative study of the oxidation of CO and CH_4 over $Au/MO_x/Al_2O_3$ catalysts. *Catalysis Today* 64, 69.

Morbidelli, M., Garvriilidis, A., and Varma, A. (2001) *Catalyst Design: Optimal Distribution of Catalyst in Pellets, Reactors and Membranes*, Cambridge University Press, Cambridge, UK.

Kinetics

Bollinger, M. and Vannice, M.A. (1996) A kinetics and Drift study of low temperature carbon monoxide oxidation over Au–TiO2 catalysts. *Applied Catalysis B: Environmental* 8, 417.

Voltz, S., Morgan, C., Liederman, D., and Jacob, S. (1973) Kinetic study of carbon monoxide and propylene oxidation on platinum catalysts. *Industrial & Engineering Chemistry Product Research and Development*, 14 (4), 294–301.

Yao, Y. (1984) The oxidation of CO and hydrocarbons over noble metal catalysts. *Journal of Catalysis*, 87, 152–162.

Kinetic Models

Duyar, M., Ramachandran, A., Wang, C., and Farrauto, R.J. (December 2015) Kinetics of CO_2 methanation over Ru/γ-Al_2O_3 and implications for energy storage applications. *Journal. of CO_2 Utilization*, (2015) 27–33.

Hinshelwood, C.N. (1940) *The Kinetics of Chemical Change*, Clarendon Press, London.

Hougen, O. and Watson, K. (1943) Solid catalysts and reaction rates general principles. *Industrial Engineering Chemistry* 35, 529.

Hurtado, P., Ordonez, S., Sastre, H., and Diez, F.V. (2004) Development of a kinetic model for the oxidation of methane over Pd/Al_2O_3 at dry and wet conditions. *Applied Catalysis B: Environmental* 51, 229–238.

Janke, C., Duyar, M., Hoskins, M., and Farrauto, R.J. (2014) Catalytic and adsorption studies for the hydrogenation of CO_2 to methane. *Applied Catalysis B: Environmental* 152–153, 184–191.

Mars, P. and van Krevelen, D. (1954) Oxidation carried out by means of vanadium oxide catalysts. *Chemical Engineering Science* 3, 41.

Rachmady, W. and Vannice, M.A. (2000a) Acetic acid hydrogenation over supported platinum catalysts. *Journal of Catalysis* 192, 322–334.

Rachmady, W. and Vannice, M.A. (2000b) Acetic acid hydrogenation over supported platinum catalysts. *Journal of Catalysis* 192, 322–334.

Vannice, M.A. (2007) An analysis of the Mars–van Krevelen rate expression. *Catalysis Today* 123 (1–4), 18–22.

Chemical and Mass Transfer Control in Heterogeneous Catalytic Reactions

Broadbelt, L. (2003) Kinetics of catalyzed reactions: heterogeneous, in *Encyclopedia of Catalysis* (ed. I. Horvath), Wiley-Interscience, Hoboken, NJ, pp. 472–494.

Heck, R., Farrauto, R., and Gulati, S. (2009) Chapter 1, in *Catalytic Air Pollution Control: Commercial Technology*, 3rd edn, pp. 18–23.

CHAPTER 2

THE PREPARATION OF CATALYTIC MATERIALS

2.1 INTRODUCTION

Active catalytic materials are located in the middle of the periodic table and are composed primarily of the precious metals (Group VIIIB), mainly Pt, Pd, Rh, and Ru, and their oxides and nonprecious metals, also referred to as base metals, and their oxides, also present in Group VIIIB, such as Fe, Co, and Ni. Group IB metals and the oxides of Cu and Ag, V (Group VB), and Cr and Mo (Group VIB) are also commonly found to be catalytically active for various reactions. Oxides with surface acidity or basicity are also catalytic and will be discussed in the context of the specific reactions they catalyze. Throughout this book, applications and the most predominant catalysts and their promoters will be discussed.

Heterogeneous catalysts, which are the dominant catalysts used in most major industrial processes and will be the focus of this book, can be classified into two general forms: supported and unsupported. This was discussed in Section 1.3. In this chapter, we will review the common process steps used to prepare these catalysts.

In the case of supported catalysts, most catalytic materials are dispersed as nanosized sites onto carriers in order to enhance their accessibility for the reactants and products. The most common carrier is γ-Al_2O_3; however, other high internal surface area carriers such as natural and synthetic carbons, TiO_2, ZrO_2, SiO_2, and alumina silicates (zeolites) are used depending on the application. Although carriers are usually believed to simply provide a surface upon which the catalytic components are dispersed, they often play a critical role in providing additional functionality that enhances reactions. The most common is surface acidity that promotes isomerization and cracking of large molecules.

It is very important to understand that most industrial catalysts are prepared, activated, and conditioned by general procedures, but many critical details of the preparations are maintained as trade secrets to protect the proprietary nature of the suppliers' products. Although many patents are held by suppliers for various preparation procedures, there is no guarantee that what is disclosed or claimed is actually practiced. This is essential in order for each supplier to maintain an advantage

Introduction to Catalysis and Industrial Catalytic Processes, First Edition. Robert J. Farrauto, Lucas Dorazio, and C.H. Bartholomew.
© 2016 John Wiley & Sons, Inc. Published 2016 by John Wiley & Sons, Inc.

over competitor's products. Consequently, what are described below are the more general, commonly used procedures.

The specific catalytic materials and their preparations will be presented in the context of their application in the chapters in this book. This chapter mainly introduces the reader to general aspects of catalyst preparation.

2.2 CARRIER MATERIALS

A carrier is usually a high surface area inorganic material containing a complex pore structure through which catalytic materials are deposited. At one time, it was thought to only provide a surface to disperse the catalytic substance to maximize the catalytic surface area. However, it is now quite clear that it can and often does play a critical role in maintaining the activity, selectivity, and durability of the finished catalyst. By far, the most common carriers are the high surface area inorganic oxides, some of which are described below.

2.2.1 Al_2O_3

Alumina is by far the most commonly used carrier in commercial applications. There are many different sources of alumina having varying surface areas, pore size distributions, surface acidic properties, composition of trace components, and crystal structures. Its properties depend on its preparation, purity, and thermal history. Various crystalline alumina hydrates are produced by precipitation from either acidic or basic solutions. It is an amphoteric oxide soluble at a pH above about 12 and below about 6. Within this broad pH range, it forms a number of different crystalline hydrates. For example, at a pH of 11, it forms a trihydrate species ($Al_2O_3 \cdot 3H_2O$) called bayerite, while at a pH of 9 it forms pseudoboehmite, which is a monohydrate crystal ($Al_2O_3 \cdot H_2O$). On the more acidic side, such as a pH of 6, the precipitate lacks any definite long-range crystal structure and is classified as amorphous. The high surface area is created by heat treating or calcining in air, typically at about 500 °C where a network is formed from Al_2O_3 particles 20–50 Å (2–5 nm) in diameter bonded together forming polymer-type chains. A scanning electron micrograph (SEM) of a specially prepared γ-Al_2O_3 crystal, magnified 80,000 times, is shown in Figure 2.1a with a surface area of about 150 m^2/g. One can see that the structure is composed of primary Al_2O_3 particles agglomerated forming highly porous networks.

Once precipitated, it is thoroughly washed to remove impurities from the precursor salts. If an acidic solution of Al^{3+} is neutralized with NaOH, the Na^+ should be removed by washing. Drying is usually performed at about 110 °C to remove excess H_2O and other salts containing volatile species such as NH_3 present from the precursor salts. Calcinations at different temperatures determine the final crystal structure, which in turn determines its chemical and physical properties.

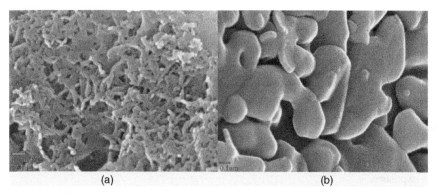

(a)   (b)

Figure 2.1   (a) SEM of γ-Al$_2$O$_3$ (80,000× magnification) and (b) SEM of α-Al$_2$O$_3$ (80,000×
magnification). (Reproduced from Chapter 2 of Heck, R.M., Farrauto, R.J., and Gulati, S.T.
(2009) *Catalytic Air Pollution Control: Commercial Technology*, 3rd edn, John Wiley &
Sons, Inc., New York.)

The monohydrate (boehmite) and trihydrate (bayerite) alumina structures
change as a function of the temperature (°C) in air.

$$
\begin{aligned}
\text{boehmite (monohydrate)} \quad &\longrightarrow \gamma\text{-Al}_2\text{O}_3 \quad (500\text{--}800\,^\circ\text{C}) \\
&\longrightarrow \delta\text{-Al}_2\text{O}_3 \quad (800\text{--}1000\,^\circ\text{C}) \\
&\longrightarrow \theta\text{-Al}_2\text{O}_3 \quad (1000\text{--}1100\,^\circ\text{C}) \qquad (2.1) \\
&\longrightarrow \alpha\text{-Al}_2\text{O}_3 \quad (>1100\,^\circ\text{C})
\end{aligned}
$$

$$
\begin{aligned}
\text{bayerite (trihydrate)} \quad &\longrightarrow \eta\text{-Al}_2\text{O}_3 \quad (300\text{--}800\,^\circ\text{C}) \\
&\longrightarrow \theta\text{-Al}_2\text{O}_3 \quad (800\text{--}1150\,^\circ\text{C}) \qquad (2.2) \\
&\longrightarrow \alpha\text{-Al}_2\text{O}_3 \quad (>1100\,^\circ\text{C})
\end{aligned}
$$

As the temperature is increased, between the ranges given above for a given
Al$_2$O$_3$ crystal structure, there is a gradual dehydration, which causes an irreversible
loss in the internal porosity or internal surface area (the material is being densified)
and a loss in its surface OH$^-$ or Brønsted acid sites. Continued heating causes a
complete transformation to another crystal structure with a continuing loss in internal
surface area and surface hydroxyl groups. For example, boehmite loses the bulk of its
water below about 300 °C, during and after which it begins to sinter or lose internal
surface area. At roughly 500 °C, it converts to γ-Al$_2$O$_3$, which typically has an internal
surface area of 100–200 m^2/g. Continuous heating causes additional sintering and/or
phase changes and loss of surface hydroxyl groups up to about 1100 °C, where it
converts to the lowest internal surface area structure (1–5 m^2/g), called α-Al$_2$O$_3$. A
SEM shown in Figure 2.1b, magnified 80,000 times, clearly shows that its morphology is much more densely packed than the γ-Al$_2$O$_3$ shown in Figure 2.1a. The internal
surface area decreases with increasing temperature. Paralleling the structural changes
that occur during heat treatment, the surface becomes progressively more dehydrated
or more hydrophobic. These transformations obey time–temperature relationships

and depend on the environment in which the material is exposed. The presence of steam, for example, greatly accelerates these transformations.

The irreversible phase transitions of Al_2O_3 result in the collapse of the internal or physical surface area of the carrier. This occludes the active catalytic species within its pore structure, resulting in a loss of accessibility by the reactants. This is a primary source of thermally induced catalyst deactivation, which will be discussed in Chapter 5. There is an advantage, however, in that the appearance of the different phases and irreversible changes in physical and chemical surface structure allow the catalyst's thermal history to be accessed. This is very important when studying the causes of catalyst deactivation in commercial installations due to excessive temperature exposure.

The presence of certain elements in the Al_2O_3 can have a profound influence on its physical surface area retention after exposure to high temperatures. Small amounts of Na_2O present in the Al_2O_3 can enhance (or catalyze) the sintering rate of the Al_2O_3 and thus act as a flux. In contrast, certain elements can act as "negative catalysts" and actually reduce the sintering rate. The presence of a few percent stabilizer such as La_2O_3 can greatly retard the sintering rate of the γ-Al_2O_3. This was of enormous importance in the development of high-temperature durable catalytic converters for the automobile in emission control.

There have been many studies offering mechanisms explaining the stabilization effects of certain metal oxides on alumina. It is generally accepted that a solid solution of the stabilizing ion in the Al_2O_3 structure decreases the mobility of the Al and O ions, resulting in a reduction in the rate of sintering and/or phase transition.

2.2.2 SiO$_2$

The inertness of SiO_2 toward reacting with sulfur oxide (SO_x) compounds in process streams makes it a suitable catalyst carrier. For example, SiO_2 is used as a carrier in the oxidation of SO_2 to SO_3 in sulfuric acid production. In contrast, Al_2O_3 is highly reactive with SO_3 and forms compounds that alter the internal surface of the carrier resulting in catalyst deactivation.

Alkaline solutions of silicate (pH >12) can be neutralized with acid, resulting in the formation of silicic acid. This polymerizes, forming a high surface area network with interconnecting pores of varying sizes.

$$SiO_4^{2-} \xrightarrow{\text{H}^+} [Si(OH)_4]_x \rightarrow SiO_2 \cdot H_2O \tag{2.3}$$

Similar to Al_2O_3, it is then washed, dried, and calcined. Surface area of SiO_2 materials can be as high as 300–$400\,m^2/g$. They have a small amount of chemically held water, giving rise to some surface acidic hydroxyl groups.

2.2.3 TiO$_2$

Titania is finding more uses as a catalyst carrier for a variety of reactions. Traditionally, because of its inertness to sulfate formation and its surface properties, TiO_2 is a preferred carrier for vanadia (V_2O_5) in selective catalytic reduction of NO_x with NH_3

from stationary sources where sulfur is present in power plant effluents. It has been shown to be a very effective support for various reduction and oxidation reactions. It is also commonly used in photocatalytic reactions. There are two crystal structures of importance: anatase and rutile. Catalytically, the anatase form is the most important in that it has the highest surface area (50–$80\,m^2/g$) and is thermally stable up to about $500\,°C$. The rutile structure has a low surface area ($<10\,m^2/g$), forming at about $550\,°C$. The phase transformation from anatase to rutile, and subsequent loss of surface area, limits the use of titania to relatively low-temperature applications ($<500\,°C$). In the case of vanadia catalysts, the phase transformation results in occlusion of the vanadia and, ultimately, deactivation.

2.2.4 Zeolites

Naturally occurring or synthetic alumina–silicate materials with well-defined crystalline structures and pore size are called *zeolites*. The Al_2O_3 and SiO_2 are bound in a tetrahedral structure with each Al and Si cation bonded to four oxygen anions. In turn, each O^{2-} is bonded to a Si^{4+} or Al^{3+} in an arrangement similar to that shown below.

To maintain charge neutrality, an extra Na^+ or H^+ must be bonded to the AlO^-, giving rise to an exchangeable cation site. These sites are acidic when the cation is H^+, and they remain on the internal or external surface, accessible by reactant molecules. The pore structure dimensions of zeolites are precise and can be between 0.3 and 0.13 nm (3 and 13 Å), which fall into the range of molecular dimensions. Any molecule with a larger cross-sectional area is prevented from entering the channel of the zeolite cage. It is for this reason that zeolites are often referred to as molecular sieves and are used in gas separations. See Figure 2.2 showing three different important zeolite structures.

Treatment in dilute acid converts this structure to the H^+-exchanged state. In the pore of a zeolite, the surface is composed of AlO^-H^+ or AlO^-M^+ (for the case of the metal-exchanged site) that provides the active sites for the desired catalytic reactions.

Zeolites are of great significance due to their well-defined crystalline structures and surface properties. Modifications to the preparation alter the SiO_2/Al_2O_3 ratio, which in turn has a profound influence on the number of surface acid or exchangeable cation sites that are the origins of the active sites. As the SiO_2/Al_2O_3 ratio increases, there are fewer Al cations within the framework and thus fewer H^+ for exchange. One might be tempted to simply increase the Al content in the zeolite to increase the number of active sites; however, thermal stability decreases as the Al_2O_3 increases. Thus, zeolites offer a great variety of properties as carriers for various catalytic metal ions depending on the reaction of interest and the conditions in which they must function.

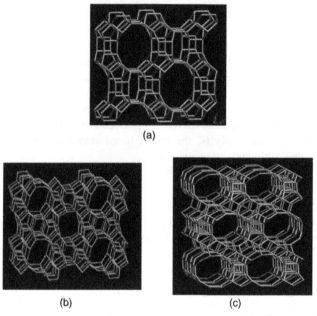

(a)

(b)   (c)

Figure 2.2 Three zeolites: (a) mordenite, (b) ZSM-5, and (c) Beta. (Reproduced from Baerlocher, Ch. and McCusker, L.B., *Database of Zeolite Structures*, International Zeolite Association, http://www.iza-structure.org/databases/.)

The most widely used zeolite is that of faujasite for fluid catalytic cracking reactions. It will be discussed in more detail in the petroleum chapter.

The zeolite mordenite (Figure 2.2a) typically has a SiO_2/Al_2O_3 ratio of 5–10 with its most important aperture (or pore size) about 6.6 Å (0.6 nm) in size. Its aperture has only one dimension into which molecules can pass. It is used as a molecular sieve used for separations by preventing molecules with larger diameters than its aperture from entering. Its acidic properties provide the active sites for the isomerization reactions (i.e., xylenes) as well as selective NO_x conversion with NH_3 in power plants in a technology called selective catalytic reduction (SCR). Synthetic zeolites are generally prepared from aqueous solutions of alkali salts of aluminum and silicon, and sometimes contain an organic amine, called a template, which aids in establishing a particular crystalline structure. The template is removed by calcination as a final step in the preparation process. Some preparations are carried out in autoclaves at temperatures between 150 and 180 °C. The zeolite is drawn as a stick figure with an O present in the midpoint of each stick, which connects two Si or a Si and Al depending on the concentration of each. It is the O ion associated with the Al ions that provide either the H^+ or the metal-exchanged cation. Mordenite has 12 member rings. Another important zeolite is ZSM-5 (Figure 2.2b), used as an additive in the FCC process to enhance gasoline production and in the methanol to gasoline process. It has a SiO_2/Al_2O_3 ratio of about 10 with one aperture 5.5 Å (0.55 nm) in diameter. Unlike mordenite, it is a three-dimensional structure with sinusoidal channels with entry from all three sides. Each ring has 10 members. Beta zeolite (Figure 2.2c) also has

12 member rings and is three dimensional. But it has two apertures, one about $6.6\,\text{Å}$ (0.66 nm) and the other $5.6\,\text{Å}$ (0.56 nm). Beta zeolite, exchanged with Fe ions, has been incorporated into some diesel oxidation catalysts as a hydrocarbon trap and also for NO_x reduction with the SCR process. Both SCR and diesel oxidation catalysts will be discussed in their respective chapters.

High surface area ceria (CeO_2) in combination with varying amounts of other metal oxides such as ZrO_2 has become a very important oxygen storage component in the automotive catalytic converters. Its ability to store and release oxygen rapidly and reversibly allows it to moderate the conversion when the gasoline engine operates near the stoichiometric air to fuel point in the three-way automobile catalytic converter. It was found to be the active component in early diesel oxidation catalysts where the liquid portion of the emissions was catalytically oxidized. It also contributes to the steam reforming reaction during fuel-rich operation. Finally, it serves as a carrier and is added to catalytic components promoting the activity for a variety of reactions including the three-way automotive catalytic converter, a promoter for oxidation reactions, and a variety of catalysts used in the generation of hydrogen. Its redox properties will be more fully described in the automotive three-way catalytic converter chapter, which is its most prominent application.

2.2.5   Carbons

Carbons are commonly used in hydrogenation reactions as carriers for precious metals such as Pd, Pt, Rh, and Ru. Their surface functionality varies considerably with their source. Many carbons are produced from natural products such as shells, trees, peat, and plants as well as coal or synthetic plastics by various heat treatments in the absence of air but often in the presence of other carbon-containing materials to add functionality. Therefore, the surface may be acidic, basic, or contain phenolic groups that can adsorb organic molecules during their conversion to other products. Another important aspect is the lack of reactivity in certain solution media. Their insolubility makes them useful for acidic or basic solutions where traditional metal oxides would dissolve. A final advantage is their combustibility. Once the catalyst is spent, recovery of the metal, especially precious metals, can be easily accomplished by simply combusting the carbon in air leaving a residue rich in the catalytic component.

2.3   INCORPORATING THE ACTIVE MATERIAL INTO THE CARRIER

2.3.1   Impregnation

The most common commercial procedure for dispersing the catalytic species within the carrier is by impregnating an aqueous solution containing a salt (precursor) of the catalytic element or elements. Most preparations simply involve soaking the carrier in the solution and allowing capillary and electrostatic forces to distribute the salt over the internal surface of the porous network. The salt generating the cations or anions, containing the catalytic element, is chosen to be compatible with the surface charge of the carrier to obtain efficient adsorption or, in some cases, ion exchange. For example, $Pt(NH_3)_2^{2+}$ salts can ion exchange with the H^+ present on the hydroxyl-containing surfaces of Al_2O_3. Anions such as $PtCl_4^{2-}$ will be electrostatically attracted to the H^+ sites. The isoelectric point of the carrier (the charge assumed by the carrier surface), which is dependent on pH, is useful in making decisions regarding salts and pH conditions for the preparation.

2.3.2   Incipient Wetness or Capillary Impregnation

The maximum water uptake by the carrier is referred to as the *water pore volume*. This is determined by slowly adding water to a carrier until it is saturated, as evident by the beading of the excess H_2O. The precursor salt is then dissolved in an amount of water equal to the water pore volume. The salt is then dispersed into the pores of the carrier as the water is drawn into the pores. Once dried and calcined (heat treatment in air), the salt decomposes leaving behind the oxidized metal. Since the amount of salt contained in the water is known, and the amount of liquid adsorbed is known, we then know that the carrier is certain to contain a precise amount of catalytic species.

As an example, consider a hypothetical preparation of 3 wt% Rh dispersed into Al_2O_3 where the water pore volume was measured to be 1.30 cm^3 H_2O/g carrier. Thus, for 100 g of carrier, the carrier will absorb 130 cm^3 liquid. To impregnate 3 g of Rh (3 wt%) into 100 g of alumina carrier, the equivalent amount of Rh salt precursor is prepared in water so that the total volume, including the volume associated with the aqueous Rh salt, is 130 cm^3. This diluted Rh salt solution is then mixed into the powdered carrier. Generally, the solution is added slowly while the powder is mechanically mixed to ensure that the liquid is dispersed onto the entire carrier mass.

2.3.3   Electrostatic Adsorption

It is customary to use a precursor salt that generates a charge opposite to that of the carrier, which is determined by the pH at a point of zero charge (PZC). In weakly alkaline solutions, the surface charge on Al_2O_3 or SiO_2 is generally negative (with respect to the PZC), so any cation should preferentially adsorb uniformly over the entire surface. Cations such as Cu^{2+}, Ni^{2+}, Pd^{2+}, Pt^{4+}, Fe^{3+}, and others derived from nitrates or oxalate salts are commonly used, while anions are generated from chloride precursor salts (e.g., $PdCl_2^{2-}$ from Na_2PdCl_4).

2.3.4   Ion Exchange

This method has the advantage of producing a highly dispersed catalytic component within the carrier. Assuming a carrier has a well-defined exchange capacity, a cation salt containing the catalytic species can exchange with the surface carrier cation. Ion exchange is commonly used for zeolite catalysts.

It is common practice to first treat the acid form of the zeolite (H^+Z^- or simply HZ) with an aqueous solution containing NH_4^+ (NH_4NO_3) to form the ammonium-exchanged zeolite (NH_4Z^-). This can then be treated with a salt solution containing a catalytic cation forming the metal-exchanged zeolite (MZ).

$$HZ + NH_4^+ \rightarrow NH_4Z + H^+ \tag{2.4}$$

$$NH_4Z + M^+ \rightarrow MZ + NH_4^+ \tag{2.5}$$

The finished exchanged zeolite is washed, dried, and calcined.

2.3.5 Fixing the Catalytic Species

Following impregnation, it is often desirable to fix the catalytic species so that subsequent processing steps such as washing, drying, and high-temperature calcination will not cause significant movement or agglomeration of the well-dispersed catalytic species.

The pH of the solution is adjusted to precipitate the catalytic species in the pores of the carrier. For example, by presoaking Al_2O_3 in a solution of NH_4OH, the addition of an acidic salt, such as $Cu(NO_3)_2$ or $Pd(NO_3)_2$, will precipitate hydrated oxide of that metal cation on the surfaces of the walls within the carrier. After all the preparation processing steps, the catalyst is treated at high temperature in air to decompose and drive off the anion salts. Sometimes a compound is added to form an insoluble compound with the catalytic component. For example, Equation 2.6 shows fixing of Rh ions as a sulfide. Then final catalyst is calcined in air at about 500–660 °C to remove the sulfur species.

$$2RhCl_3 + 3H_2S \rightarrow Rh_2S_3 + 6HCl \tag{2.6}$$

$$Rh_2S_3 + 3O_2 \rightarrow 2Rh + 3SO_2 \tag{2.7}$$

One can also precipitate the catalyst-containing species forming a water-insoluble compound that is washed to remove the $NaNO_3$ (Equation 2.8). The precipitated species is then decomposed (Equation 2.9). If necessary, reduction to the metallic state using H_2 can be done as required by the application (Equation 2.10).

$$Cu(NO_3)_2 + 2NaOH \rightarrow Cu(OH)_2 + 2NaNO_3 \tag{2.8}$$

$$Cu(OH)_2 \rightarrow CuO + H_2O \tag{2.9}$$

Direct reduction can also be used to reduce and fix the catalytic component (Pd) as shown below.

$$HCOOH + Pd^{2+} \rightarrow Pd + 2H^+ + CO_2 \tag{2.10}$$

This method is particularly effective for the precious metals because they are easily reduced to their metallic states. The advantage of the reducing agents mentioned above is that, upon subsequent heat treatment, they leave no residue.

2.3.6 Drying and Calcination

All of the dispersion methods discussed in the prior sections involved aqueous-based processes. A catalyst containing salt (referred to as the precursor salt) is dissolved in water and impregnated into the carrier such as in a simple wet impregnation. The excess water and some volatile species are removed by air drying at about 110 °C yielding a dried powder. After drying, it is calcined in air to about 400–600 °C. During calcining at these elevated temperatures, the metal salts are decomposed producing a metal oxide on the surface. The optimal temperature for air calcining depends highly on the chemistry of the catalytic species on the surface and will vary with the metal and type of salt precursor used. The metal oxide formed on the surface is now insoluble and immobilized.

2.4 FORMING THE FINAL SHAPE OF THE CATALYST

The nature of the industrial process will dictate the shape and size of the catalyst material. When fixed bed reactors are used, pressure drop will be a concern and relatively large diameter catalyst particles are desired. When the reaction is to be conducted in a fixed bed reactor and is limited by pore diffusion, we will want catalyst shapes that minimize the path length for diffusion, such as structures with "holes" integrated into the geometry or small particles. For reactions controlled by bulk mass transfer, particles with a high geometric surface area, such as a large number of small particles or carriers with irregular external shapes (i.e., triangular or star shaped), are ideal to induce turbulent flow around the particles and maximize the geometric surface area available for mass transfer. Similarly, when the reaction is limited by heat transfer, the catalyst shape and composition will be designed to minimize the resistance for heat transfer (i.e., metal support or some other high heat transfer material). When the reaction is conducted in the liquid phase (slurry phases) or a fluidized bed, very fine (tens of micrometers in cross section) catalyst powders are used where the size of the powder is optimized based on the fluid dynamics of the fluid bed or slurry reactor. Some typical particulate catalysts were shown in Figure 1.6.

Once the optimal catalyst geometry is determined, there are a variety of different solid processing techniques available to form the optimal shape. With the optimal shape for each chemistry process being different, a wide variety of different shapes and structures have been used in industrial processes, some of which are illustrated in Figure 2.3. The following sections will describe the common methods used to form catalyst shapes.

A monolith design has become the state of the art for environmental applications mostly notably for stationary and automotive emission control. A variety of monoliths are shown in Figure 2.3.

2.4.1 Powders

Here we consider the catalytic component dispersed on a suitable carrier material in a powder for a reaction carried out in a slurry phase or fluidized bed. For these reactors, the catalyst must be suspended in a liquid circulating in a stirred tank reactor (slurry phase) or fluidized in a fluidized bed reactor. In these applications, powders comprised of particles with a diameter of less than 100 μm are desired. Powders

Figure 2.3   Ceramic and metallic (center image) monoliths of different shapes and cell geometries. (Reproduced with permission from BASF.)

of this particle size are generally formed through one of the two possible routes: milling and sieving, or spray drying.

2.4.1.1 *Milling and Sieving*   For milling, larger particles or an agglomerated calcined powder is reduced in size using a variety of available techniques. The optimal technique is a function of the process involved and the size of the desired product. When milling particles, two variables are of interest: mean particle size and the distribution of particle size. The dynamics of milling do not allow precise control over the milled size. Instead, a distribution of sizes will result where there will exist a mean size. The milling technique used will affect both the minimum mean size and the width of the particle size distribution.

As a slurry, particles can be milled using a media mill or ball mill. This type of mill is packed with very hard spherical or cylindrical media (tens of millimeters in cross section), and then filled with slurry containing the particles to be milled. Mechanical motion is induced causing the media to impact one another, catching the much smaller catalyst particles in between the impacting media resulting in particle breakage and size reduction. To an extent, the size of the final powder is a function of the residence time in the mill.

For dried particulates, a variety of milling techniques are available. For larger particulate or agglomerates, hammer mills are very effective to reduce particle size. In general, hammer mills consist of multiple metal fixtures (hammers) rotating at high speed that impact a particle. In one design, the metal "hammers" rotate vertically within a housing where material is fed in through the top. On the bottom of the milling housing is a screen. The size of the screen dictates the size of the material that can "escape" the milling chamber, where finer screen sizes yield finer products. The mill is designed to allow the screens to be easily replaced, so one mill can be used to create a variety of different particle sizes. The screen acts as an internal mechanical "classifier" that controls particle size. An alternative to the mechanically classified mill is a design that integrates an air classifier into the hammer mill. This mill design uses hammers mounted onto a horizontal spinning disk. The material to be milled is fed from the top and dropped into the spinning hammers. The mill is designed to allow air to be drawn from the bottom up around the spinning hammers into a separate device integrated into the mill design called an "air classifier." The air classifier controls the size of the particle allowed to escape the milling chamber, similar to the mechanical screen. Rejected particles are directed back to the hammers for additional milling. The advantage of the air-classified hammer mill is tighter control over particle size and particle size distribution.

For fine particle grinding of dry powders (<10 μm), "jet" milling is generally preferred. While multiple designs exist, the general concept involves using opposing jets of compressed gas to create the energy required for breaking particles. As an example of one design, compressed gas is introduced at high velocity through nozzles at four equidistant points around a circular milling chamber. The jets of compressed gas are directed to converge at a common point in the center of the circular milling chamber. The material to be milled is fed above the jets and allowed to fall down into the converging jets. Particles are entrained in the high-velocity jets and impact other particles where the compressed jets converge. Particles break due to particle-to-particle collisions. The use of the high-velocity jets allows greater energy for milling compared with hammer milling and therefore smaller particle sizes are achievable. However, generally the feed to the jet mill needs to be of a relatively small particle size, compared with a hammer mill that can accept relatively large particles. Often, fine powder sizing is achieved through a combination of hammer milling followed by milling in a jet mill.

Often milling results in a particle size distribution that is unacceptably wide and additional particle classification is required. This is generally done using either mechanical (sieving) or air classification. Mechanical sieving simply involves passing the material over screens where particles smaller than the screen opening fall through and larger particles do not. For particles larger than 100 μm, this method is ideal. However, for particle sizes much smaller than 100 mm, air classification offers an advantage albeit at a cost of increased system complexity and associated cost. Many different designs for air classifiers are available; however, they are based on the same general operating principle. In one common design, a vanned wheel rotates at high speed inside a housing. Air is drawn through the housing and through the rotating wheel. The material to be classified is fed into the housing and entrained into the air flowing toward the rotating wheel. The rotating wheel creates rotating air currents with velocities on the same magnitude as the rotating vanes, yet air is still being drawn through the wheel. As particles enter this region, a centrifugal force acts on the particles. The size of the particle dictates whether it is thrown outward or passes through the rotating wheel. This method is more suitable for classifying fine powders.

2.4.1.2 Spray Drying Spray drying is a method for producing dry powders from a liquid slurry by rapid drying with a hot gas. One advantage of spray drying is the ability to produce very uniformly sized spherical dried aggregates, which is the primary reason this method is used for catalyst production. In this method, the desired solids are dispersed into a liquid to form a slurry. The drying medium is typically air, unless a flammable solvent is used where an inert medium such as nitrogen is required. The air is heated and flows through a drying chamber. The slurry is atomized into fine droplets inside the drying chamber using a nozzle or high-speed rotating wheel. The droplets rapidly dry leaving behind a solid spherical aggregate of solid powder initially dispersed into the liquid. Spray drying is a continuous process and the dried solids are then conveyed in the drying medium out of the drying chamber. Typically, a cyclone is placed downstream of the drying chamber to separate the solids from the flowing gas. Spray drying can also be used to disperse metal salts onto a solid by dissolving the metal salt into the slurry being fed to the drying chamber. During the drying process, the volatile components of the slurry will evaporate leaving behind the solids and dissolved salts.

2.4.2   Pellets, Pills, and Rings

Small shapes on the order of millimeters in cross section can be formed into the desired geometry by pressing the dry finished catalyst powder in a die at pressures ranging between 100 and 4000 atm. While a cylinder is the common geometry formed, a wide variety of geometrical shapes as well as formation of channels through the pellet is possible using sophisticated die designs. The pelleting process occurs with a machine usually containing multiple dies and has the ability to produce pellets at a high rate. The pelleting process and the equipment are very similar to those used in the pharmaceutical industry to produce pills. The ability of the material to form strong pellets depends on the properties of the catalyst powder including its tensile strength, mesoporosity, and moisture content. For materials that do not naturally form strong pellets, additional materials to act as a binder must be added. The advantage of pelleting over extrusion, which is the topic of the next section, is the ability to produce precisely shaped, high-strength particles.

2.4.3   Extrudates

Extrusion is another process often used to form catalyst particles. In this process, the catalyst powder is mixed with a liquid, most often water, to form an "extrudable" paste. Sometimes for powders with poor rheology, it is necessary to add lubricating aids such as alcohols to decrease resistance to extrusion through the die. The consistency of the paste must be such that it is fluid enough to move through the extruder, yet "dry" enough to allow the extrude to maintain its shape once compressed through the extrusion die. In these systems, the extrusion paste is contained in a hopper and fed into a screw drive, which forces the material through a die. As the material moves through the die, the material is compressed into the desired shape. As the extrudate comes out of the die, a knife is used to cut the shapes off the die at the desired length. The shape of the die determines the cross-sectional shape of the extrudate. A wide variety of shapes are possible including circles, rings, ovals, lobed circles, or stars. The extrudates are then dried to form the final product. Often extrusion is performed before the calcination step described above to take advantage of the good binding properties of the uncalcined catalyst. In this case, the extrudates would then be calcined to form the desired product. Compared with the pelleter described in the previous section, extrudates will tend to be weaker and have higher porosity and less regular shapes; however, they are often less expensive to produce.

2.4.4 Granules

Granules are formed by milling/sieving larger catalyst particles formed from some other process or formed directly as the result of the synthesis process, such as the case for the iron-based ammonia catalyst (Section 7.1). For ammonia synthesis, the iron catalyst is prepared by pouring molten iron ore into a cold bath to produce a particle "shot." While cost effective, the irregular shaped granules have the potential to pack tightly in the reactor, reducing bed porosity and increasing pressure drop. Higher pressure drop is precisely the issue with the iron-based ammonia catalyst granules and is the reason why ammonia reactors are designed to provide radial flow in order to decrease pressure drop through the densely packed granules.

2.4.5 Monoliths

The catalyst support for environmental applications is almost exclusively ceramic or metal monolithic structures with the catalyst washcoated onto the walls of the channels. They have varying numbers of parallel channels upon which a thin catalyst/carrier (called a washcoat) is fixed. The gases diffuse into the channels and diffuse into the washcoat reacting with the catalytic sites. Figure 2.4 shows a double-coated washcoated ceramic monolith.

The advantages of monoliths will be discussed in more detail in the environmental chapters but relative to particulates they offer excellent mechanical integrity in applications where attrition is a problem, such as in catalytic converters for mobile

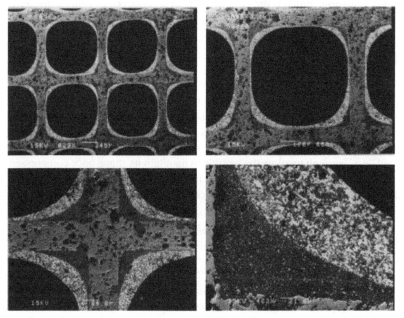

Figure 2.4 Ceramic washcoated monoliths. (Reproduced from Chapter 2 of Heck, R.M., Farrauto, R.J., and Gulati, S.T. (2009) *Catalytic Air Pollution Control: Commercial Technology*, 3rd edn, John Wiley & Sons, Inc., New York.)

devices such as cars and trucks. An important feature of ceramic cordierite ($2MgO–2Al_2O_3–5SiO_2$) as the monolith composition is its low expansion during temperature gradients during the driving cycle. They also offer low pressure drop since they are open-channel structures (70–90% open frontal area) with thin catalyzed washcoats (<200 μm) offering little or no pore diffusion resistance. Bulk mass transfer is an issue since the parallel channel monoliths support laminar flow. Reactor designs and new monoliths have addressed the fluid dynamics.

Washcoats are prepared from aqueous slurries (30–40% solids) containing the catalyst/carrier powder. The catalyzed powder is milled to <10 μm to be compatible with the porosity of the ceramic monolith walls. The monolith is dipped into the slurry where it coats the walls. The channels are then blown free of excess slurry with an air purge, dried, and calcined to establish an adherent bond. For metal monoliths, the procedure is the same but usually the metal surface must be thermally or chemically treated to generate a roughened surface. Often an aluminum-containing alloy is used, which when treated in air at >800 °C generates a roughened "skin" of Al_2O_3 that permits greater bonding between the metal wall and the washcoat. This step is necessary to ensure good adhesion because the metal expands to a greater extent than the washcoat during thermal transients in a vehicle.

2.5 CATALYST PHYSICAL STRUCTURE AND ITS RELATIONSHIP TO PERFORMANCE

It is critical to understand the rate-limiting steps in any process so that the catalyst and reactor design can be optimized. For a reaction known to be limited by chemical kinetics, it is necessary to have the proper loading (concentration) of catalytic components on the carrier since maximizing the number of sites will enhance the reaction kinetics. For reactions controlled by bulk mass transfer, the geometric surface area and the particle size must be optimized for the reaction. For the reactions limited by pore diffusion, the structure should be sufficiently small (short diffusion path) with large pores to enhance transport through the porous network of the carrier to the active sites. Of course, the reactor and process are designed with all of these factors in mind.

Throughout the application section, the key parameters for catalyst composition and structure will be discussed in the context of the application. This approach will demonstrate the design aspects of the catalyst and the reactor and process conditions.

2.6 NOMENCLATURE FOR DISPERSED CATALYSTS

The combination of the catalytic species supported on a carrier is presented by stating the amount and specific catalytic material, followed by the name of the carrier separated by a slash, that is, 0.5% Pt/SiO_2, 1% Pd/Al_2O_3, 10% Ru/C, 3% V_2O_5/TiO_2, 20% Ni/Al_2O_3, and so on. The catalytic component is always shown as weight percent. It must be clearly understood that this only describes the general composition of the catalyst and does not describe the nature or chemistry of the active sites responsible for the particular catalytic reaction. These are often not known in real processes.

QUESTIONS

1. Why is high surface area an important characteristic of a good catalyst support?

2. What does it mean for the active material to be "poorly dispersed" and why is this undesirable?

3. From the scientific literature, find an article providing a detailed description of a heterogeneous catalyst. Summarize the chemical composition, important phases, structures, and properties.

4. Why is alumina a better high-temperature support ($>500\,°C$) than titania?

5. What is shape selectivity? What feature of zeolite structures explains shape selectivity?

6. Using the Internet or any suitable reference, find the following zeolites and their physical and chemical properties such as aperture size, Si/Al ratio, degree of acidity (related to Al), and number of tetrahedral sites per ring. What is the significance of some of these properties?

   a. Faujasite (FAU).

   b. Mordenite (MOR).

   c. ZSM-5 (MFI).

7. Compare ceramic and metal monoliths as supports for catalyzed washcoats. When would you use each? Compare porous particulates and monoliths. State the advantages and disadvantages.

8. What are the performance consequences of changing the following monolith properties?

   a. Smaller channel diameter.

   b. Thinner monolith wall thickness.

   c. Decreasing cells/in.^{2} with increased channel diameter.

9. For fixed bed reactors, what is the primary advantage of larger formed particulate shapes over granular particulate catalyst?

10. Compare and contrast pelletizing versus extrusion for forming particulate shapes.

11. From the scientific literature, find an article providing a detailed description of the synthesis of a catalyst. Summarize the process used and the reason for each step.

BIBLIOGRAPHY

Auerbach, S.M., Carrado, K.A., and Dutta, P.K. (eds) (2003) *Handbook of Zeolite Science and Technology*, Dekker, New York.

Bartholomew, C. and Farrauto, R.J. (2006) *Fundamentals of Industrial Catalytic Processes*, 2nd edn, John Wiley & Sons, Inc., New York.

Beguin, B., Garbowski, E., and Primet, M. (1991) Stabilization of alumina toward thermal sintering by silicon addition. *Journal of Catalysis* 127, 595–604.

Emitec Product Literature (1992) *The new generation of metallic catalytic converter substrates*. Lohmar, Germany.

Harris, G. (1993) A review of precious metal refining, in *Precious Metals 1993: 17th International Precious Metals Conference, Newport, RI* (ed. R. Mishra), pp. 351–374.

Johnson, M.J. (1990) Surface area stability of aluminas. *Journal of Catalysis* 123, 245–259.

Kato, A., Yamashita, H., Kawagoshi, H., and Matsuda, S. (1987) Preparation of lanthanum beta alumina with high surface area by co-precipitation. *Communications of the American Ceramic Society* 70 (7), C157–C161.

Kolb, W.B., Papadimitriou, A.A., Cerro, R., Leavitt, D.D., and Summers, J. (1993) The ins and outs of coating monolithic structures. *Journal of Chemical Engineering Progress* 89 (2), 61–67.

Komiyama, M. (1985) Design and preparation of impregnated catalysts. *Catalysis Reviews: Science and Engineering* 27 (2), 342–372.

Lachman, I. and Williams, J. (1992) Extruded monolithic catalyst supports. *Catalysis Today* 14, 317–329.

LePage, J.F. (1997) Preparation of solid catalysts, in *Handbook of Heterogeneous Catalysis*, Vol. 1 (eds G. Ertl, H. Knozinger, and J. Weitkamp), VCH, Weinheim, Germany, pp. 49–72.

Machida, M., Eguchi, K., and Arai, H. (1988) Preparation and characterization of large surface area $BaO \cdot 6Al_2O_3$. *Bulletin of the Chemical Society of Japan* 61, 3659–3665.

Mishra, R. (1993) A review of platinum group metals recovery from automobile catalytic converters, in *Precious Metals 1993: 17th International Precious Metal Conference, Newport, RI* (ed. R. Mishra), pp. 449–474.

Park, J. and Regalbutto, J. (1995) A simple accurate determination of oxide PZC and the strong buffering effect of oxide surfaces at incipient wetness. *Journal of Colloid and Interfacial Science* 175, 239–252.

Regalbutto, J. (ed.) (2006) *Catalyst Preparation*, Taylor & Francis, New York.

Schuth, F. and Unger, K. (1997) Precipitation and co-precipitation, in *Handbook of Heterogeneous Catalysis*, Vol. 1 (eds G. Ertl, H. Knozinger, and J. Weitkamp), VCH, Weinheim, Germany, pp. 72–86.

Stiles, A. (1983) *Catalyst Manufacture: Laboratory and Commercial Preparations*, Dekker, New York.

Thomas, A. and Brundrett, C. (1980) Catalyst development: lab to commercial scale. *Chemical Engineering Progress* 76 (6), 41–45.

Trimm, D. (1980) *Design of Industrial Catalysts*, Elsevier, Amsterdam.

Wan, C. and Dettling, J. (1986) *High temperature catalyst and compositions for internal combustion engines*. U.S. Patent 4,624,940.

Wefers, K. and Misra, C. (1987) *Oxides and Hydroxides of Aluminum*, Alcoa Laboratories, East Saint Louis, IL.

Worstell, J.H. (1992) Succeed at catalyst upgrading. *Chemical Engineering Progress* 88 (6), 33–39.

CHAPTER 3

CATALYST CHARACTERIZATION

3.1 INTRODUCTION

The characterization of a heterogeneous catalyst is the quantitative measure of its physical and chemical properties assumed to be responsible for its performance in a given reaction. These measurements have value in the preparation and optimization of a catalyst, and even more importantly in elucidating mechanisms of deactivation and subsequent catalyst design to minimize such deactivation. Physical properties such as pore size, surface area, and morphology of the carrier, as well as the geometry and strength of the catalyst support, must be well defined for the given end-use application. Similarly, determining the composition, structure, and nature of the carrier and the active catalytic components and their changes during the catalysis process is a critical goal in characterization.

The importance of the physical and chemical properties can be appreciated when we restate the fundamental steps involved in a heterogeneous catalytic process. (These were stated in more detail in Chapter 1.) The example used refers to any supported catalyst material.

1. *Bulk mass transport of the reactants from the bulk fluid to the outside surface of the catalyst.* This is strongly influenced by the geometric surface area of the catalyst, its shape, and the flow turbulence.

2. *Diffusion or transport of reactant(s) to active sites through the pore structure of the catalyst.* This depends on the pore size and diameter of the catalyst. A well-optimized catalyst will have a sufficiently large pore size and a small diameter (small diffusion path) to permit easy access of reactant and product molecules to and from the active sites, respectively.

3. *Chemisorption of reactants(s) onto the catalytic active sites.* This is dependent on the nature of the active site and the chemistry of the adsorbing molecule(s).

4. *Chemical conversion of the chemisorbed species to products.* This step depends on the number and nature of the activated complexes formed by the chemisorbed molecule and the catalytic site.

5. *Desorption of products(s) from active sites.* This step depends on the number of adsorbed species and the strength of their bonds to the active surface.

Introduction to Catalysis and Industrial Catalytic Processes, First Edition. Robert J. Farrauto, Lucas Dorazio, and C.H. Bartholomew.
© 2016 John Wiley & Sons, Inc. Published 2016 by John Wiley & Sons, Inc.

6. *Diffusion or transport of product(s) through the pore structure.* This step is affected mostly by the size and shape of the diffusing molecule and the pore and small diffusion path of the particulate catalyst.

7. *Bulk diffusion of the products from the outside surface of the particulate to the bulk fluid.* The process is influenced by the particulate size and shape and the turbulence of the flow.

During the conversion process, a number of physical and chemical properties have to be well defined to produce an optimized catalytic system. The most important properties are discussed below.

3.2 PHYSICAL PROPERTIES OF CATALYSTS

3.2.1 Surface Area and Pore Size

Surface area, pore size, pore size distribution, pore structure, and pore volume of the carrier are among the most fundamentally important properties in catalysis because the active sites are present or dispersed throughout the internal surface through which reactants and products are transported. The size and number of pores determine the internal surface area. It is usually advantageous to have high surface area (large number of small pores) to maximize the dispersion of catalytic components. However, if the pore size is too small, diffusional resistance becomes a problem.

3.2.1.1 Nitrogen Porosimetry A standardized procedure for determining the internal surface area of a porous material with surface areas greater than 1 or 2 m^2/g is based on the adsorption of N_2 at liquid N_2 temperature onto the internal surfaces of the carrier.

Each adsorbed molecule occupies an area of the surface comparable to its cross-sectional area (16.2 Å^2). By measuring the number of N_2 molecules adsorbed at monolayer coverage, one can calculate the internal surface area. This plot is shown in Figure 3.1a. The adsorption of N_2 first rapidly rises with pressure and then flattens in the general region where monolayer coverage is occurring. At high relative N_2 partial pressures, coverage beyond a monolayer occurs, as does condensation of liquid N_2 in the pores, giving rise to the large increase in volume adsorbed. The Brunauer–Emmett–Teller (BET) equation (Equation 3.1) describes the relationship between volume adsorbed at a given partial pressure and the volume adsorbed at monolayer coverage:

$$\frac{P}{V(P_o - P)} = \frac{1}{V_m C} + \frac{P(C-1)}{CP_o V_m} \tag{3.1}$$

This equation can be stated in the linear form $Y = sx + b$, where P/P_o is x and the term on the left-hand side is Y as plotted in Figure 3.1b. The intercept b is equal to $1/V_m C$, and the slope s is $(C-1)/V_m C$. The most reliable results are obtained at relative pressures (P/P_o) between 0.05 and 0.3.

The same equipment can be used to determine the pore size distribution of porous materials with diameters less than 100 Å, except that high relative pressures

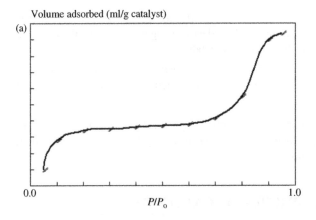

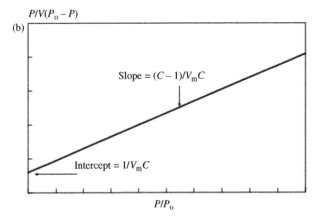

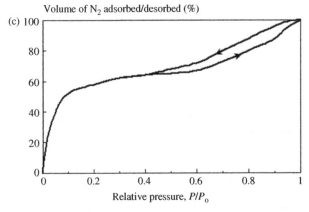

Figure 3.1   (a) Adsorption isotherm for nitrogen for BET surface area measurement.
(b) Linear plot of the BET equation for surface area measurement. (c) Nitrogen adsorption/
desorption isotherm for pore size measurement. (Reproduced from Chapter 3 of Heck,
R.M., Farrauto, R.J., and Gulati, S.T. (2009) *Catalytic Air Pollution Control: Commercial
Technology*, 3rd edn, John Wiley & Sons, Inc., New York.)

are used to condense N_2 in the catalyst pores. The procedure involves measuring the volume adsorbed in either the ascending or descending branch of the BET plot at relative pressures close to 1.0. Figure 3.1c shows the volume of N_2 adsorbed and desorbed as the pressure is increased and decreased, respectively. Capillary condensation occurs in the pores in accordance with the Kelvin equation (Equation 3.2).

$$\ln\left(\frac{P}{P_o}\right) = \frac{-2\sigma V_m \cos\theta}{r_p RT} \tag{3.2}$$

Hysteresis between the adsorption–desorption isotherms in Figure 3.1c at relative pressures of 0.6–0.9 is observed with carriers having a significant volume of mesopores (diameters between 20 and 500 Å). It results because pores that fill at a given pressure during adsorption require a lower pressure to empty during desorption. The form of the Kelvin equation given above describes the desorption isotherm, and it is the preferred one for calculations of the pore size distribution.

3.2.1.2 Pore Size by Mercury Intrusion
For materials with pore diameters greater than about 30 Å, the mercury intrusion method is preferred (ASTM D4284-83, 1988). The penetration of mercury into the pores of a material is a function of applied pressure. At low pressures, mercury penetrates the large pores, whereas at higher pressures the smaller pores are progressively filled. Due to the non-wetting nature of mercury on oxide carriers, penetration is met with resistance. The Washburn equation (see Equation 3.3) relates the pore diameter, d_p, with the applied pressure, P.

$$d_p = \frac{-4\gamma \cos\theta_c}{P} \tag{3.3}$$

The wetting or contact angle θ_c between mercury and the solid is usually 130°, and the surface tension of the mercury (γ) is 0.48 N/m. Pressure is expressed in atmospheres and d in nanometers (10 Å). This technique is satisfactory for pores down to 30 Å diameter; however, this is a function of the instrument capability. Maximum diameters measured are usually 10^6 Å. Typical pore size distribution data are shown in Figure 3.2, where the integral penetration of mercury into the pores is plotted as a function of applied pressure. The differential curve (Figure 3.3) clearly shows a bimodal pore distribution with mean pore diameters at 10,000 and 1000 Å. A nitrogen desorption isotherm is required to obtain an accurate measure in the region below 100 Å (see Section 3.2.2).

3.2.2 Particle Size Distribution of Particulate Catalyst

3.2.2.1 Particle Size Distribution
The size of the powdered particles that make up the carrier must be compatible with the reaction conditions for a slurry phase or fluid bed process. The carrier powders are ball milled to produce the desired particle size consistent with the applications that include separation and recovery of the catalyst from the product for its reuse.

Sieves of various mesh sizes have been standardized; thus, one can determine particle size ranges by noting the percentage of material, usually based on weight, that

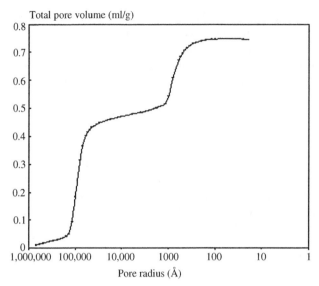

Figure 3.2   Mercury penetration as a function of pore size of catalyst. (Reproduced from Chapter 3 of Heck, R.M., Farrauto, R.J., and Gulati, S.T. (2009) *Catalytic Air Pollution Control: Commercial Technology*, 3rd edn, John Wiley & Sons, Inc., New York.)

passes through one mesh size but is retained on the next finer screen. Sieves are stacked with the coarsest on top and the underlying screens progressively finer. A precise weight of catalyst material is added to the top screen. The stack of sieves is vibrated, allowing the finer particles to pass through coarser screens until retained by those screens finer in opening than the particle size of the material of interest. Each fraction is then weighed, and a distribution determined.

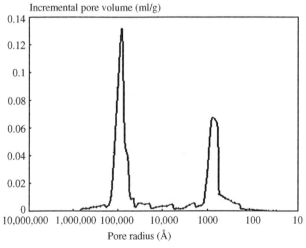

Figure 3.3   Differential porosimetry for a porous catalyst. (Reproduced from Chapter 3 of Heck, R.M., Farrauto, R.J., and Gulati, S.T. (2009) *Catalytic Air Pollution Control: Commercial Technology*, 3rd edn, John Wiley & Sons, Inc., New York.)

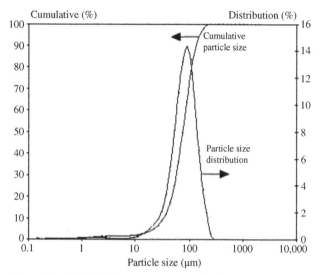

Figure 3.4   Particle size measurement using laser light scattering analysis. (Reproduced from Chapter 3 of Heck, R.M., Farrauto, R.J., and Gulati, S.T. (2009) *Catalytic Air Pollution Control: Commercial Technology*, 3rd edn, John Wiley & Sons, Inc., New York.)

This method is reliable only for particles larger than about 40 μm. Below this, sieving is slow and charging effects influence measured values. Sophisticated instrumentation is available for measuring the distribution of finer particles. Methods include electronic counting, light scattering, image analysis, sedimentation, centrifugation, and volume exclusion. In this method, the dried powder from the slurry is suspended in an electrolyte and pumped through a tube containing a small orifice. An electric current passes through this tube, and as individual particles pass through the orifice, a fraction of the current is interrupted. This fractional change in current is proportional to the particle size. The magnitude of the change in current flow is subdivided over the range of sizes limited by the size of the orifice. The particle diameters are calculated on the equivalent sphere of the excluded volume and, assuming constant density for the particles, the results are commonly recorded as weight percentage as a function of incremental particle size.

More recently, laser techniques have been found effective. A He–Ne laser beam is passed through an aqueous suspension of particles and it is diffracted in proportion to the radius of the particle. A distribution for an Al_2O_3 carrier is illustrated in Figure 3.4. The values on the left axis indicate the cumulative percentage of particles the size of which is displayed on the *x*-axis. The right axis shows the percentage distribution for a given particle size.

3.2.2.2 *Mechanical Strength*

Strength measurement for catalyst materials is quite simple. A single unit representative of the lot is placed between parallel plates of a device capable of exerting compressive stress, and the force necessary to crush the material is noted. The particulate (or monolith) can be placed within the plates so as to measure axial or radial crush strength. A sufficient number of units must be tested to obtain proper statistics.

The resistance to attrition is determined by vigorously shaking the particulates together simulating their interaction in a slurry phase or fluid bed reactor. Passing high-velocity liquid or air through the catalyst bed and collecting the attrited fragments is a reasonable measure of resistance to mechanical losses.

3.2.3 Physical Properties of Environmental Washcoated Monolith Catalysts

3.2.3.1 Washcoat Thickness Optical microscopy is the method used most frequently to obtain washcoat thickness directly. A portion of washcoated monolith is mounted in epoxy and sliced to obtain a cross section. The contrast between washcoat and monolith is sufficient to permit optical thickness measurements. A typical cross section of a washcoat on a ceramic auto exhaust monolith is shown in Figure 2.4.

3.2.3.2 Washcoat Adhesion Washcoated monolithic catalysts are subjected to high gaseous flow rates and rapid temperature fluctuations. As a result, adhesion loss is a likely cause for concern. The most common technique is to subject the washcoated monolith to a jet of gas that simulates the linear velocities anticipated in use and to note the weight change of the catalyst due to erosion. One can also collect the attrited particles and correlate their weight with the weight of washcoat initially bound to the monolith.

Monolithic materials are frequently subjected to substantial thermal gradients (thermal shocks) in start-up and shutdown, as in the exhaust of any combustion process. Frequent thermal shocks cause the washcoat to delaminate due to the expansion difference between it and the monolith. This is most pronounced when the monolith is metallic because of its greater thermal expansion compared with an oxide washcoat.

One can evaluate this phenomenon by cycling the catalyzed monolith between temperature extremes anticipated in service and noting weight losses. Monoliths can be mounted on a rotating "carousel" that moves in and out of streams of heated gas.

3.3 CHEMICAL AND PHYSICAL MORPHOLOGY STRUCTURES OF CATALYTIC MATERIALS

3.3.1 Elemental Analysis

The proper combination of chemical elements is essential in catalysts for optimum performance. More often than not, small amounts of promoter oxides intentionally added (often less than 0.1%) can influence activity, selectivity, and life.

Catalyst suppliers must meet many manufacturing specifications that apply to the proper chemical analysis of a catalyst. Carriers are derived from raw materials, which contain various impurities. Some can be detrimental to catalytic performance and therefore must be removed. Impurities such as alkali and alkaline earth compounds, if used in excess, act as fluxes, causing sintering or loss of surface area of carriers such as Al_2O_3. When added in the proper amount, the same impurities can enhance stability against sintering or, in some cases, improve selectivity. Hence, it is not just the presence of impurities, but the manner in which they have been introduced that is important. Obviously, one has greater control when starting with a relatively pure material to which predetermined amounts of promoters can be added.

The quantitative procedures used to analyze catalysts are no different from those for any other chemical material. Special procedures are often needed to dissolve catalysts in preparation for analysis—particularly refractory materials such as certain noble metals and ceramics.

3.3.2 Thermal Gravimetric Analysis and Differential Thermal Analysis

Thermal gravimetric analysis (TGA) measures the weight change of a material upon heating in a highly sensitive balance. The energy change (heat adsorbed or liberated) associated with any transformation is measured with the differential thermal analysis (DTA) mode. It is common practice for both to be measured simultaneously in a combined TGA/DTA unit. A few milligrams of catalyst are loaded into a quartz pan suspended in the microbalance. A controlled gas flow and temperature ramp is initiated and a profile of weight and energy change versus temperature is recorded. Frequently, TGA units are equipped with a mass spectrometer, so the off-gases from the catalyst can be measured as a function of temperature.

The following is just one example of how TGA/DTA can be used to study catalyst chemistry. During catalyst synthesis, aqueous solutions of catalyst precursor salts are impregnated into the porous carrier. After deposition onto the carrier, the combination is calcined in air to a temperature sufficient to decompose the salt.

$$Ba(CH_3COO)_2 + 2O_2 \rightarrow BaO + 2CO_2 + 3H_2O$$

TGA/DTA can be used to study this decomposition process, as shown in Figure 3.5, where barium acetate is decomposed after it has been impregnated into high surface area carrier such as cerium oxide.

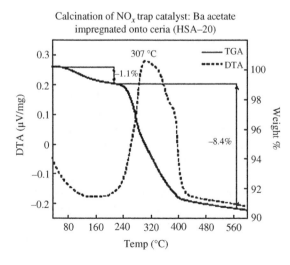

Calcination of NO$_x$ trap catalyst: Ba acetate impregnated onto ceria (HSA–20)

Figure 3.5   Thermal gravimetric analysis and differential thermal analysis of the decomposition of barium acetate on ceria. (Reproduced from Chapter 3 of Heck, R.M., Farrauto, R.J., and Gulati, S.T. (2009) *Catalytic Air Pollution Control: Commercial Technology*, 3rd edn, John Wiley & Sons, Inc., New York.)

The solid line shows the weight change associated with heating the ceria after impregnation with the acetate salt of barium. A small weight loss (measured on the right axis) is observed between 80 and 240 °C due to the evaporation of H_2O. The dotted line (measured on the left axis) shows the DTA profile with a negative slope consistent with the endothermic event of water loss. Decomposition of the acetate begins around 250 °C where the weight change (solid line) is seen to sharply decrease until about 400 °C where it plateaus. Associated with this exothermic decomposition event (positive slope) is the heat liberated in the DTA profile showing a maximum at 307 °C. The ceria was preconditioned to be stable during this procedure. This simple test allows manufacturing personnel to establish the optimized conditions for drying and calcining the catalyst during its large-scale production. This particular material, BaO/CeO_2, is being used for adsorbing or trapping NO_x in a lean burn engine exhaust such as a diesel.

This technique is also used to establish the temperature and environmental conditions under which a metal will oxidize (temperature-programmed oxidation (TPO)) or a metal oxide is reduced in environment such as H_2 or CO (temperature-programmed reduction (TPR)).

In Chapter 5, other examples of how these useful tools can be used to understand catalyst deactivation and regeneration modes will be described.

3.3.3 The Morphology of Catalytic Materials by Scanning Electron Microscopy

In Chapter 2, the reader was introduced to a variety of carrier materials upon which the catalytic components are dispersed to maximize the number of sites available for reactants to chemisorb. Clearly, $\gamma\text{-}Al_2O_3$ is the most common material for industrial application. It has a surface area of over 150–200 m^2/g with a highly porous structure as shown in the scanning electron micrograph (SEM) of Figure 2.1a. Its pores range from about 20 to 200 Å (2–20 nm). As it experiences elevated temperatures, it slowly transforms to other lower porosity and more crystalline intermediary structures terminating with $\alpha\text{-}Al_2O_3$, the lowest surface area and the most crystalline structure as shown in Figure 2.1b. The structural transformations of $\gamma\text{-}Al_2O_3$ will be further discussed in Chapter 5 because they represent one of the most commonly occurring deactivation modes for supported catalysts.

Catalysts contain a variety of catalytic metals and metal oxide promoters in addition to the Al_2O_3 carrier. Each has a specific function and must have its proper location within the final catalyst particle. Some components must be in intimate contact with each other to contribute to electronic promoting effects but could poison other components and cause deactivation. Such is the case with certain metals that form undesirable alloys or compounds. In the scanning mode, the electron beam focused on the sample is scanned by a set of deflection coils. Backscattered electrons or secondary electrons emitted from the sample are detected. As the electron beam passes over the surface of the sample, variations in composition and topology produce variations in the intensity of the secondary electrons. The raster of the electron beam is synchronized with that of a cathode ray tube, and the detected signal then produces an image on the tube. Spot or area analysis is also possible when an electron microscope

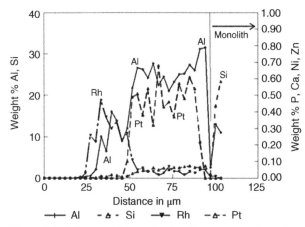

Figure 3.6   Electron microprobe showing a two-washcoat-layer monolith catalyst. The top layer is Rh on Al_2O_3 and the bottom layer is Pt on Al_2O_3. (Reproduced from Chapter 3 of Heck, R.M., Farrauto, R.J., and Gulati, S.T. (2009) *Catalytic Air Pollution Control: Commercial Technology*, 3rd edn, John Wiley & Sons, Inc., New York.)

is equipped with an energy dispersive analyzer (EDX) or wavelength dispersive analyzer (WDS). The bombardment of a sample with electrons generates X-rays characteristic of the elements present. Thus, the EDX can determine the composition of any portion of the sample, while the WDS permits the mapping of the location of species present. This is particularly important when foreign matter is present either from a contamination problem in manufacturing or by poisoning during a catalytic process. The electron microprobe is another form of electron microscopy that is extremely important for metal location studies requiring high resolution. It is similar to the scanning electron microscope; however, its primary function is to detect characteristic X-rays produced by the electron beam interaction with the specimen. The X-ray emissions can be used to determine the elemental composition of the specimen quantitatively, and also to detect the location of a particular element within the morphology or topological structure of the specimen.

Figure 3.6 shows a microprobe line profile of a ceramic monolith upon which has been deposited two different washcoat layers. The top layer from roughly 50 to 100 μm contains Rh on Al_2O_3 while the second or bottom layer from 50 to 100 μm contains Pt on Al_2O_3. The total thickness of the two layers is 100 μm as evidenced by the decrease and the absence of Pt at depths greater than 100 μm and the appearance of Si, which is a component of the monolith substance.

3.3.4   Structural Analysis by X-Ray Diffraction

Provided a material is sufficiently crystalline to diffract X-rays and is present in an amount greater than about 1%, X-ray diffraction (XRD) can be used for qualitative and quantitative analyses. Crystal structures possess planes made by repetitive arrangements of atoms, which are capable of diffracting X-rays. The angles of diffraction differ for the various planes within the crystal. Thus, every compound

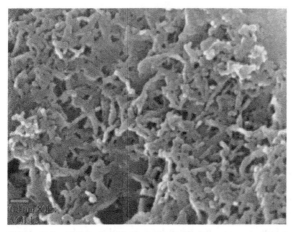

Figure 3.7 SEM of γ-Al₂O₃ with its highly porous network. (Reproduced from Chapter 3 of Heck, R.M., Farrauto, R.J., and Gulati, S.T. (2009) *Catalytic Air Pollution Control: Commercial Technology*, 3rd edn, John Wiley & Sons, Inc., New York.)

or element has its own somewhat unique diffraction pattern. Comparing the patterns allows differentiation of various structures.

3.3.5 Structure and Morphology of Al₂O₃ Carriers

Figure 3.7 shows the highly porous SEM micrograph of γ-Al₂O₃. Its highly porous network is utilized in dispersing catalytic components throughout its structure.

Figure 3.8 shows the XRD patterns of two Al₂O₃ structures, amorphous (low-crystallinity) γ-Al₂O₃ and highly crystalline α-Al₂O₃. Crystalline means a high degree of structural order while amorphous means little long-range structural order. γ-Al₂O₃ is the high surface area, lower temperature structure, whereas α-Al₂O₃ is produced at high temperatures and has low surface area. Below crystallite sizes of 50 Å, a well-defined X-ray pattern will not be obtained. Materials with crystallites smaller than this are more precisely called amorphous since they do not diffract X-rays and their patterns are broad and diffuse. Structures in this class, which are quite common for freshly prepared catalysts, must be characterized by other techniques such as those listed below. Crystalline materials such as α-Al₂O₃ have sharp and well-defined peaks.

3.3.6 Dispersion or Crystallite Size of Catalytic Species

3.3.6.1 Chemisorption One of the most frustrating facts facing the catalytic scientist is that often when a structure has a definite XRD pattern and can be well characterized, it usually has less than optimum activity. This is because most catalytic reactions are favored by either amorphous materials or extremely small crystallites. Their sizes typically fall into what we commonly call nanosize range (<200 nm or <20 Å). Small crystals can agglomerate or grow and produce large crystallites that

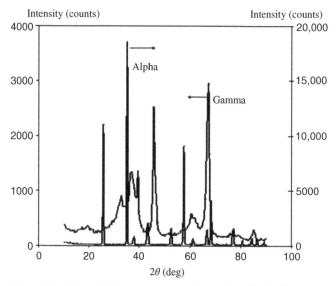

Figure 3.8   X-ray diffraction patterns of γ- and α-Al$_2$O$_3$. (Reproduced from Chapter 3 of Heck, R.M., Farrauto, R.J., and Gulati, S.T. (2009) *Catalytic Air Pollution Control: Commercial Technology*, 3rd edn, John Wiley & Sons, Inc., New York.)

diffract X-rays and thus generate easily read patterns; however, the atoms of the small crystals are buried within the larger crystal, making them inaccessible to reactant molecules. The purpose of the preparation technique is to disperse the catalytic components in such a way as to maximize their availability to reactants.

$$\% \text{ dispersion} = \frac{\text{number of catalytic sites on surface}}{\text{theoretical number of sites present as atoms}} \times 100 \qquad (3.4)$$

When this is done effectively, only small metal catalytic sites lacking long-range structure are present and the diffraction of X-rays is almost nonexistent. As the crystals get smaller and smaller, the XRD peaks get broader and broader and eventually are undetectable above the background. However, it is these "X-ray-amorphous" species that are often the most active for a given catalytic reaction.

Standardized techniques exist for obtaining information regarding the distribution and number of catalytic components dispersed within or on the carrier. Selective chemisorption can be used to measure the accessible catalytic component on the surface by noting the amount of gas selectively adsorbed per unit weight of catalyst. The stoichiometry of the chemisorption process must be known to estimate the available catalytic surface area. One assumes that the catalytic surface area is proportional to the number of active sites. A gas that will selectively chemisorb only onto the metal and not the support is used under predetermined conditions. Hydrogen and carbon monoxide are most commonly used as selective adsorbents for many supported metals. There are reports in the literature of instances in which gases such as NO and O$_2$ have been used to measure catalytic areas of metal oxides; however, due to difficulty in interpretation, they are of limited use.

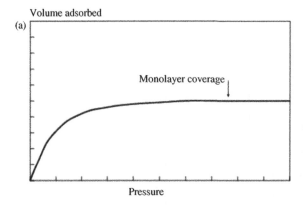

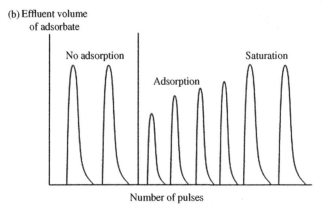

Figure 3.9   (a) Chemisorption isotherm for determining surface area of the catalytic component. (b) Pulse chemisorption profiles for the dynamic chemisorption method. (Reproduced from Chapter 3 of Heck, R.M., Farrauto, R.J., and Gulati, S.T. (2009) *Catalytic Air Pollution Control: Commercial Technology*, 3rd edn, John Wiley & Sons, Inc., New York.)

The measurements are usually carried out in a static vacuum system similar to that used for BET surface area measurements. The pressure of gas above the sample is increased and the amount adsorbed measured at equilibrium as shown in Figure 3.9a. When there is no further adsorption with increasing pressure (shown in the flat portion of the figure), the catalytic surface is saturated with a monolayer of adsorbate. Noting the number of molecules of gas adsorbed and knowing its stoichiometry with the surface site (i.e., one H per metal site), one can determine the catalytic surface area by multiplying molecules adsorbed by cross-sectional area of the site and dividing by the weight of catalyst used in the measurement. For example, the cross-sectional area of Pt is $8.9\,\mathring{A}^2$ and that of Ni is $6.5\,\mathring{A}^2$.

The static vacuum technique is time consuming, so alternative methods have been devised. A dynamic pulse technique has been used in which a pulse of adsorbate such as H_2 or CO is injected into a stream of inert gas and passed through a bed of catalyst. Gas adsorption is measured by comparing the amount injected with the amount passing through the bed unadsorbed.

As shown from left to right in Figure 3.9b, the first two pulses are used for calibration and bypass the catalyst sample. The second set of pulses that pass through the catalyst are first diminished due to adsorption. Once saturation or monolayer coverage is reached, no further adsorption from the gas phase occurs. The amount adsorbed is found by the difference in areas under the peaks compared with those under the calibration pulses. The major difference between dynamic and static methods is that the former measures only those species that are strongly adsorbed, whereas the latter, performed under equilibrium conditions, measures strong and weakly chemisorbed species. Thus, static techniques usually give better dispersion results.

3.3.6.2 Transmission Electron Microscopy

In transmission electron microscopy (TEM), a thin sample, usually prepared by a microtome, is subjected to a beam of electrons. The dark spots on the positive of the detecting film correspond to dense areas in the sample that inhibit electron transmission. These dark spots form the outline of metal particles or crystallites and hence their sizes can be determined directly.

Figure 3.10 shows a TEM of sintered Pt, about 500 Å in size, dispersed on a TiO_2 carrier. Crystallites of Pt about 10 nm (100 Å) in size, dispersed on CeO_2, are shown in the TEM of Figure 3.11. Using image analysis, a size distribution and average crystallite size can be calculated. Assuming a spherical shape for the crystallites, the percent dispersion (i.e., ratio of surface atoms to total atoms in the crystallite × 100) can be calculated. It should be understood that this technique measures only a small fraction of the catalyst, so obtaining data representative of the entire sample is quite difficult. However, it does give a direct measure of the catalytic components.

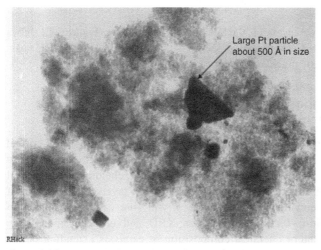

Large Pt particle about 500 Å in size

Figure 3.10   Transmission electron micrograph of Pt on TiO_2. (Reproduced from Chapter 3 of Heck, R.M., Farrauto, R.J., and Gulati, S.T. (2009) *Catalytic Air Pollution Control: Commercial Technology*, 3rd edn, John Wiley & Sons, Inc., New York.)

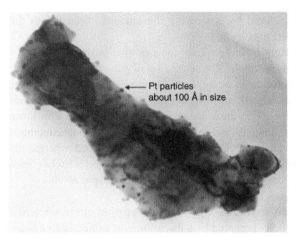

Figure 3.11 Transmission electron micrograph of Pt on CeO_2. (Reproduced from Chapter 3 of Heck, R.M., Farrauto, R.J., and Gulati, S.T. (2009) *Catalytic Air Pollution Control: Commercial Technology*, 3rd edn, John Wiley & Sons, Inc., New York.)

3.3.7 X-Ray Diffraction

The larger the crystals of a given component, the sharper the peaks on the XRD pattern for each crystal plane. The Scherer equation relates the breadth, B, at half-peak height of an XRD line due to a specific crystalline plane to the size of the crystallites, L.

$$B = k\lambda/L\cos\theta \tag{3.5}$$

where λ is the X-ray wavelength, θ is the diffraction angle, and k is a constant usually equal to 1.

As the crystallite size increases, the line breadth B decreases. Figure 3.12 shows the sharp XRD pattern of CeO_2 treated at 1500 °C. The CeO_2 treated at 800 °C has a broad profile indicative of much smaller crystallite size.

3.3.8 Surface Composition of Catalysts by X-Ray Photoelectron Spectroscopy

The composition of the catalyst surface, as opposed to its bulk, is of critical importance since this is where the reactants and products interact. It is on these surfaces that the active sites exist and where chemisorption, chemical reaction, and desorption take place. Furthermore, poisons deposit in the layers of the surface of the catalyst and thus knowing their concentration will give valuable insight into activity and deactivation. Techniques such as XRD and electron microscopy measure the structure and/or chemical composition of catalysts extending below the catalytic surface. The composition of the surface is usually different from that of the bulk, and thus its analysis must be carried out by techniques specific to the surface.

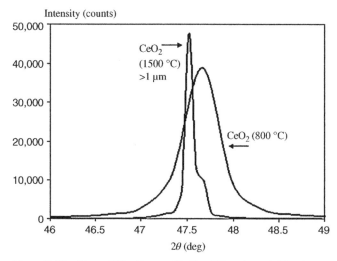

Figure 3.12   X-ray diffraction profile for different crystallite sizes of CeO$_2$. (Reproduced from Chapter 3 of Heck, R.M., Farrauto, R.J., and Gulati, S.T. (2009) *Catalytic Air Pollution Control: Commercial Technology*, 3rd edn, John Wiley & Sons, Inc., New York.)

Tools for fundamental research for surface composition characterization (X-ray photoelectron spectroscopy (XPS), Auger electron spectroscopy (AES), ion scattering spectroscopy (ISS), and secondary ion mass spectroscopy (SIMS)) are available. XPS is used more widely than the others for studying the surface composition and oxidation states of industrial catalysts, and thus its application will be discussed in some detail.

The acronym XPS refers to the technique of bombarding the surface with X-ray photons to produce the emission of characteristic electrons. These are measured as a function of electron energy. Because of the low energy of the characteristic electrons, the depth to which the analysis is made is only about 40 Å. The composition of this thin layer as a function of depth can be determined by removing or sputtering away top layers and analyzing the underlying surfaces. A number of important catalytic properties have been studied by this technique, including oxidation state of the active species, interaction of a metal with an oxide carrier, and the nature of chemisorbed poisons and other impurities.

If a small amount of a gas-phase impurity, that is, S, Cl, or P, deposits on the surface of the catalyst, its concentration is not likely to be detected by bulk chemical analysis. XPS allows only the top few monolayers to be analyzed and thus allows insight into its effect on the performance. Another example is the oxidation state of the surface catalytic component. Palladium can exist in three oxidation states, all of which have different activities toward specific reactions. For example, Pd metal is the active component in hydrogenation reactions while higher oxidation states of Pd are active for hydrocarbon oxidations. XPS allows us to determine the respective oxidation states. In Figure 3.13, the Pd is found in three oxidation states: 25% as Pd(0), 56% as Pd(+2), and 19% as Pd(+4).

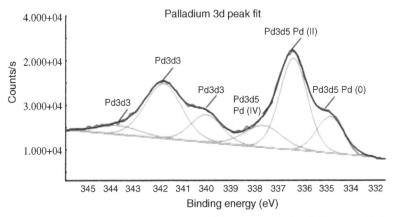

Figure 3.13   An XPS spectrum of various oxidation states of palladium on Al_2O_3. (Reproduced from Chapter 3 of Heck, R.M., Farrauto, R.J., and Gulati, S.T. (2009) *Catalytic Air Pollution Control: Commercial Technology*, 3rd edn, John Wiley & Sons, Inc., New York.)

3.3.9   The Bonding Environment of Metal Oxides by Nuclear Magnetic Resonance

Zeolites have long been of great importance in the chemical and petroleum industries for their unique pore size and acidity, both of which have a strong influence on catalytic activity. They are now finding greater use in environmental applications as adsorbents for hydrocarbons during cold exhaust conditions and as carriers for metal cations that generate high activities for special reactions such as the selective NO_x reduction.

The zeolites with varying ratios of SiO_2 and Al_2O_3 are bound together through an oxygen bridge forming a tetrahedral structure. A nuclear magnetic resonance (NMR) profile for faujasite, the main catalyst in catalytic cracking of heavy oils to gasoline range, is shown in Figure 3.14.

The first major peak occurs at -106 ppm indicating no bridges Si (O—Al), while the second peak at -100 ppm reflects the number of single Si—O—Al bridges Si (1Al). The peaks at progressively lower ppm show fewer bridges (Si(2Al), Si(3Al), and Si(4Al)) as evident by their low intensity. Thus, the distribution of tetrahedral Si—O—Al sites that represent the active sites can be determined. The higher the Si/Al ratio in the zeolite, the more tetrahedral bridges are expected to exist. When most zeolites undergo high-temperature exposure, they deactivate and the tetrahedral Si—O—Al bridge is broken. One can follow deactivation by the decrease in peak intensities. The importance of this will become clear when deactivation of zeolites is discussed in Chapter 5.

Atomic nuclei spin in much the same way as electrons. The movement of an electric charge generates a magnetic field. The silicon isotope (^{29}Si) with an uneven number of protons and neutrons is present in significant amounts in all compounds including zeolites. In such a case, a dipole (separation of charge) exists that can interact with an external magnetic field. The dipole can exist in different spin states

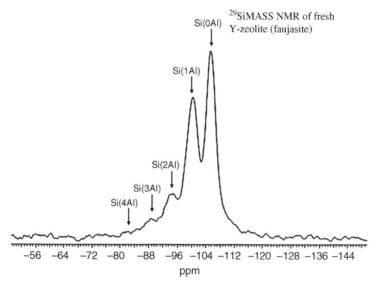

Figure 3.14   NMR profile of a Y faujasite zeolite. (Reproduced from Chapter 3 of Heck, R. M., Farrauto, R.J., and Gulati, S.T. (2009) *Catalytic Air Pollution Control: Commercial Technology*, 3rd edn, John Wiley & Sons, Inc., New York.)

and at a particular frequency of the imposed magnetic field it can be elevated to a higher spin state during "resonance." Upon relaxation of the external field, the dipole decays to a lower energy state and emits energy that is characteristic of that material. The frequency at which "resonance" occurs is influenced by the elements to which the isotope is bonded. So the number of Si—O—Al bridges will alter the frequency of the resonant energy of ^{29}Si creating the profile shown in Figure 3.14.

3.4 SPECTROSCOPY

This chapter is intended to show the methodology commonly used for the characterization of real commercial catalysts. There are many techniques designed to obtain more fundamental characterization that are beyond the scope of this book. The techniques described in this chapter are performed on catalysts extracted from the process after aging, that is, *ex situ*. It is very desirable, but also extremely difficult, to measure catalyst properties *in situ* during an actual process. Once removed from its catalytic environment, the nature of the adsorbed surface species will be changed, and what is measured will significantly differ from the actual catalytic surface. Nevertheless, many of its other properties (e.g., surface area, pore size, crystalline structure, crystallite size, and chemical composition) are essential data in determining which factors influenced its performance. Infrared DRIFT spectroscopy allows the nature of the adsorbed species to be observed before, during, and after catalytic reaction. In Figure 3.15, CO adsorption is studied on the surface of an automobile catalyst containing Pt, Pd, and Rh prior to the addition of O_2. The nature of the bonding of the

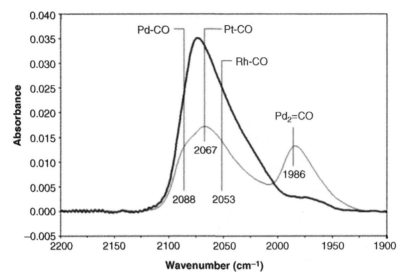

Figure 3.15 DRIFT spectra of CO chemisorbed on different precious metal particles of catalysts prepared in different ways. The CO chemisorption followed by FT-IR measurements was performed at room temperature after the catalysts were treated at 400 °C for 1 h with 7% H_2 in Ar gas. (Reproduced from Chapter 3 of Heck, R.M., Farrauto, R.J., and Gulati, S.T. (2009) *Catalytic Air Pollution Control: Commercial Technology*, 3rd edn, John Wiley & Sons, Inc., New York.)

CO to each element is shown. The addition of O_2 will remove the CO from the most active sites.

Catalytic scientists have created an exciting new approach toward the study of fundamental catalytic reactions. So-called operando spectroscopy permits the characterization of the surface of a catalyst during real catalysis.

QUESTIONS

1. Why are BET surface areas and Hg measured pore sizes important in heterogeneous catalysis?

2. What do X-ray diffraction patterns tell us about the components of the catalyst?

3. How may a TGA/DTA be used in

   a. Catalyst preparation?

   b. Activation of a catalyst?

   c. Burn off of hydrocarbon oils deposited on the catalyst?

4. How is NMR used in determining the thermal history of zeolites?

5. What can we learn from CO chemisorption measurements of metal-supported catalysts?

6. What can we learn from CO chemisorption measurements of metal-supported catalysts?

7. A chemist is experimenting with a new Pt/Al_2O_3 catalyst synthesis procedure that involves precipitation of Pt in liquid slurry onto the surface of the solid catalyst. The chemist suspects that not all the platinum precipitated onto the catalyst. What characterization test(s) could be used on the catalyst to determine how much of the Pt has been deposited onto the catalyst?

8. Following the above-mentioned scenario, after calcining the catalyst, the chemist now wants to know how well dispersed the platinum is on the surface of the catalyst. What characterization test(s) could be performed to measure how well dispersed the active crystallites are on the surface?

9. Another chemist was trying to convert γ-alumina into α-alumina by heating the sample in air slowly up to 1000 °C overnight. In the morning, they realized that the furnace shut off unexpectedly at some temperature less than 1000 °C. They are curious about roughly what temperature the furnace shut down. What characterization test could they perform on the alumina sample to roughly determine the maximum temperature it saw? Explain how they would use the result to estimate temperature.

10. A chemist recently prepared a $Pt–Re/Al_2O_3$ reforming catalyst. As part of the characterization he performed to study the new catalyst, XRF and XPS composition measurements were conducted. The XRF results agreed with their synthesis procedure and indicated equal amounts of Pt and Re contained in the catalyst. However, the XPS result suggested that there was significantly more Pt and Re. Explain why the tests would yield differing results. Provide an educated guess as to what may be happening on the catalyst surface to explain the difference.

11. It is suspected that an automotive catalyst has experienced a high temperature and the catalytic metal crystallites have sintered. How will the dispersion of metal crystallites change as a result of sintering? What characterization tests could be used to confirm this theory?

12. From the scientific literature, find an article where characterization was performed to describe a catalyst. Summarize the different characterization tests performed and the reason for performing each.

BIBLIOGRAPHY

Anderson, R. and Dawson, P. (1976) *Experimental Methods in Catalytic Research*, Academic Press, New York.

ASTM (1988a) *Committee D-32 on Catalysis*, 2nd edn, ASTM, Philadelphia, PA.

ASTM D3663-85 (1988b) *Standard test method for surface area of catalysts*. ASTM, pp. 3–6.

ASTM D4641-87 (1988c) *Standard test method for determination of nitrogen adsorption/desorption isotherm for pore size measurements*. ASTM, pp. 41–44.

ASTM D4284-83 (1988d) *Standard test method for determination of pore volume distribution by mercury porosimetry*. ASTM, pp. 26–29

ASTM D4513-85 (1988e) *Standard test method of particle size of catalytic materials by sieving*. ASTM, pp. 71–72.

ASTM D4438-85 (1988f) *Standard test method for distribution of catalytic materials by electronic counting*. ASTM, pp. 63–65.

ASTM D4464-85 (1988g) *Standard test method for particle size by laser light scattering*. ASTM, pp. 66–67.

ASTM D4642-86 (1988h) *Standard test method for chemical analysis of Pt on Al_2O_3 catalysts*. ASTM, pp. 14–17.

ASTM D3908-82 (1988i) *Standard test method for hydrogen chemisorption of supported Pt on Al₂O₃ catalyst by volumetric method.* ASTM, pp. 13–16.

Banares, M. (2005) Operando methodology: combination of in-situ spectroscopy and simultaneous activity measurements under catalytic reaction conditions. *Catalysis Today* 100, 71.

Bartholmew, C. and Farrauto, R.J. (2006) Chapter 3, in *Fundamentals of Industrial Catalytic Processes*, 2nd edn, John Wiley & Sons, Inc., New York.

Delannay, F. (1984) *Characterization of Heterogeneous Catalysts*, Dekker, New York.

Deviney, M.L. and Gland, J. (1985) *Catalyst Characterization Science: Surface and Solid State Chemistry*, ACS Symposium Series No. 288, American Chemical Society.

Farrauto, R.J. and Hobson, M.C. (1992) Catalyst characterization, in *Encyclopedia of Physical Science and Technology*, Vol. 2 (ed. R.A. Meyers), Academic Press, New York, pp. 735–761.

Knozinger, H. (2003) Catalyst characterization, in *Encyclopedia of Catalysis*, Vol. 2 (ed. I. Horvath), Wiley-Interscience, Hoboken, NJ, pp. 142–182.

CHAPTER 4

REACTION RATE IN CATALYTIC REACTORS

4.1 INTRODUCTION

In the design and operation of any catalytic reactor, a thorough understanding of kinetic parameters and the variables that will influence rate is critical. At any point in the reactor, the rate of reaction will be limited by one of the three phenomena described in Section 1.6 and illustrated in Figure 4.1: (1) "bulk" mass transfer rate between the bulk phase and the external catalyst surface, (2) "pore" diffusion rate between the external catalyst surface and the catalytic sites within the pore structure, or (3) surface reaction kinetics at the sites. It is possible and even common for the rate-limiting process to vary over the length of the reactor as reaction conditions change. Regardless of the industrial process, fundamental questions that arise during the reactor design will almost always include (1) how much catalyst is needed to achieve a particular conversion, (2) how will fluctuations in reaction conditions affect conversion and reactor operation, and (3) what design parameters would need to change to decrease the amount of required catalyst (i.e., increase reaction rate). Beyond reactor design, in the development of new catalyst formulations, it is important to ensure that rate measurements are conducted at conditions where it is the surface kinetic reaction limiting rate and not mass transfer (diffusion) of reactants to the catalyst surface. Answering these questions requires an understanding of the relationship between reaction rate and reaction conditions. The purpose of this chapter is to describe these relationships.

4.2 SPACE VELOCITY, SPACE TIME, AND RESIDENCE TIME

Before proceeding to the main discussion of this chapter, some basic terms and equations commonly used in reactor engineering should be reviewed. The first is the concept of space time, which is defined in Equation 4.1. In physical terms, space time represents the time necessary to process one reactor volume of reaction fluid at the entrance conditions. The volume of the reactor (V_{rxtr}) for catalytic systems is the volume of the catalyst. A related term is mean residence time, which is the mean time

Introduction to Catalysis and Industrial Catalytic Processes, First Edition. Robert J. Farrauto, Lucas Dorazio, and C.H. Bartholomew.
© 2016 John Wiley & Sons, Inc. Published 2016 by John Wiley & Sons, Inc.

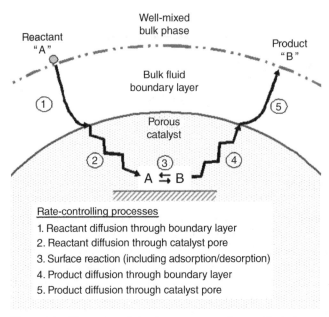

Rate-controlling processes
1. Reactant diffusion through boundary layer
2. Reactant diffusion through catalyst pore
3. Surface reaction (including adsorption/desorption)
4. Product diffusion through boundary layer
5. Product diffusion through catalyst pore

Figure 4.1   Illustration of the three processes that can limit the reaction rate during heterogeneous catalysis.

each molecule spends in the reactor. Mean residence time will equal space time when the reaction proceeds at constant temperature, pressure, and density, and volumetric flow is calculated at the reactor inlet condition (i.e., volumetric flow is constant along the reactor length). There are many catalytic applications where these conditions are met. Where they are not, space time is still an easily calculated term approximating the time molecules must spend inside the reactor to achieve a particular conversion.

$$\tau = \frac{V_{rxtr}}{V_{flow}^{inlet}} \tag{4.1}$$

Another useful term used in reactor engineering is space velocity (SV), which is defined as the number of reactor volumes processed per unit time, almost always reported in units of h^{-1}. The general equation for space velocity is given in Equation 4.2.

$$SV = \frac{V_{flow}}{V_{rxtr}} \tag{4.2}$$

Most often, the volumetric flow (V_{flow}) is that measured at standard temperature and pressure (STP). However, it is not unreasonable to define space velocity at other conditions that are more physically meaningful to the user; thus, it is important to know the conditions used to define space velocity when using data reported in the literature or elsewhere. While it may seem that space velocity and space time are the inverse of each other, as we have defined them, the volumetric flow is different in each expression due to the different defining condition used. The concept of space velocity is independent of process scale. Thus, if a chemist develops a catalyst capable of operating at a SV of $50,000\,h^{-1}$ in his laboratory-scale reactor, the engineer easily knows how this will translate into his large-scale reactor.

The volumetric flow used in Equation 4.2 is defined differently for liquid- and gas-phase reactions. For liquid-phase reactions, we define volumetric flow as liquid flow rate at STP and specify space velocity as liquid hourly space velocity (LHSV). For gas-phase reactions, we define volumetric flow as gaseous flow rate at STP and specify space velocity as gas hourly space velocity (GHSV). When species are present that are gaseous at reaction conditions, but liquid at STP (e.g., water), the general convention is to calculate the volumetric flow of the condensable at STP as an ideal gas. For catalytic systems, it is the mass of catalyst present that is critical. Thus, a more physically meaning definition of SV for catalytic systems is weight hourly space velocity (WHSV), which is defined as mass flow at STP divided by the weight of catalyst contained in the reactor (Equation 4.3).

$$\text{WHSV} = \frac{M_{\text{flow}}}{W_{\text{cat}}} \tag{4.3}$$

4.3 DEFINITION OF REACTION RATE

Consider the hypothetical reaction below in Equation 4.4 where a, b and c are the stoichiometric coefficients.

$$a\text{A} + b\text{B} \rightarrow c\text{C} \tag{4.4}$$

By definition, the rate of a chemical reaction is defined as the number of moles of reactant or product that is consumed or produced per unit time per unit reactor volume/length. Mathematically, the definition of rate is derived from a material balance around the reactor yielding the expression in Equation 4.5, which assumes steady-state plug flow through a tubular reactor. Equation 4.5 can be easily rewritten into more meaningful terms. In catalytic reactions, it is the amount of catalyst that is critical, not the volume of the reactor and Equation 4.5 can be written in terms of catalyst mass (Equation 4.6). The different notation for the rate, r', only indicates that the units for rate are different (per mass) from those in Equation 4.5 (per volume). The two rate terms are simply related by the catalyst bulk density, $r' = r/\rho$.

$$\frac{dF}{dV} = -r \tag{4.5}$$

$$\frac{dF}{dW} = -r' \tag{4.6}$$

The above rate expressions are written in terms of the disappearance of reactants, which is why the "negative" sign precedes the rate term, r. If the expression were written in terms of the rate of appearance of a product species, the negative sign would not be used. Regardless of the physical processes limiting the rate of reaction, these expressions always apply. They define what rate is and how it can be physically measured. The rate-limiting process, which is discussed in detail in the following sections, will define the expression used to represent the right-hand side of Equations 4.5 and 4.6. If the reaction is limited by surface kinetics (at the reaction sites), the rate term could be represented by a kinetic rate expression, such as the power rate law that will be discussed in the following section. If the rate of bulk mass transfer limits the reaction, the rate term will be represented with an expression representing the rate of diffusion to the external catalyst surface. The remainder of this chapter will discuss how rate is expressed for each limiting process.

4.4 RATE OF SURFACE KINETICS

When the rate of reaction is dictated by the rate of chemical reaction on the surface of the catalyst, the reaction is said to be "kinetically controlled." This will generally be the case at low temperatures (i.e., low conversions) due to the higher activation energy associated with chemical reactions compared with diffusion. There are many instances in industrial processes where this will be the case. For example, the syntheses of ammonia or methanol are both exothermic reactions and must be conducted at relatively low temperatures to avoid chemical equilibrium limitations. For specialty chemical reactions, such as selective oxidations, with multiple possible product paths, high conversions cause a loss of selectivity and therefore low conversions with feed recycle are necessary. When designing the reactor, we need to know how reaction rate and conversion will change with reaction conditions such as temperature and reactant partial pressure. Additionally, we need to know whether to expect the rate-limiting process to change during the operation of the reactor and where this transition is likely to occur. For example, in the environmental catalytic application of the three-way catalytic converter used in automobile exhaust for emission control, the reactor is initially kinetically limited and then becomes diffusion limited as the reaction temperature increases. Answering these questions requires an understanding of surface kinetics and the development of an expression to predict it. The objective of the next section is to present an expression for the kinetic rate of reaction that can be used for reactor design and discuss how the parameters in this expression are determined.

4.4.1 Empirical Power Rate Expressions

It is always preferred to know the precise mechanism by which a catalytic reaction occurs allowing a kinetic model to be developed as was shown in Sections 1.4.2.1–1.4.2.3. Often this is not feasible for complicated reactions where the kinetic rate-limiting step (related to mechanism) is not obvious. For these cases, we can develop an empirical power model that measures reaction orders of reactants and products, which allows concentrations to be adjusted to optimize the rate. For example, a positive order for a reactant means the rate will increase as its concentration is increased, while the reverse is true if it is a negative order. If a product has a negative order, its increase in production will further inhibit the forward reaction and thus it is advisable to remove it during the process. Examples will be given in the following sections.

For the hypothetical catalytic reaction in Equation 4.7, the rate of the heterogeneous surface reaction can be modeled using a pseudo-homogeneous power rate law expression. Using this approach, all the elementary steps of the heterogeneous reaction, including adsorption and desorption, are fit to a power law with the general form of Equation 4.8.

$$aA + bB \rightleftharpoons cC + dD \tag{4.7}$$

$$r = k_f[A]^w[B]^x - k_r[C]^y[D]^z \tag{4.8}$$

where k_f is the rate constant for the forward reaction and k_r is the rate constant for the reverse reaction, which are defined by the Arrhenius expression given in Equation 4.9. The exponents w, x, y, and z represent the reaction orders or concentration dependence for each species. This indicates that the rate begins to decrease toward zero as products C and D are being produced and reactants A and B are consumed. In other words, equilibrium is approached.

$$k = k_o \exp\left(-\frac{E_A}{RT}\right) \tag{4.9}$$

4.4.2 Experimental Measurement of Empirical Kinetic Parameters

When measuring chemical kinetics of the catalyst and determining kinetic parameters, it is critical to operate the reactor under "differential reaction conditions." Differential conditions are those where there is little change in reaction temperature and reactant concentrations across the reactor length. Thus, the conversion is kept very small by operating the experiment at high space velocity. Oftentimes, the catalyst is diluted with inert material to further prevent accumulation of heat in the bed. To measure chemical kinetics of a reaction in the laboratory, it is common to maintain conversions below about 20%. For highly exothermic reactions, that is, $\Delta H > 200$ kJ/mol, measurements should be made at conversions no greater than 5–10% to maintain the temperature as close as possible to that desired. When measuring catalyst kinetics, it is equally important to ensure that the reaction is far from equilibrium. Under ideal conditions when only reactants are fed into the reactor and differential conditions are maintained, chemical equilibrium limitations are generally not an issue. However, when measurements are being conducted under industrially realistic feed conditions where a mixture of reactants and products is present, the effect of chemical equilibrium on the net rate of reaction must be considered. Finally, the experimenter should ensure that chemical kinetics will be controlling the rate.

By performing simple mathematical manipulations to the power rate law, we can easily determine the kinetic parameters in Equations 4.8 and 4.9: pre-exponential factor, activation energy, and concentration dependencies (reaction orders). The stoichiometric terms (a, b, c, and d) describe the reaction material balance but do not describe the reaction orders (w, x, y, and z) in Equation 4.8, which are kinetic terms. When the reaction is operated sufficiently far from chemical equilibrium, the reverse reaction can be neglected, leaving Equation 4.10 as our starting point for determining kinetic parameters.

$$r = k_f[A]^\alpha[B]^\beta \tag{4.10}$$

When applying the power rate law to the global reaction, the pre-exponential factor becomes more or less a mathematical fitting parameter and much of its physical meaning is lost. Most often, we are interested in determining the activation energy and concentration dependencies, both of which give us valuable insight into the chemistry and required process variables of importance. As such, this section will focus on determining only these parameters. Neglecting the calculation of the pre-exponential factor also allows a simplification that makes determining the activation energy easier.

We start with determining the concentration dependencies. In general, the procedure involves "linearizing" the power rate law in order to express the desired concentration dependence as the slope of the linear equation. Using this approach allows us to plot the experimental rate measurements and extract the concentration dependence as the slope of the resulting plot.

For example, to determine the concentration dependence α in Equation 4.10, we would linearize Equation 4.10 by taking the natural logarithm of both sides. We use the log operation so that mathematical manipulation of Equation 4.10 yields a linear equation, where α exists as the slope of the line. For this case, the resulting equation is given in Equation 4.11, where the first two terms on the product side are essentially constant; α is slope of the line.

$$\ln(r) = \ln k_r + \beta \ln[B] + \alpha \ln[A] \tag{4.11}$$

[A] is to be varied while always maintaining [B] 5–10 times that of [A]. Since [B] is in large excess, it is essentially constant during the experiment. The temperature is also maintained constant, so k_f is also constant. A series of experiments are conducted where concentration of A is varied and reaction rate determined. A ln–ln plot of rate and [A] is generated and the slope determined, which is α. This process is described in Figure 4.2. The same process is followed to determine β, except that Equation 4.11 is rearranged so that β is slope of the line on the ln–ln plot of rate and [B].

It should be understood that this reaction order is empirical and corresponds to the specific range of concentrations and temperature where the experiments were

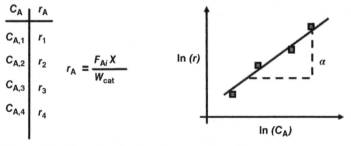

Figure 4.2   Illustration showing how experimental rate measurements can be plotted in order to determine the concentration dependence used in the power rate law.

conducted. For other concentration ranges, the reaction order may be different and thus should be measured in the new range.

Reaction orders reflect (but may not be identical to) the coefficients of reactants and products in the rate-limiting step. They should not be confused with the stoichiometric coefficients in the overall chemical reaction. Often power law rate expressions show fractional orders and not the expected whole numbers. This is because there are often more complicated terms in real rate-limiting steps such as the case for the Langmuir–Hinshelwood kinetics (see Equation 1.25) where depending on the conditions of the experiment there can be positive and negative terms for the same reactant.

A similar mathematical process is followed to determine the activation energy where the Arrhenius expression is used as the starting point.

$$k_f = k_o \exp\left(-\frac{E_A}{RT}\right) \tag{4.9}$$

The Arrhenius expression is linearized into the form given in Equation 4.12.

$$\ln k_f = \ln k_o - \left(\frac{E_A}{R}\right)\left(\frac{1}{T}\right) \tag{4.12}$$

In Equation 4.9, the activation energy still represents the energy barrier of the rate-limiting step in the reaction sequence and the exponential term is still a fraction representing the number of species with energy greater than this barrier.

A plot of ln(rate) versus $1/T$ yields a straight line with the slope of E_a/R. It is clear that the activation energy is effectively the temperature dependence of the reaction. The pre-exponential factor (k_o), however, becomes more or less a fitted parameter representing the frequency of some event that drives the reaction forward.

Experimentally, a series of experiments are conducted at differential reactor conditions where the concentrations are maintained constant and temperature is varied. If we are not interested in determining the pre-exponential factor, we can simply plot the natural log (ln) of rate versus inverse temperature. When using this simplified approach, the y-axis intercept no longer represents the natural log of A. However, since we are most often only interested in determining the activation energy, this simplification is very useful. The experimental process of determining activation energy is described in Figure 4.3.

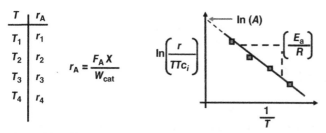

Figure 4.3   Illustration showing how experimental rate measurements can be plotted in order to determine the activation energy and pre-exponential factor used in the Arrhenius expression.

For the special case of a first-order (or pseudo-first-order) reaction, one can use the expression shown in Equation 4.13. This case is a common occurrence in environmental pollution oxidation reactions where a small amount of a pollutant is present in a large excess of air. One example is the oxidation of ethane in a large excess of air (O_2). The O_2 concentration does not change during the reaction and thus it is treated as pseudo-zero order. Furthermore, the equilibrium constant is sufficiently large that the reverse reaction can be ignored. Since CO_2 and H_2O are the products, one assumes that there are no inhibition effects due to their production.

$$-\frac{dC}{dt} = r = kC \tag{4.13}$$

The integration of this equation from the residence time $t = 0$ to $t = t$ with the initial concentration of the pollutant designated as C_{pi} and its outlet concentration designated as C_{po} gives

$$\ln\left(\frac{C_{pi}}{C_{po}}\right) = kt \tag{4.14}$$

If we make the simplifying assumption that $t \sim (GHSV)^{-1}$, then

$$\ln\left(\frac{C_{pi}}{C_{po}}\right) = \frac{k}{GHSV} \tag{4.15}$$

One makes a conversion versus temperature plot by varying the GHSV. From the change in the ratio of inlet to outlet concentration, one can determine the rate constant at several temperatures as shown in Figure 4.4.

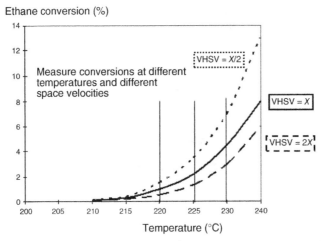

Ethane conversion (%)

Figure 4.4 Conversion versus temperature at different space velocities. Experiment is performed to determine the rate constant at various temperatures. (Reproduced from Chapter 4 of Heck, R.M., Farrauto, R.J., and Gulati, S.T. (2009) *Catalytic Air Pollution Control: Commercial Technology*, 3rd edn, John Wiley & Sons, Inc., New York.)

4.4.3 Accounting for Chemical Equilibrium in Empirical Rate Expression

While kinetic experiments are most often conducted at conditions far from chemical equilibrium, realistic industrial processes (i.e., NH_3 and methanol syntheses) are often conducted at integral conditions where chemical equilibrium is a major factor influencing the observed rate. When sizing the reactor, engineering calculations must capture the effect of chemical equilibrium.

The chemical equilibrium constant is

$$K_{eq} = \frac{[C]_{eq}^c [D]_{eq}^d}{[A]_{eq}^a [B]_{eq}^b} = e^{(-\Delta G_{rxn}(T)/RT)} \tag{4.16}$$

The temperature dependence of free energy and hence the equilibrium constant is reflected in Equation 4.17.

$$\Delta G_{rxn}(T) = G_D(T) + G_C(T) - G_A(T) - G_B(T) \tag{4.17}$$

One approach to account for the effect of chemical equilibrium on the rate of reaction is to include one additional term in the power rate law: $(1 - \beta)$.

$$r = k_f [A]^w [B]^x (1 - \beta) \tag{4.18}$$

β, which is described in Equation 4.19, varies from 0 (far from equilibrium) to 1 (at equilibrium). Thus, the term $(1 - \beta)$ approaches zero as chemical equilibrium is achieved, which drives the forward rate to zero.

$$\beta = \frac{\dfrac{[C]^c [D]^d}{[A]^a [B]^b}}{\dfrac{[C]_{eq}^c [D]_{eq}^d}{[A]_{eq}^a [B]_{eq}^b}} = \frac{[C]^c [D]^d}{[A]^a [B]^b} \frac{1}{K_{eq}} \tag{4.19}$$

The term $(1 - \beta)$ is the "approach to equilibrium." This reflects the fact that chemical reactions are reversible and reaction in both directions occurs simultaneously.

4.4.4 Special Case for First-Order Isothermal Reaction

For the special case of a first-order isothermal reaction, one can develop the following reactor sizing expressions. This special case is often encountered in environmental abatement applications where the pollutant exists in small concentrations in a large excess of oxygen (pseudo-zero order), and where temperature is effectively constant due to the small reaction enthalpy resulting from the conversion of the diluted pollutant.

One can estimate the volume of the reactor (V) needed for a given amount of conversion by substituting space velocity for time in the integrated first-order reaction rate expression:

$$\ln\left(\frac{C_i}{C_o}\right) = kt = \frac{k}{GHSV} = k\frac{V}{v_o} = k\frac{AZ}{v_o} \tag{4.20}$$

If 85% conversion is desired, then $C_i/C_o = 1/0.15$.

$$\ln(6.66) = k\frac{V}{v_o} = k\frac{AZ}{v_o} \tag{4.21}$$

Assuming that cross-sectional area (A) is known, v_o is the volumetric flow rate at STP and the length of the reactor (Z) required to achieve the desired conversion can be determined. A word of caution is necessary. Because catalysts deactivate (k likely decreases) during use, this calculation only serves as a minimum guide for reactor sizing. Furthermore, it assumes that no mass transfer limitations exist. Nevertheless, it does show how approximate calculations can be made.

4.5 RATE OF BULK MASS TRANSFER

4.5.1 Overview of Bulk Mass Transfer Rate

As described in Section 4.1, there are three physical processes occurring in the catalytic reactor: reaction of the catalyst surface, diffusion of reactants and products through the boundary layer surrounding the catalyst particles, collectively called "bulk mass transfer," and diffusion of reactants and products through the pores of the catalyst, collectively called "pore diffusion." There is a rate of each of these three processes and either one may be rate limiting. In the previous section, we discussed the rate of the reaction on the catalyst surface. In this section, we discuss the rate of bulk mass transfer.

When the rate of bulk mass transfer limits the overall reaction, the reaction rate is dictated by Equation 4.22. Later in this section, we will discuss the origin of this expression in more detail; however, we will start with a brief discussion of its physical interpretation.

$$r = k_{MT}a_s C_{A,bulk} \tag{4.22}$$

The bulk mass transfer coefficient is a function of the diffusivity of reactant A through the reaction mixture and the thickness of the boundary layer (i.e., resistance to diffusion). While diffusivity is effectively given for a particular reaction, the thickness of the boundary layer is strongly influenced by reactor design. Catalyst particle size, shape, surface roughness, and most importantly the linear velocity of the gas traveling through the catalyst bed all influence the mass transfer coefficient. Specifically, higher fluid velocity and smaller irregular shaped particles both yield higher mass transfer coefficients (enhanced turbulence) and higher reaction rates when bulk mass transfer is rate controlling. From Equation 4.22, another critical parameter is the geometric surface area (a_s), which is the external surface area (m^2) of the particle or surface to which reactants must flow. Larger geometric surface area provides greater area for external diffusion to occur, thus resulting in higher rates of bulk mass transfer. Geometric surface area is dictated by the design of the catalyst shape and is a major consideration in the design of catalytic reactors. The downside of a higher mass transfer coefficient and higher geometric surface area is higher pressure drop through the bed. All of the parameters that we have discussed that increase diffusion will also result in higher pressure drop.

4.5.2   Origin of Bulk Mass Transfer Rate Expression

Bulk mass transfer limitations are introduced as the chemical species diffuse through the boundary layer on their way to the catalyst surface (bulk mass transfer). As a result of frictional forces between the catalyst solid and the bulk flowing fluid, an ultrathin layer of near-stagnant fluid will form against the outer skin of the catalyst, which is referred to as the hydrodynamic boundary layer. The thickness of this fluid layer is defined in terms of fluid velocity, which ranges from zero at the surface of the catalyst to free stream velocity some distance away from the catalyst surface. In the bulk fluid, outside of the boundary layer, fluid mixing is rapid and no concentration gradient exists. However, within the boundary layer, convective mixing diminishes significantly and mass transfer occurs primarily by molecular diffusion. Just as the hydrodynamic boundary layer is defined by the point where the boundary layer velocity approaches the free stream velocity, the mass transfer boundary layer is defined by the point where the boundary layer concentration approaches the free stream concentration. Although the mass transfer boundary layer forms as a result of the hydrodynamic layer, it does not necessarily mean that the thicknesses of the two will be equal. The relative thickness of the two layers is a function of momentum diffusivity (hydrodynamic boundary layer thickness) and mass diffusivity (mass transfer boundary layer thickness), which is defined as the Schmidt number (Equation 4.23). When $Sc > 1$, the thickness of the hydrodynamic boundary layer is said to be larger than the mass transfer boundary layer.

$$Sc = \frac{\text{momentum diffusivity}}{\text{mass diffusivity}} = \frac{\nu}{D_{AB}} \tag{4.23}$$

To estimate the rate of mass transfer through the boundary layer, it is reasonable to treat the boundary layer as a static film with linear concentration profile bounded by the concentrations at the surface and bulk phase. Fick's law (Equation 4.24) can then be applied to calculate the mass transfer rate through the boundary layer. In Equation 4.24, the concentration difference is the driving force for mass transfer and the resistance to mass transfer is the ratio of the diffusion distance (δ_m) and mass diffusivity ($D_{i,\text{mix}}$). The inverse of mass transfer resistance is defined to be the mass transfer coefficient, k_c. Equation 4.25 then provides the expression for the rate of mass transfer through the boundary layer, where a_s is the ratio of particle surface area to particle volume that transforms mass flux into a rate per unit volume.

$$J_i \sim D_{i,\text{mix}} \frac{(C_{i,\text{bulk}} - C_{i,\text{surface}})}{\delta_M} \tag{4.24}$$

$$r_{\text{MT},i} \sim \frac{D_{i,\text{mix}}}{\delta_M} a_s (C_{i,\text{bulk}} - C_{i,\text{surface}}) \sim k_{\text{MT}} a_s (C_{i,\text{bulk}} - C_{i,\text{surface}}) \tag{4.25}$$

In calculating the mass transfer rate, the challenge is proper estimation of the mass transfer coefficient. Analytical solutions are possible by simultaneous solution of the continuity and momentum equations. For a single sphere in a flowing fluid, this approach would be challenging, but relatively straightforward. However, for a packed bed of spheres, this approach is unrealistic. Instead, the mass transfer coefficient is determined empirically. This problem has been rigorously studied and experimental correlations for the mass transfer coefficient can be found in the literature for a variety of geometries at a variety of different hydrodynamic conditions. These experiments typically involve measuring the rate of dissolution or evaporation of a substance as a function of geometry and fluid dynamics. In the case of gas-phase mass transfer, the rate of evaporation for a given geometry and flow condition is used to measure the mass transfer coefficient. Many measurements for a given geometry are taken for a range of flow conditions. The results are then fitted to Equation 4.26, where Re is the Reynolds number (Equation 4.27), Sc is the Schmidt number (Equation 4.28), and Sh is the Sherwood number (Equation 4.29). The terms a_{Sh} and x_{Sh} are the geometry-dependent fitted parameters. Thus, by knowing the fluid properties and characteristic length of the catalyst particle, these correlations can be used to estimate the mass transfer coefficient.

$$Sh = a_{Sh}Re^{x_{Sh}}Sc^{1/3} \tag{4.26}$$

$$Re = \frac{\text{fluid inertia}}{\text{momentum diffusivity}} = \frac{d_p u}{\nu} \tag{4.27}$$

$$Sc = \frac{\text{momentum diffusivity}}{\text{mass diffusivity}} = \frac{\nu}{D_{A,mix}} \tag{4.28}$$

$$Sh = \frac{\text{convective mass transfer}}{\text{diffusive mass transfer}} = \frac{k_{mt}d_p}{D_{A,mix}} \tag{4.29}$$

4.6   RATE OF PORE DIFFUSION

4.6.1   Overview of Pore Diffusion

In this chapter, we have discussed two of the three possible processes that can limit the rate of reaction: kinetics of surface reaction and bulk mass transfer. The remaining possible rate-limiting process is the rate at which reactant or products diffuse through the network of pores contained within the catalyst, collectively called "pore diffusion." This section will focus on the factors that influence the rate of pore diffusion.

As will be described in more detail in the next section, the rate of pore diffusion is influenced by the diffusion path length (distance a molecule must travel to reach the innermost portion of the catalyst) and the physical characteristics of the catalyst pore structure. When the reaction rate is limited by pore diffusion, any change made to catalyst geometry that decreases the distance for diffusion will increase the rate of reaction. For example, decreasing the diameter of a catalyst pellet, reducing the washcoat thickness on a monolithic support, or incorporating holes into the design of a catalyst pellet are all changes that would increase the rate of reaction. Another possible modification to the catalyst would be to design a catalyst pellet where the active material is concentrated near the exterior surface of the pellet. The advantage of this design is that we are able to increase pore diffusion rate (by decreasing the distance a molecule must diffuse to reach the innermost catalytic site) without affecting pressure drop.

4.6.2 Pore Diffusion Theory

The internal structure of a catalyst particle or washcoat is comprised of vast network of pores varying in size. The rate at which a species diffuses through this porous network will be a function of the nature of the diffusing species, concentration gradient, diffusion path length, and the structure of pore (Equation 4.30). A given species will diffuse more slowly through a pore characterized by a tortuous path than it would through a more ordered pore network. These characteristics of the catalyst are captured by three factors: the pore channel tortuosity (τ_p), constriction (σ_p), and overall particle porosity (ε_p). The channel tortuosity captures effect of the diffusion distance between abrupt changes in direction. A species diffuses faster through a straight channel than one with frequent changes in direction. The constriction factor captures the effect of varying pore diameter of the length of the pore. The overall porosity captures the combined effect of pore diameter and the overall volume of the pore network. Together, these parameters are factored into the diffusivity of the diffusing species to yield the effective diffusivity, D_e (Equation 4.31).

$$r_{pore} = a_s D_e \frac{(C_{surface} - C_{center})}{r_{pellet}} \tag{4.30}$$

$$D_e = \frac{\varepsilon_p \sigma_p}{\tau_p} D_{AB} \tag{4.31}$$

In the case of catalyst particles, the greatest path length for diffusion is the pellet radius. The actual path length will be longer, but this fact is captured within the effective diffusivity term. To determine whether pore diffusion is rate limiting, the diffusion rate must calculated based on the greatest diffusion path and it is for this reason the radius of the catalyst pellet is in the denominator of Equation 4.30.

As the pore diameter approaches the mean path length of the diffusing molecule, the mechanism for diffusion will change from bulk diffusion (particle–particle-driven diffusion) to Knudsen diffusion (particle–wall-driven diffusion). The mean path length (Equation 4.32) is the average distance a molecule will travel in between collisions with other molecules. When the pore diameter approaches the mean path length of the diffusing species, the rate of molecular diffusion will be influenced more by collisions with the pore wall than collisions with other molecules. The Knudsen diffusion coefficient, D_k (Equation 4.33), of a gaseous molecule moving through a straight channel of radius R_{pore} is given by the product of the mean molecule velocity, v_{rms}, and the average distance the molecule travels before contacting a pore wall, which is taken to be twice the pore radius. As a rough approximation, the effective diffusivity can be taken as one-quarter of the Knudsen diffusivity (Equation 4.34).

$$\lambda = \frac{RT}{PN_A \pi \sqrt{2}} \frac{1}{d_m^2} \tag{4.32}$$

$$D_k = \frac{2}{3} R_{pore} \sqrt{\frac{3RT}{MW_i}} \tag{4.33}$$

$$D_e \sim \frac{1}{4} D_k \tag{4.34}$$

The reaction becomes pore diffusion limited when the rate of diffusion into the pore is slower than the reaction occurring on catalytic sites. When this occurs, the reactant is being consumed before it has time to diffuse to the innermost portions of the catalyst pellet. The end result is a concentration gradient that will exist within the catalyst pellet, with the highest concentration near the exterior surface and zero at the core. Experimentally, the measured reaction rate will be on the same order as the calculated diffusion rate. This leads to the Weisz–Prater criterion, C_{WP} (Equation 4.35), which is the ratio of the observed reaction rate to the rate of pore diffusion. When C_{WP} is less than 1, the reaction is not pore diffusion limited and would be kinetically controlled. When C_{WP} is approximately 1, the rate of reaction is limited by pore diffusion. The difference $(C_{WP} - 1)$ is then a relative measure of the concentration gradient in the pellet. As this difference becomes large, the concentration gradient in the pellet becomes more severe and less of the catalyst in the core is utilized.

$$C_{WP} = \frac{r_{rxn}(obs)}{r_{pore}} = \frac{r_{rxn}\rho_{cat}R_{pellet}}{D_e\left(\dfrac{C_{i,s} - 0}{R_{pellet}}\right)} \tag{4.35}$$

4.7 APPARENT ACTIVATION ENERGY AND THE RATE-LIMITING PROCESS

When conducting laboratory experiments, it is essential to understand which physical process is limiting reaction rate: kinetics, bulk mass transfer, or pore diffusion. One way to determine the controlling process is to measure the activation energy for kinetic control and the "apparent" activation energy for processes limited by mass transfer. We use the term "apparent" activation energy because mass transfer is a physical phenomenon distinct from activated chemical processes where bonds are being made and broken. As discussed above, the activation energy is a measure of the temperature dependence of the reaction. When kinetics controls the reaction rate, the reaction rate will behave according to the Arrhenius equation and rate will respond exponentially to changes in temperature. Thus, the activation energy should be relatively high. In contrast, when the reaction is controlled by bulk mass transfer, the temperature dependence is relatively weak. Thus, the "apparent" activation energy will be relatively low. If we measure the rate or rate constant over a wide temperature range where all controlling processes could be observed, the resulting plot of $\ln k$ versus $1/T$ (Figure 4.5) shows three distinctly different slopes, each related to a different rate-limiting step. The largest slope is for a reaction controlled by chemical kinetics, intermediate slope for pore diffusion, and the smallest slope for bulk mass transfer control.

When the reaction is controlled by one of the chemical steps, mass transfer of reactants is relatively fast. The concentration of reactants within the catalyst/carrier is essentially uniform. With pore diffusion control, the concentration of reactants decreases from the outer periphery of the catalytic surface toward the center. Finally, with bulk mass transfer control, the concentration of reactants approaches zero at the boundary layer near the outside surface of the catalyst.

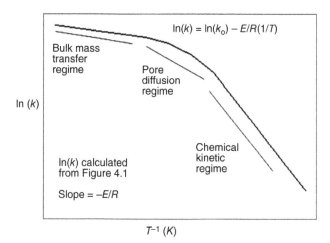

Figure 4.5   Arrhenius plot for determining activation energies. (Reproduced from Chapter 4 of Heck, R.M., Farrauto, R.J. and Gulati, S.T. (2009) *Catalytic Air Pollution Control: Commercial Technology*, John Wiley & Sons, Inc., New York.)

4.8 REACTOR BED PRESSURE DROP

When reactant enters a bed of catalyst particulates, its flow is opposed by friction and the presence of a volume of catalyst material. This creates a backpressure that must be overcome by increased head pressure. The additional pressure must be provided by a compressor or blower, which adds to the process expense. In all applications, there are maximum permitted pressure drops as part of the overall system engineering and economics.

The basic equation for pressure drop can be derived from the energy balance and results in the following expression:

$$-\frac{1}{\rho_g}\frac{dP}{dZ} = \frac{2fu^2}{g_c d_p} \tag{4.36}$$

The change in pressure drop (dP) as a function of distance (dZ) divided by the gas density (ρ_g) is shown in the left-hand side of Equation 4.36. The right-hand side of the equation shows the pressure drop with the linear velocity (u) (calculated using the volumetric flow rate (v_o) divided by the cross-sectional area of bed (A)), the particle diameter (d_p), the friction factor (f), and the gravitational constant g_c.

$$u = \frac{v_o}{\rho_g A \epsilon} \tag{4.37}$$

The linear velocity (u) is an extremely important design parameter for reducing pressure drop in the reactor bed and for increasing turbulence enhancing mass transfer rates. It is effectively a measure of the turbulent flow in the reactor. By designing the reactor to have a large cross-sectional area (A), the linear velocity is decreased reducing the pressure drop. The void fraction (ε) is the space unoccupied by mass. A smaller catalyst particle decreases ε since the void volume is decreased due to closer packing. This leads to a higher linear velocity. If additional mass transfer conversion is necessary, an increase in linear velocity enhances turbulence and conversion but at the expense of increased pressure drop.

4.9 SUMMARY

A generalized approach for determining kinetics and diffusional resistances has been developed in this chapter. Methods to determine empirical power rate law parameters have been presented. Additionally, variables affecting pressure drop have been reviewed. This chapter has provided important tools required to understand catalyst kinetics and reactor designs presented in the following chapters that discuss various catalyst applications in detail.

QUESTIONS

1. What temperature is required for a catalyst with an activation energy of 15,000 cal/mol to operate with the same volume as one with an activation energy of $E = 12,000$ cal/mol (operation temperature is 227 °C) with the same conversion? Assume k_o values are the same for simplicity.

2. You are employed as an environmental engineer for a major chemical company that uses alcohols as liquid carriers for pigments to be used for decorating. Currently to meet local environmental regulations, the harmful emissions (1500 vppm) resulting from spraying and drying the coatings are thermally incinerated. However, this requires a thermal burner, which is expensive, consumes large amounts of fuel, and generates some NO_x (NO, NO_2, and/or N_2O). The plant manager, knowing you had a course in catalysis, is seeking an alternative to thermal incineration. You are asked to evaluate some fixed bed catalysts that will abate the alcohol emissions. You contact a catalyst company and they suggest you use a monolithic support with benefits of a lower low-pressure drop than a particulate bed of catalyst. They offer two catalysts: an inexpensive one with an activation energy (E) of 21,200 cal/mol (21.2 kcal/mol) and a second more expensive one with an activation energy of 12,000 cal/mol (12 kcal/mol). Compare the volumes needed for each assuming you want to achieve 90% conversion at 227 °C. Assume both catalysts have the same k_o (pre-exponential function).

3. Compare the volumes needed for two catalysts, one with $E = 48$ kJ/mol (12 kcal/mol) and the other with 60 kJ/mol (15 kcal/mol).

4. What must the activation energy (E) be for a catalyst to operate at 275 °C (548 K), where one with an $E = 30$ kcal/mol operates at 550 °C (823 K). Assume both will give the same conversion and both have the same k and k_o. (This is not likely since k_o includes the number of active sites and other more fundamental parameters including collision frequencies that govern conversion for each catalyst.)

5. Determine the activation energy for a fixed bed catalyst where the reaction conditions are adjusted so that percentage conversion for each temperature is low and about the same for the following three rates:

Rate $(400\,K) = 10$.

Rate $(410\,K) = 20$.

Rate $(420\,K) = 40$.

6. Why is it useful to know the reaction orders for the reactants and products? Consider the empirical equation where A and B are reactants and C and D products and a, b, c, and d are reaction orders.

$$\text{Net rate} = [A]^a [B]^b [C]^c [D]^d$$

7. Consider the oxidation reaction where methane, which is a strong greenhouse gas, is catalytically oxidized to CO_2 and H_2O by Pd/γ-Al_2O_3 catalyst.

$$CH_4 + 2O_2 = 2H_2O + CO_2$$

How would you determine the activation energy where the reaction conditions are adjusted (very high SV) so that the percentage conversion for each reactant is less than 5% for the following three measured relative rates:

Rate $(312\,°C) = 99.9$.

Rate $(297\,°C) = 87.2$.

Rate $(282\,°C) = 58.2$.

8. The first oxidation catalyst used in the United States in 1975 for abating CO and HC emissions from the automobile exhaust (gasoline) was a combination of 2% Pt and 1% Pd dispersed on γ-Al_2O_3//monolith. In Brazil, where ethanol (derived from sugarcane) is often used as a fuel, the catalyst companies first tried the same catalyst. The first laboratory tests conducted used 2% ethanol in excess air and a conversion versus temperature plot was made. The ethanol began converting at $150\,°C$; however, the product distribution (selectivity), especially at low temperatures, showed undesirable acetaldehyde (CH_3CHO) formation along with the desired CO_2 and H_2O.

a. Write the general rate equations for the formation of each product. Ignore the reverse reaction since the K_e is very large. The formation of acetaldehyde was determined to be first order in ethanol while for the CO_2 reaction path the ethanol order was 0.25. In both cases, the O_2 was pseudo-zero order because it was present in large excess. What process variable could you change in the system to enhance CO_2 production and minimize acetaldehyde?

b. The CO_2 path has an activation energy of 80 kJ/mol while the undesired path has an activation energy of 45 kJ/mol. In the exhaust of the vehicle, what process parameters could you change and how could you optimize the system design to minimize the undesired reaction?

c. Given that the Pt/Pd catalyst worked for gasoline engines but has not performed so well for ethanol fuel vehicles, suggest one or two simple catalyst preparations to minimize or eliminate acetaldehyde production completely?

9. You are requested to perform a laboratory evaluation of a 1.5% Pd/ZrO$_2$//monolith catalyst used in a coal-fired power plant. The coal contains 1000 ppm of sulfur compounds and 2% ash. The catalyst is expected to abate CO and HC emissions. After 2 years of operation, an inspection by the local environmental agency claims it is not meeting regulations for either CO or HCs.

   a. Develop a short laboratory plan to perform the catalyst tests to determine its performance compared with the fresh catalyst.

   b. Assume you observe the following in the conversion versus temperature profile: The fresh catalyst lights off (begins to oxidize) the HC at 250 °C and reaches 50% conversion at 350 °C and 90% at 450 °C. The used catalyst lights off at 400 °C and reaches 70% conversion at 450 °C and 90% at 550 °C. Sketch the profiles and explain what occurred using some of the characterization tools we have studied.

10. An extruded "cylindrical" catalyst is used for VOC abatement. Over time, the conversion is observed to increase by 10% despite the feed conditions and reaction temperature being the same. When the catalyst is unloaded, it is observed that the extrudates broke into smaller pieces over time. What rate-controlling regime was the reactor likely operating in? Why would the conversion increase?

11. A Pt/alumina monolithic catalyst is used for abatement of hydrocarbons from a natural gas-fired turbine. The turbine system temporarily fires at a much higher rate than normal, resulting in the catalyst temperature increasing well above the temperature where sintering of the Pt is expected. Yet, there is no change in conversion. What rate-controlling regime was the reactor likely operating in? Why would the conversion not change if the Pt sintered?

12. A catalytic reactor currently operates with a cylindrical shaped catalyst. The reactor is used for a very endothermic reaction (steam reforming, $\Delta H_{rxn} = 200$ kJ/mol). The shape is changed such that a hole is incorporated into the catalyst shape (see figure below). However, the amount of catalyst contained in each catalyst has been maintained to be the same. This is accomplished by increasing the concentration of catalyst in the design using the hole. How will conversion change if the reaction is limited by the following processes? Explain your answer.

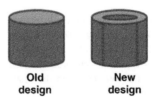

Old   **New**
design   **design**

   a. Kinetics.

   b. Pore diffusion.

   c. Bulk mass transfer.

   d. Heat transfer.

13. Explain the shape of the plot below. For each section, describe the rate-limiting process and explain the shape relative to the neighboring sections (i.e., why is the curve slope steep, shallow, or transitioning between).

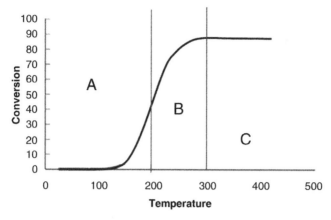

14. A Pt/Al$_2$O$_3$ catalyst is 1000 times more active than a Ni/Al$_2$O$_3$ catalyst. Which catalyst will yield a higher conversion when the reaction is controlled by bulk mass transfer? Explain.

15. The Ni concentration of a Ni/Al$_2$O$_3$ methanation catalyst is increased from 3 to 4%. You can assume that the Ni dispersion increases equally as much. The conversion is observed to increase significantly from 10 to 20%. Explain this observation.

16. The conversion is initially 70% for a catalytic reaction. The reaction temperature is increased 20 °C. The conversion does not change. Explain this observation.

17. How is space velocity maintained constant if the reactor diameter is decreased? Why would one do this? How will the pressure drop change if the reactor diameter is further decreased?

18. A pellet catalyst is used in a fixed bed reactor. The synthesis of catalyst A uses an impregnation technique that uniformly disperses the active metal through the catalyst pellet. Catalyst B synthesis uses an "eggshell" impregnation technique that concentrates the metal in the outer third of the pellet volume. Otherwise, the pellet geometry is identical. The conversion for the two catalysts is identical. What is the rate-limiting process?

19. A tubular packed bed reactor is loaded with cylindrical catalyst shapes. To increase production, the volumetric flow and catalyst volume are increased proportionally (space velocity is maintained constant). The conversion is observed to increase. What is the rate-limiting process?

BIBLIOGRAPHY

Bird, R., Stewart, W., and Lightfoot, E. (1970) *Transport Phenomena*, John Wiley & Sons, Inc., New York.

Bonacci, J., Farrauto, R., and Heck, R. (1989) Catalytic incineration of hazardous wastes, in *Encyclopedia of Environmental Control Technology, Vol. I: Thermal Treatment of Hazardous Wastes* (ed P. N. Cheremisinoff), Gulf Publishing Company, Houston, TX.

Broadbelt, L. (2003) Kinetics of catalyzed reactions—heterogeneous, in *Encyclopedia of Catalysis*, Vol. 4 (ed I. Horvath), Wiley-Interscience, Hoboken, NJ, pp. 472–490.

Fogler, S.H. (1992) *Elements of Chemical Reaction Engineering*, 2nd edn, Prentice Hall.

Hodgman, C.D. (ed.) (1960) *Handbook of Chemistry and Physics* Chemical Rubber Publishing Company, Cleveland, OH.

Lachman, I. and McNally, R. (1985) Monolithic honeycomb supports for catalysis. *Chemical Engineering Progress* 84 (1), 29–31.

CHAPTER 5

CATALYST DEACTIVATION

5.1 INTRODUCTION

Catalysts are subject to the conditions in which they must operate. The two most frequent deactivation modes are sintering of the catalytic components and carrier due to high temperature and exposure to poisoning from feed contaminants. Temperatures above 900 °C are experienced in applications such as nitric acid and hydrocyanic acid production as well as automobile catalytic converters where temperature close to 1000 °C are experienced periodically during the lifetime of the vehicle. Other sources of deactivation such as poisoning can occur due to process contaminants adsorbing onto or blocking active catalytic sites. It is essential to understand the modes of poisoning in order to develop resistant materials and methods of regeneration when possible. Deactivation by attrition and erosion of particulate materials or powdered catalysts must also be considered.

Throughout the application sections of this book, reference will be made to common sources of deactivation specific to a particular application, methods of regeneration, and design considerations to minimize these effects. A convenient tool for studying deactivation and regeneration is the model reaction. Here one takes a representative model compound and conditions under which the catalyst is expected to function and develop a conversion versus temperature curve as shown in Figure 1.5. Shifts in the profile give great insight into the mechanism of deactivation. This theme will be further developed throughout this chapter.

This chapter will serve as a convenient working guide for determining the mechanism of deactivation and commonly used industrial diagnostic tools to monitor these changes. It will be particularly useful when attempting to perform technical service on a catalyst that is not performing in accordance to specifications.

5.2 THERMALLY INDUCED DEACTIVATION

It is the objective of the catalyst manufacturer to maximize accessibility of the reactants to the active sites by depositing the catalytic components often on a carrier. A perfectly dispersed (100% dispersion) catalyst is one in which every atom (or molecule) of active component is available to the reactants. This is shown

Introduction to Catalysis and Industrial Catalytic Processes, First Edition. Robert J. Farrauto, Lucas Dorazio, and C.H. Bartholomew.

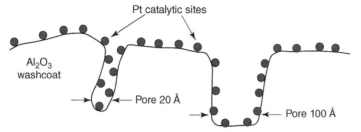

Figure 5.1   Idealized cartoon of perfectly dispersed Pt on a high-surface γ-Al$_2$O$_3$. (Reproduced from Chapter 5 of Heck, R.M., Farrauto, R.J., and Gulati, S.T. (2009) *Catalytic Air Pollution Control: Commercial Technology*, 3rd edn, John Wiley & Sons, Inc., New York.)

as a cartoon in Figure 5.1 where the dots represent the catalytic component (indicated as Pt in the figure) dispersed on a high surface area γ-Al$_2$O$_3$. It should be understood that surface of the Al$_2$O$_3$ contains hydroxide sites, not shown for convenience, which are in direct contact with the catalytic components and play an important role in immobilizing them. The carrier can be thought of as an inorganic sponge possessing small pores of varying size and shapes into which reactants can flow and interact with catalytic components dispersed on the surface. The products formed pass through the porous network out to the bulk fluid.

Some catalysts are made in this highly active state, but are highly unstable and thermal effects cause crystal growth resulting in a loss of catalytic surface area. Additionally, the carrier with a large internal surface network of pores tends to undergo sintering with a consequent loss in internal surface area. Not uncommon are reactions of the catalytically active species with the carrier resulting in the formation of a less catalytically active species. All of these processes are influenced by the nature of the catalytic species, the carrier, and the process gas environment, but mostly by high temperatures.

5.2.1   Sintering of the Catalytic Species

It is common for a highly dispersed catalytic species in the nanosize to undergo growth to structured crystals as a consequence of their high surface/volume ratio. As this process proceeds, the sites grow larger decreasing the surface/volume ratio with fewer catalytic atoms or molecules on the surface of the crystal available to the reactants. In other words, many active sites are buried within the crystal and with fewer sites participating in the reaction, a decline in performance is most frequently noted.

This phenomenon is represented in Figure 5.2 by a simple model in which individual active catalytic components are designated as dots on a carrier. Pt is shown, but this applies to any metal or metal oxide supported on a carrier.

Initially, the sites are well dispersed but undergo coalescence or crystal growth induced thermally. As catalytic components undergo coalescence, the number of surface sites and the reaction rate decrease.

The driving force for catalytic sintering can be explained by the high surface/volume energy ratio possessed by small crystallites in the nanometer range. Thermodynamically, this is an unstable state and crystal growth, or sintering, occurs to minimize the free energy.

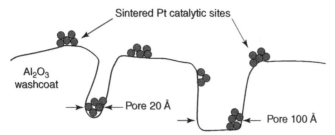

Figure 5.2 Conceptual diagram of sintering of the catalytic component on a carrier. (Reproduced from Chapter 5 of Heck, R.M., Farrauto, R.J., and Gulati, S.T. (2009) *Catalytic Air Pollution Control: Commercial Technology*, 3rd edn, John Wiley & Sons, Inc., New York.)

The extent of sintering can be measured by selective chemisorption techniques in which a thermally aged catalyst selectively adsorbs much less adsorbate than when it was fresh. The growth in crystal structures can also be observed by the XRD pattern; however, for most industrial metal-supported catalysts, the catalytic component is present in such a dilute concentration (often in the nanosize range) that X-rays are not diffracted. High-resolution transmission electron microscopy (TEM) permits the most direct method of observation of the growth process, but requires a great number of micrographs for statistical significance and although occasionally used, it is not considered a routine test in commercial practice. An excellent example of a sintered precious metal catalyst is shown in the TEM micrograph of Figure 5.3. Here the Pt is seen to have grown from 2 to $-20\,nm$ (0.2–2 Å)

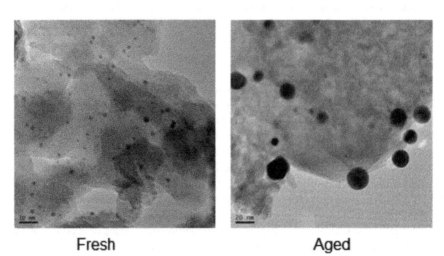

Fresh   Aged

Figure 5.3 TEM of fresh and sintered Pt on Al_2O_3 in an automobile catalytic converter application. "Black dots" are platinum crystallites. The size difference in crystallites between the two pictures is the result of sintering. (Reproduced from Chapter 5 of Heck, R.M., Farrauto, R.J., and Gulati, S.T. (2009) *Catalytic Air Pollution Control: Commercial Technology*, 3rd edn, John Wiley & Sons, Inc., New York.)

Catalytic scientists have found preparation and compositional additions to decrease the rate of sintering by using the proper carriers and/or stabilizers, but with time all materials reach a quasi-steady state as they seek to approach minimum free energy as dictated by thermodynamics.

The loss of performance due to active metal or metal oxide sintering is so important for high-temperature applications that research in learning to stabilize high dispersions has and will be conducted for years to come. The addition of certain stabilizers to the catalyst formulation has been a fruitful approach. Certain rare earth oxides such as CeO_2 and La_2O_3 have been effective in reducing sintering rates of Pt in the automobile exhaust catalytic converter. Once again the precise mechanism of stabilization is not clearly understood, but obviously the stabilizers fix the catalytic components to the surface-minimizing mobility and crystal growth. In some cases, the sintering can occur simply by migration over the surface as a function of the wetting angle of the metal or metal oxide to the carrier surface. Catalytic components sinter more readily on SiO_2 that has a lower concentration of hydroxide species on its surface compared to gamma γ-Al_2O_3.

In the idealized conversion versus temperature plot of Figure 5.4, the conversion of the sintered catalyst is displaced to higher temperatures relative to the fresh catalysts since fewer sites are available. It should be noted that the activation energy, as evidenced by the parallel profile, is unchanged since the kinetic control is still the operative rate-limiting step.

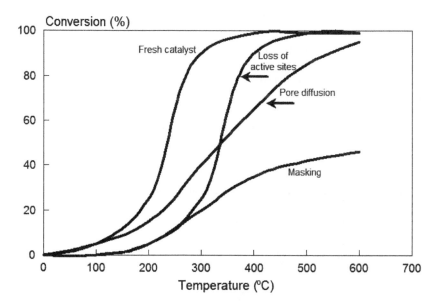

Loss of active sites = sintering of sites and carrier
Pore diffusion = smaller effective pores
Masking = some surface sites covered and pores blocked

Figure 5.4 Idealized conversion versus temperature for various aging phenomena. (Reproduced from Chapter 5 of Heck, R.M., Farrauto, R.J., and Gulati, S.T. (2009) *Catalytic Air Pollution Control: Commercial Technology*, 3rd edn, John Wiley & Sons, Inc., New York.)

5.2.2 Sintering of Carrier

Within a given structure, such as γ-Al_2O_3, the loss of internal (BET) surface area is associated with the loss of surface hydroxyl groups and a gradual loss of the internal pore structure network, leading to occlusion of the some of the catalytic sites. This is shown in the simple cartoon of Figure 5.5. The surface hydroxyl groups are not shown, but the sintering process results in the expulsion of a surface H_2O molecule and the bonding of adjacent Al sites to the remaining O ion.

$$\begin{array}{cc} OH & OH \\ | & | \\ Al & Al \end{array} \longrightarrow \quad \begin{array}{c} O \\ \diagup \diagdown \\ Al \quad Al \end{array} + H_2O$$

As the sintering phenomena occur, some pore openings become progressively smaller, introducing greater pore diffusion resistance. Thus, a chemically controlled reaction may gradually become limited by pore diffusion. The occurrence of these phenomena can be suspected when the apparent activation energy of the catalytic reaction is observed. In the plot of conversion versus temperature of Figure 5.4, the slope of the curve becomes progressively lower. In the extreme case, the pores completely close masking the catalytic sites within that pore rendering them inaccessible to the reactants. The 2θ peaks in the X-ray diffraction pattern become sharper relative to the less sintered material since diffraction occurs on more well-defined crystal planes as the crystal becomes relative to the amorphous-like state of the fresh and well-dispersed material.

The second mechanism for the loss of internal surface area involves an irreversible conversion to a new crystal structure. Each conversion results in a decrease in the porosity and surface area of the Al_2O_3. At about 800 °C, γ-Al_2O_3 converts to delta (δ)-Al_2O_3, with a surface area about 50–60% that of γ, with a more XRD crystalline structure. At about 1000 °C, δ-Al_2O_3 converts to theta θ-Al_2O_3 with a 75–95% reduction in its surface area. The X-ray diffraction peaks become even sharper with greater crystallinity. The final high-temperature stable structure at about 1100 °C is alpha (α)-Al_2O_3, which is the most crystalline of all Al_2O_3 structures with virtually no porosity on surface area (i.e., 1–2% γ). It is highly compact as was shown in Figure 2.1. In a similar manner, TiO_2, used as a carrier for supporting V_2O_5 for selective reduction of NO_x with NH_3, is irreversibly converted from its high surface

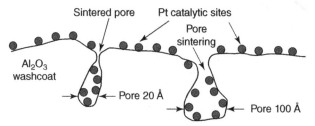

Figure 5.5   Illustration of the sintering of the catalyst carrier occluding the catalytic component. (Reproduced from Chapter 5 of Heck, R.M., Farrauto, R.J., and Gulati, S.T. (2009) *Catalytic Air Pollution Control: Commercial Technology*, 3rd edn, John Wiley & Sons, Inc., New York.)

(a)   (b)

Figure 5.6   Microscopy images of low surface area rutile (a) and high surface area anatase (b). Each set of four photos show the structure at increasing magnification. (Reproduced from Chapter 5 of Heck, R.M., Farrauto, R.J., and Gulati, S.T. (2009) *Catalytic Air Pollution Control: Commercial Technology*, 3rd edn, John Wiley & Sons, Inc., New York.)

area anatase structure ($\sim 80 \, m^2/g$) to its rutile structure with a surface area $<10 \, m^2/g$ at about 550 °C. Figure 5.6 shows the morphology change associated with this transformation. The X-ray diffraction pattern has new 2θ values that allow differentiation of the structures present. The peaks positions are sharper and characteristic of low surface area rutile.

In the conversion versus temperature (*X–T*) profile of Figure 5.4, carrier sintering results in a shift to higher temperatures since some active sites are no longer accessible to the reactants. When the pores get smaller, the rate-limiting step changes from kinetic control to pore diffusion and the slope becomes lower since the activation energy for pore diffusion is lower than that for kinetic control.

The occurrence of either mechanism is primarily detected by N_2 surface area measurements, that is, BET, XRD, and to a lesser extent, pore size distribution. Conversion from one phase to another will generate a completely different XRD pattern and allow a semiquantitative estimate of crystal sizes. The X-ray diffraction patterns for high surface area and low surface area structures were shown in Figure 3.6. The BET surface area measurement is also commonly used; however, other mechanisms, that is, masking or fouling, can lead to a decline in apparent internal surface area without sintering having occurred. Similarly, pore size measurements are useful, but do not present a complete picture since pore blockage due to masking also cause an apparent change in the pore size distribution.

The presence of specific amounts of stabilizers, such as BaO, La_2O_3, SiO_2, and ZrO_2 (see Chapter 2), can retard the rate of sintering in certain carriers. These were discovered during the early studies of the development of the catalytic converter where sintering occurred due to the extremely high exhaust temperatures experienced.

When a zeolite experiences high temperature, the Si–O–Al bridges undergo a process called dealumination, where Al is extracted from the framework of its tetrahedral structure. Since Al is the site for active catalytic sites such as H^+ or metal cations, they also leave the framework and the electronic environment within the pore of the zeolite is altered. Naturally, this changes the activity/selectivity of the catalyst. The Al-containing species form new structures with penta- and octahedral coordination sites, none of which have well-defined pore sizes like the zeolite from which they are derived. Thus, the crystallinity of the zeolite is decreased. This is shown in Figure 5.7a where the fresh zeolite loses Si–O–Al bridges, that is, Si(3Al), Si (2Al), and Si(Al), after high-temperature exposure. The terminology describes the

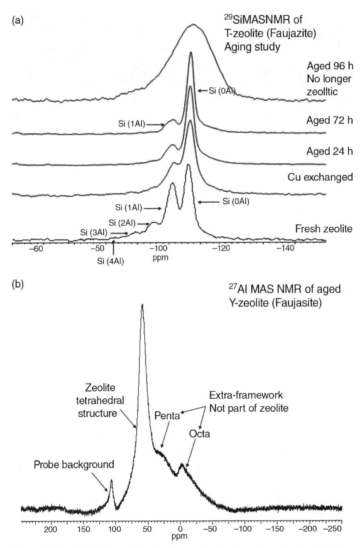

Figure 5.7   (a) NMR profile of a thermally aged zeolite showing the loss of the Si–O–Al bridges. Si(3Al), Si(2Al), and Si(Al) are seen to decrease in intensity with the progressively more severe thermal aging. (b) Growth of penta- and octahedral coordination sites in a thermally deactivated zeolite. (Reproduced from Chapter 5 of Heck, R.M., Farrauto, R.J., and Gulati, S.T. (2009) *Catalytic Air Pollution Control: Commercial Technology*, 3rd edn, John Wiley & Sons, Inc., New York.)

number of Si–O–Al bridges present. The formation of undesirable penta- and octahedral sites are shown in Figure 5.7b as a function of various prolonged thermal exposures.

5.2.3 Catalytic Species–Carrier Interactions

Reaction of the active catalytic component with the carrier can be a source of deactivation if the product is less active than the initially dispersed species. For example, Rh_2O_3 reacts with a high surface area γ-Al_2O_3 forming an inactive compound during high-temperature lean conditions (called fuel shut-off) in the automobile exhaust. This is a particularly important mechanism for deactivation of NO_x reduction activity. The reaction believed to be occurring is conceptually shown in Equation 5.1. Fortunately, this reaction is reversible when exposed to rich fuel conditions, that is, H_2 is generated when it returns to its stoichiometric operating condition:

$$Rh_2O_3 + Al_2O_3 \underset{H_2}{\overset{800\ °C\ (air)}{\rightleftharpoons}} Rh_2Al_2O_4)$$ (5.1)

One of the reasons base metal oxides such as Cu, Ni, and Co were excluded from use as autoexhaust oxidation catalysts was their reaction with Υ-Al_2O_3 forming inactive materials.

$$CuO + \Upsilon\text{-}Al_{23} \xrightarrow{800\ °C} CuAl_2O_4$$ (5.2)

The conversion versus temperature plot (Figure 5.4) shifts to higher temperatures due to the loss of active sites. The slope may or may not change depending on the catalytic activity and activation energy of the new structure formed. This undesirable reaction has led to the development of carriers such as SiO_2, ZrO_2, TiO_2, and their combinations that are less reactive with Rh_2O_3 than Al_2O_3. The interaction problem can be solved by these alternative carriers, but often they are not as stable against sintering and, thus, there is a trade-off in performance that must be factored into the new catalyst formulations.

Rh also has an undesirable reaction with the oxygen storage component (CeO_2-containing compounds), an important component in the three-way catalytic converters washcoat. Separating these components by incorporating them in different washcoat layers on the monolith is one approach toward minimizing deactivation.

Reactions between components that form new compounds can be monitored by XRD provided the amount present is large enough to be XRD visible. The Rh_2O_3/Al_2O_3 compound formed cannot be easily studied by XRD due to the relatively small amount of Rh present. Selective chemisorption may be considered the most useful technique, but its measurement requires a reduction of the Rh to the metallic state that decomposes the Rh–Al compound and consequently regenerates the Rh. Temperature-programmed reduction in which the amount and temperature at which H_2 decomposes the Rh_2O_3–Al_2O_3 complex is a convenient technique for studying these interactions. One measures the amount of H_2 consumed in the reduction of the Rh ion in the compound. X-ray photoelectron spectroscopy (XPS) has become useful in elucidating the oxidation state changes of the various catalytic components.

5.3   POISONING

A common cause of catalyst deactivation results from contaminants present in the feedstock or from the process equipment depositing onto the catalyst surface. There are two basic mechanisms by which poisoning occurs: (i) selective poisoning, in which an undesirable contaminant directly reacts with the active site or the carrier, rendering it less active and (ii) nonselective poisoning such as deposition of fouling agents onto or into the catalyst carrier, masking sites and pores resulting in a loss in performance due to a decrease in accessibility of reactants to active sites.

5.3.1   Selective Poisoning

This is a discriminating process, by which a poison directly reacts with an active site, decreasing its activity or selectivity for a given reaction, as the cartoon shows in Figure 5.8. This mechanism applies to many metallic catalysts such as Ni, Cu, and Pd, especially when the poison is sulfur oxide or H_2S.

Some poisons chemically react with the catalytic component such as Pb, Hg, and Cd forming catalytically inactive alloys. The activity of Pt is dramatically reduced when a Pt–Pb alloy forms by reaction of the Pb compounds, previously present in gasoline with Pt. For this reason, Pb compounds have been removed from today's gasoline. Some poisons merely adsorb (chemisorb) onto sites, that is, SO_2 onto a metal site (i.e., Pd, CuO, Ni, etc.), and block that site from further reaction. Some mechanisms are reversible in that heat treatment, washing, or simply removing the poison from the process stream will desorb the poison from the catalytic site restoring its catalytic activity. When some of the active sites are directly poisoned, there is a shift in the conversion versus temperature profile to higher temperatures. However, no change in the slope occurs since the remaining sites can function as before with no change in the overall process activation energy. The conversion versus temperature diagram will look similar to that for catalytic sintering with a loss in active sites and a parallel shift to higher temperatures. Chemisorption and XPS are common methods for analysis; however, XPS is conducted in vacuum that could remove poisons unintentionally during the measurement. The pretreatment for chemisorption is a high-temperature reduction that can also remove the poisons; so care must be exercised in data interpretation.

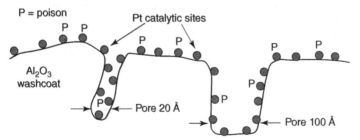

Figure 5.8   Conceptual cartoon showing selective poisoning of the catalytic sites. (Reproduced from Chapter 5 of Heck, R.M., Farrauto, R.J., and Gulati, S.T. (2009) *Catalytic Air Pollution Control: Commercial Technology*, 3rd edn, John Wiley & Sons, Inc., New York.)

When the carrier reacts with a constituent in the gas stream to form a new compound, as in the case of $Al_2(SO_4)_3$, pores are generally partially blocked resulting in increased diffusional resistance. This will cause a decrease in the activation energy and thus the conversion versus temperature curve will shift to higher temperatures with a lower slope, as shown in Figure 5.4. XRD is a useful tool for monitoring this mode of deactivation.

5.3.2 Nonselective Poisoning or Masking

In some applications, aerosols or high molecular weight material from upstream equipment can physically deposit onto the surface of the catalyst clogging pores and blocking access of reactants to the catalytic sites. This mechanism of deactivation is referred to as fouling or masking. Reactor scale, that is, Fe, Ni, Cr, and so on, corrosion, silica- and alumina-containing dusts, phosphorous from lubricating oils, and so on, are frequently found on catalysts. A cartoon sketch of these phenomena is shown in Figure 5.9. These phenomena are nonselective and independent of the catalytic component or carrier since it is physical deposition.

These mechanisms are nondiscriminating in that they deposit physically on the outer surface of the catalyst. The conversion versus temperature curve will show a substantial stepwise drop in the maximum conversion due to the loss in geometric area that impacts the bulk mass transfer area (see Figure 5.4). Those poisons that penetrate into the porous network will coat the inside of the pores leading to enhanced pore diffusion resistance. The slope of the conversion versus temperature plot will be decreased due to the decrease in activation energy. Methods for regeneration will be discussed in Chapter 12.

Surface chemical analysis by SEM or XPS is a commonly used method for detecting the nature of the masking agent. Accompanying masking, there usually is a decline in surface area due to pore blockage. An XPS spectrum of poisoned Pt/Al_2O_3 catalysts used to abate emissions from a sulfur-containing fuel-rich burning engine shows a heavy concentration of poisons, as shown in Figures 5.10. The surface has large concentrations of carbon (6.6%), Fe (2.7%), and S (2.4%). Also shown are the fresh components of the catalysts: Al at 35.6%, Pt at 0.2%, and O at 52.5%. Fe was found to be derived from reactor scale corrosion.

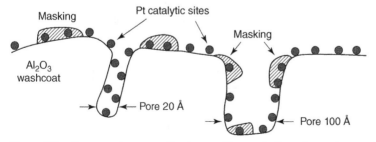

Figure 5.9   Conceptual cartoon showing masking or fouling of a catalyst washcoat. (Reproduced from Chapter 5 of Heck, R.M., Farrauto, R.J., and Gulati, S.T. (2009) *Catalytic Air Pollution Control: Commercial Technology*, 3rd edn, John Wiley & Sons, Inc., New York.)

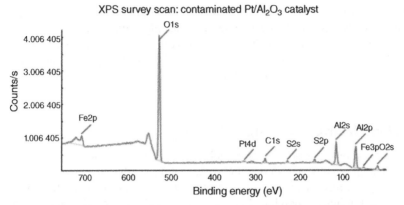

Figure 5.10   XPS spectrum of the surface of a contaminated Pt on Al_2O_3 catalyst.

A very special type of masking or fouling is coking in which a carbon-rich hydrogen-deficient material is formed on and in the porous network, which is discussed below. Nonselective poisoning, such as fouling, is a physical phenomenon and contaminants usually accumulate at the inlet of the catalyst bed or in the case of the monolith about 1–3 cm axially downstream from the inlet face. Furthermore, they tend to concentrate on the outer periphery of the catalyst particle, usually no more than about 30 μm deep from the gas–surface interface. Selective poisons, such as sulfur oxides, are more discriminating and will deposit on the specific sites and thus may penetrate into the depths of the catalyst.

These two mechanisms can be seen in Figure 5.11. Aerosols of phosphorous, calcium, and zinc, originating from the detergent package in lubricating oils, have

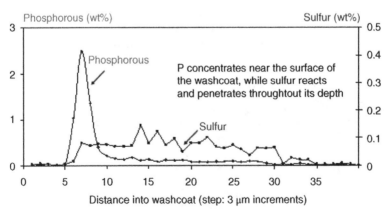

Figure 5.11   Electron microprobe showing the deposition location of the poisons within the washcoat of a monolith catalyst used in an automobile catalytic converter. The X-ray beam is scanned perpendicular to the axial direction through thickness of the washcoat.

deposited within about 30 μm depth of the washcoat thickness. (The units on the x-axis require multiplication by 3.) The sulfur, existing in the gas phase as SO_2/SO_3, forms $Al_2(SO_4)_3$ and is present more uniformly throughout the depths of the washcoat.

5.4   COKE FORMATION AND CATALYST REGENERATION

Coke is a common source of catalyst deactivation in petroleum and chemical processes. It can be thought of as a highly unsaturated species that deposits both chemically and physically on and within the catalyst particle that can lead to selective poisoning as well as masking. Surface acid sites catalyze coke formation by forming high molecular weight polymers that accumulate over time. Some metals catalyze coke formation to varying degrees. For example, Ni is an excellent catalyst for dehydrogenating molecules by forming olefins, such as ethylene, known to be a powerful coke precursor. In contrast, Rh has a lower activity toward hydrocarbon dehydrogenation and thus is considered more coke tolerant; however, it still forms coke depending on the temperature and the other reactants present. For example, in the steam reforming process to produce H_2 and CO, excess steam decreases coke formation for all catalysts, especially for Rh. This mechanism of deactivation will be discussed in more detail as it relates to petroleum processing in Chapter 10.

Thermal analysis (TGA/DTA) is an excellent tool to establish conditions for regenerating coked and poisoned catalysts. By measuring the weight and energy change of a representative sample of contaminated catalyst in a microbalance, the temperature needed for coke burn-off and regeneration can be determined. Coke deposition and its burn-off are measured by TGA/DTA (temperature-programmed oxidation). Coke forms mainly in petroleum and many chemical processes, but can also occur in the rich mode of an exhaust catalyst. This is shown in Figure 5.12 for a Rh-containing hydrocarbon steam reforming catalyst. The first minor weight loss below 200 °C is due to volatile components. Note that the weight loss is endothermic indicating volatility. At about 425 °C, the coke begins to combust. The weight loss is accompanied by the exothermic heat of combustion that peaks at 556 °C.

In Figure 5.13, a TGA/DTA profile for endothermic desorption of sulfur from a Pd-containing/Al_2O_3 oxidation catalyst is shown. Desulfation from the metal is complete around 800 °C after which the $Al_2(SO_4)_3$ begins to decompose. The high temperature for its decomposition indicates it is a very stable compound and therefore is not a preferred carrier for high sulfur-containing gas streams. The $Al_2(SO_4)_3$ is a large-volume compound that blocks pores and causes enhanced diffusional resistance for the reactants decreasing activity and reducing its life.

Core from deactivated (coked) TWC obtained from full-scale engine test

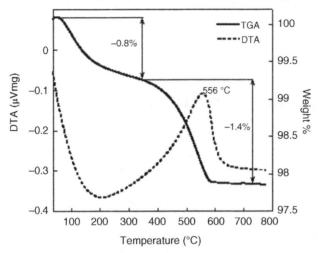

Figure 5.12 TGA/DTA in air of coke burn-off from a catalyst. (Reproduced from Chapter 5 of Heck, R.M., Farrauto, R.J., and Gulati, S.T. (2009) *Catalytic Air Pollution Control: Commercial Technology*, 3rd edn, John Wiley & Sons, Inc., New York.)

Desulfation of sulfated PtPd/MgO-ZrO$_2$/alumina Doc catalyst

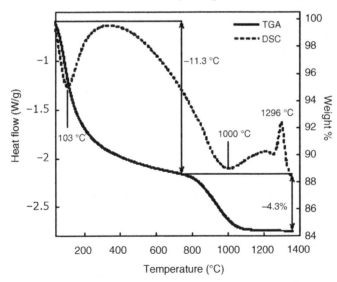

Figure 5.13 TGA/DTA profile for desulfation of Pd on Al$_2$O$_3$ catalyst. (Reproduced from Chapter 5 of Heck, R.M., Farrauto, R.J., and Gulati, S.T. (2009) *Catalytic Air Pollution Control: Commercial Technology*, 3rd edn, John Wiley & Sons, Inc., New York.)

QUESTIONS

1. How can you use TGA/DTA for catalysis preparation, deactivation, and regeneration?

2. Describe three different deactivation mechanisms and provide examples for each. Provide at least one analytical test that could be used to confirm the decay mechanism.

3. A Pt/Al_2O_3//monolith catalyst is returned from the field because the customer is complaining about poor performance. You extract a sample for laboratory testing to determine deactivation modes. Suggest the causes of deactivation for each set of results.

   a. The conversion–temperature profile is shifted to the right of the fresh catalyst, but the slope is the same. The chemisorption of CO is decreased by 75%, but the BET surface area is unchanged. The Pt crystallite size was examined by microscopy (TEM) and there appears to be no change in size of the Pt.

   b. A Pd/TiO_2//monolithic catalyst used for combusting solvent hydrocarbons emitted from an automobile paint factory shows a small decrease in BET surface area and the XRD shows that the TiO_2 washcoat has sharper peaks. The CO chemisorption is the same, but the slope of the conversion–temperature profile has decreased. What are the possible causes of deactivation?

4. A Pd/Al_2O_3//monolith ozone decomposition catalyst is removed from the air intake of a Boeing 777 wide-body commercial jet after 10,000 h of operation. In the laboratory, it is noted that the conversion–temperature profile is shifted to the right with a reduction in slope. It never reaches its maximum fresh activity. What is the deactivation mode?

5. A customer complains that the catalyst is deactivating in the process stream, but when you remove it and test it in the laboratory with a model gas stream containing the main components to be converted, there is no sign of deactivation. What are some possible explanations besides the customer being crazy?

6. A $2\%Pt/Al_2O_3$ catalyst, as a tablet, is to be used to abate emissions from a coal-fired power plant. What characterization methods would be used to determine the various modes of deactivation? Also sketch the conversion–temperature profile from a sample of deactivated catalyst using a model reaction in the laboratory. In some cases, it is possible that there is more than one deactivation mode. Consider each of them.

   a. The data recorded overnight indicated the catalyst saw an excessively high temperature due to a failed heat exchanger upstream.

   b. A compressor fails upstream and oil leaks into the exit gas to be abated. Consider (a).

   c. What process and catalyst changes would you suggest to minimize the possible deactivation modes?

   d. Secondary air is necessary to maintain a lean (excess air) environment for the catalyst. If the filter on the injected air stream gets clogged with dust, what might happen to the catalyst?

   e. How might you recover the activity for (d)?

7. Consider a highly exothermic hydrocarbon oxidation reaction proceeding down the length of a Pd/TiO_2//monolith. The feed stream contains a high concentration of sulfur dioxide, which is known to selectively poison only the Pd sites reducing its activity to zero. Assuming you have a series of thermocouple probes inserted at different axial lengths within the monolith, how would the temperature profile (conversion) change with time?

8. **a.** An inexpensive motorcycle operates fuel-rich two-cycle engine in which the oil is mixed directly with the fuel. Its exhaust therefore contains significant amounts of inorganic components (ash) from the oil such as (oxides of P, Zn, and Ca) added as a lubricant. This exhaust contains a high concentration of CO and additional air must be added. A Pt/ZrO$_2$//monolith is located in the engine exhaust to abate the CO. The ash nonselectively deposits a few millimeter within the inlet of the monolithic catalyst. Assuming you have a series of thermocouple probes inserted at different axial lengths within the monolith, how would the temperature profile (conversion) change with time?

   b. Can you suggest a solution to preventing ash from decreasing the activity of the catalyst?

9. A Cu-exchanged zeolite is used as a selective NO$_x$ reduction catalyst to elemental nitrogen in a desulfurized natural gas-fueled power plant operating in an air-rich mode. Ammonia is injected as the selective reductant. The outlet NO$_x$ levels are very low until there is a large-temperature spike in the catalyst caused by a leak of some natural gas and its oxidation in the catalyst bed. A large breakthrough of NO$_x$ appears and never decreases. The catalyst is replaced with a new one, but the plant manager wants to understand what changes occurred in the catalyst. What characterization techniques would you use to determine the cause?

10. A Ni/Al$_2$O$_3$ steam reforming catalyst has been observed to be slowly losing activity. It is well known that the Ni catalysts tend to produce and accumulate significant quantities of coke. So, naturally you suspect coke accumulation is the cause of deactivation. Suggest the characterization test that would be best to confirm the presence of coke and roughly quantify the amount on the catalyst surface. Explain in a few concise sentences how this test would be performed and how you would interpret the results determine the amount of coke on the catalyst surface.

11. Following from Question 9, you perform your test and realize there is no measurable amount of coke on the Ni/Al$_2$O$_3$ catalyst. It is also well known that sulfur species will poison the catalyst surface. The sulfur species will strongly adsorb onto the active site and can be detected using the appropriate catalyst characterization technique. So, now you suspect sulfur poison to be the cause of deactivation. What is the best test or combination of tests to determine whether sulfur species are on the surface of the catalyst? List only *one* test (or combination of tests) and explain your choice.

12. A Rh/γAl$_2$O$_3$ LPG steam reforming catalyst is returned from the field due to deteriorating performance. A laboratory-scale activity test of a small sample revealed the conversion–temperature profile has shifted to the right, the slope has decreased, and the maximum conversion has decreased. Surface characterization revealed a significant decrease in BET (nitrogen adsorption) surface area and a significantly lower CO chemisorption uptake. XRD did not detect the presence of high-temperature alumina phases. *What is most likely cause of deactivation. What is the best follow-up test to confirm your diagnosis?*

13. A Ni/Al$_2$O$_3$ catalyst is used for slurry-phase hydrogenation. The catalyst is observed to deactivate over time. A catalyst sample is withdrawn for characterization. Its XRD pattern is shown to have decreased in half-width for the Al$_2$O$_3$ and indicates the presence of Ni, which was not detected in the fresh sample. From this single result, *what deactivation mechanism is indicated?* A laboratory-scale activity test is planned. *What shape for the conversion–temperature plot do you suspect?*

14. A Pt/Al_2O_3 oxidation catalyst was observed to deactivate. The deactivation was observed after short temperature spike. A small sample of the deactivated catalyst was evaluated in the laboratory. The light-off temperature increased and the slope of the $X–T$ plot decreased. XPS results did not reveal foreign surface species. TEM analysis indicated the Pt crystallite size was roughly equal to the fresh catalyst. Nitrogen BET surface area measurements indicated a significant decrease in catalyst surface area. *What cause of deactivation is consistent with these results? What follow-up testing should be performed to confirm your diagnosis?*

15. A Pt/Ce_2O_3 catalyst is used for water gas shift reaction. The catalyst is observed to deactivate. A sample is evaluated in the laboratory. The $X–T$ plot is observed to have a higher light-off temperature, but otherwise the slope and maximum conversion is unchanged. TEM reveals no change in Pt crystallite size. CO chemisorption indicates lower CO uptake. N_2 adsorption indicates no change in pore size or internal surface area. Attempts to regenerate the catalyst by heating in dilute oxygen were not successful. *Which cause of deactivation is consistent with these results? What follow-up testing should be performed to confirm your diagnosis?*

BIBLIOGRAPHY

Bartholomew, C. (2003) Catalyst deactivation/regeneration, in *Encyclopedia of Catalysis*, Vol. 2 (ed I. Horvath), Wiley–Interscience, Hoboken, NJ, pp. 182–316.

Bartholomew, C. and Farrauto, R. (2006) *Fundamentals of Industrial Catalytic Processes*, 2nd edn, John Wiley & Sons, Inc., Hoboken, NJ.

Chen, J., Heck, R.M., and Farrauto, R.J. (1992) Deactivation, regeneration and poison resistant catalysts: commercial experience in stationary pollution abatement. *Catalysis Today* 11 (4), 517–546.

Forzatti, P., Buzzi-Ferraris, G., Morbidelli, M., and Carra, S. (1984) Deactivation of Catalysts: 1. Chemical and Kinetic Aspects. *International Chemical Engineering* 24 (1), 60–73.

Heck, R., Farrauto, R., and Lee, H. (1992) Commercial development and experience with catalytic ozone abatement in jet aircraft. *Catalysis Today* 13, 43.

Hegedus, L. and Baron, L. (1978) Phosphorus accumulation in automotive catalyst. *Journal of Catalysis* 54, 115.

Klimish, R., Summers, J., and Schlatter, J. (1975) The chemical degradation of automobile emission control catalysts, in *Catalysts for the Control of Automotive Pollutants* ACS Series 143 (ed J. McEvoy), American Chemical Society, Washington, DC, p. 103.

Lampert, J., Kazi, S., and Farrauto, R. (1997) Palladium catalyst performance for methane emissions abatement for lean burn natural gas engines. *Applied Catalysis B: Environmental* 14, 211.

Oudet, F., Vejux, A., and Courtine, P. (1989) Evolution during thermal treatment of pure and lanthanum doped Pt/Al_2O_3 and $Pt-Rh/Al_2O_3$ automotive exhaust catalysts. *Applied Catalysis* 50, 79–86.

Trimm, D. (1997) Deactivation and regeneration, in *Handbook of Heterogeneous Catalysis*, Vol. 3, Wiley-VCH Verlag GmbH, Weinheim, Germany, pp. 1263–1282.

CHAPTER 6

GENERATING HYDROGEN AND SYNTHESIS GAS BY CATALYTIC HYDROCARBON STEAM REFORMING

6.1 INTRODUCTION

6.1.1 Why Steam Reforming with Hydrocarbons?

Hydrogen is an essential element for many industrial processes and therefore its generation is critical for commerce. The largest concentration of hydrogen in our universe is in water (H_2O). The equilibrium constant for the formation of H_2O at room temperature (298 K) is 10^{40} or 10^{-40} for its dissociation. To dissociate about 50% of the steam into H_2 and O_2 ($K \sim 2$), a temperature of about 3100 °C is required. This temperature in principle could be provided by concentrated solar energy.

$$H_2 + O_2 \rightleftharpoons H_2O \tag{6.1}$$

Alternatively, electrolysis can split H_2O, at room temperature, with an applied voltage well above the thermodynamic value of 1.23 V. This energy can be provided via solar energy coupled with photovoltaic, nuclear, and wind energy, among others, all of which are being investigated as alternatives to fossil fuel. The industrial chemical process currently uses catalytic steam reforming (SR) of natural gas (e.g., methane), which permits the endothermic reaction to be performed at about 800–900 °C for essentially complete conversion of the H_2O and CH_4 to $3H_2$ and CO. The equilibrium constant for this reaction at 900 °C is about 1.4×10^3. So reaction with CH_4 reduces the external thermal energy necessary to split H_2 from steam making the process viable chemically. The presence of the proper catalyst (usually supported Ni) allows this reaction to occur at reasonable rates at 700–900 °C.

$$CH_4 + H_2O \rightleftharpoons 3H_2 + CO \tag{6.2}$$

Introduction to Catalysis and Industrial Catalytic Processes, First Edition. Robert J. Farrauto, Lucas Dorazio, and C.H. Bartholomew.
© 2016 John Wiley & Sons, Inc. Published 2016 by John Wiley & Sons, Inc.

6.2 LARGE-SCALE INDUSTRIAL PROCESS FOR HYDROGEN GENERATION

6.2.1 General Overview

Large-scale hydrogen production by catalytic reforming of fossil fuels has been practiced for many years by the chemical and petroleum industries. Industrial hydrogen generation involves a collection of catalytic steps that are illustrated in Figure 6.1. While any hydrocarbon can be used as the feedstock for steam reforming, the massive infrastructure of natural gas is the most common hydrocarbon source. Natural gas is mostly methane, but depending on its source typically also contains other C_2–C_5 hydrocarbons. Natural gas is odorless and small amounts of sulfur (<5 ppm)-containing compounds (mercaptans and thiophenes) are intentionally added to give it an easily detectable odor for safety since it is widely used in homes, schools, and businesses. However, these sulfur compounds are known poisons for catalysts and downstream processes and thus must be removed prior to the steam reforming process. The most common method for sulfur removal in industrial hydrogen processes is hydrodesulfurization (HDS). HDS involves a two-step process where the sulfur compounds react over a catalyst with hydrogen to form hydrogen sulfide, which is then adsorbed by zinc oxide in a downstream adsorption step. Typically, these processes are carried out at close to 30 atm.

With the sulfur in the hydrocarbon feedstock removed, steam is then mixed with the hydrocarbon feed and fed into the catalytic steam reforming reactor. In this step, the steam partially oxidizes the carbon in the fuel to CO while the hydrogen in the water and fuel is liberated as diatomic hydrogen. The reaction conditions are set to yield equilibrium concentrations with about >99% CH_4 conversion.

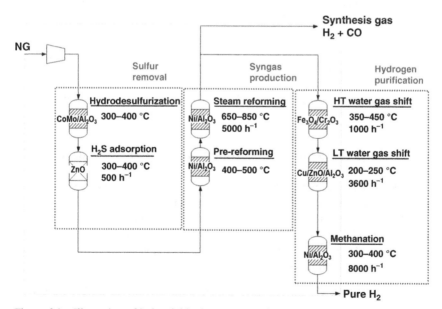

Figure 6.1   Illustration of industrial hydrogen generation process.

The effluent from the steam reformer is called synthesis gas, which is a name given to reflect the various uses of the mixture of CO and H_2. Depending on the final use of the synthesis gas, the CO concentration often must be reduced from the percent to ppm concentration level. This is accomplished in several process steps. Almost always, the water gas shift (WGS) reaction is used, which has the dual advantage of CO reduction while producing additional H_2. When ppm levels of CO concentrations are required, additional steps beyond the water gas shift reaction are required. Several technologies are available for this step including CO adsorption through a technology called pressure swing adsorption (PSA), methanation, and the preferential oxidation of CO. The method that is chosen is a function of the particular circumstances of the hydrogen plant. These will be discussed in this chapter.

6.2.2 Hydrodesulfurization

The HDS process operates at about 400 °C (673 K) and 30 atm of H_2, over a packed bed of particulate catalysts composed of 3% Co and 15% Mo (S) supported on Al_2O_3 (1.5–3 mm tablets). It is common practice to presulfide the catalyst to avoid high initial activity that consumes large amounts of H_2 while producing undesired light gases. The organic sulfur compound is designated as R-S while the hydrogen-rich product formed is designated as R-H.

$$\text{Hydrodesulfurization:} \quad \text{R-S} + 1.5H_2 \rightarrow \text{R-H} + H_2S$$
$$\text{(R-S = organic sulfur compounds)} \tag{6.3}$$

The sulfur-free CH_4 passes through the reactor unconverted and is not shown in Equation 6.3. The H_2S generated is reacted with ZnO pellets at about 400 °C consistent with the exit temperatures from HDS. This temperature provides reasonable equilibrium reaction of the H_2S with ZnO ($K_{673\,K} = 6.6 \times 10^5$) and sufficient kinetics.

$$\text{H}_2\text{S adsorption:} \quad H_2S + ZnO \rightarrow ZnS + H_2O \tag{6.4}$$

It should be noted that no steam was added to the HDS process since the reaction of H_2S with ZnO is favored by low steam concentrations as is evident from the equilibrium. The hydrogen used for HDS is recycled from the final product of the reforming process.

6.2.3 Hydrogen via Steam Reforming and Partial Oxidation

6.2.3.1 Steam Reforming The sulfur-free natural gas (methane) at 30 atm is mixed with pressurized steam and the mixture preheated to 800–900 °C (1073–1173 K), where it is fed to a Ni-containing packed bed catalytic reactor system. The H_2O/CH_4 ratio is between 2.5 and 3 to promote the equilibrium toward more H_2 and to decrease the tendency for coke generation on the catalyst surface. The excess water is then used in the downstream water gas shift reaction.

$$CH_4 + H_2O \rightleftharpoons 3H_2 + CO, \quad \Delta H° = +201\,\text{kJ/mol}, \quad \Delta G° = +206\,\text{kJ/mol} \tag{6.5}$$

Note that the standard free energy is larger than the reaction enthalpy due to the expansion of gaseous products increasing the reaction entropy. The equilibrium constant is unfavorable at 298 °C (571 K) ($K_{571\,K} = 9 \times 10^{-10}$), so temperatures up to 900 °C (1173 K) are required ($K_{1173\,K} = 1.4 \times 10^3$) to achieve acceptable conversions. Le Chatelier's principle indicates that the reaction is favored at low pressure due to the expansion of product gases relative to the reactants. However, the final H_2 produced is preferred to be at high pressure for storage and other downstream processes. The penalty in H_2 yield is offset by the higher volume product not having to be further compressed.

On the surface of the catalyst, methane steam reforming occurs as a sequence of five basic steps: (1) the methane and water adsorb onto the catalyst surface, (2) the adsorbed methane decomposes to form hydrocarbon fragments and carbon atoms, (3) the adsorbed H_2O dissociates to form H and OH species, (4) the hydrocarbon fragments and OH species recombine to form CO and H_2, and (5) CO and H_2 desorb from the catalyst surface. The overall rate of methane steam reforming is limited by the rate of the initial activation of the C—H bond in CH_4 (step 2). As such, the intrinsic rate of methane steam reforming is generally observed to be first order in CH_4 and pseudo-zero order in H_2O. The zero order of water is a consequence of the excess of water typical in commercial processes, but also due to the fact that water can be activated (H_2O dissociation) much more easily than methane (initial hydrogen extraction) on the catalyst surface. Overall, the activation energies vary from about 60 to 80 kJ/mol.

The steam reforming catalyst contains two primary components: (1) the active metal and (2) the metal oxide carrier. The active metal serves primarily as the site for hydrocarbon adsorption and decomposition. The appropriate metal for steam reforming is one capable of adsorbing the hydrocarbon, but not too strongly as to inhibit decomposition. Numerous metals have been observed to have activity for steam reforming, including Rh, Ru, Pd, Pt, and Ni. Due to cost considerations, Ni is almost always the active metal used in large-scale industrial processes. The carrier, α-Al_2O_3, can serve a number of roles. First, the carrier is the porous support for dispersing the active Ni metal. However, equally important, the metal oxide carrier may also provide sites for water activation to form H and OH species. The α-Al_2O_3 also has a relatively high heat conductivity that is important for heat transfer to all the catalyst in the reactor. An ideal candidate for a carrier material will be one that will maintain the stability of the dispersed metal, activate water, and transport the activated water to the adsorbed hydrocarbon. The α-Al_2O_3 carrier must also be mechanically strong to resist attrition during loading and the severe reforming operations of high temperature and pressure in the presence of steam.

The highly endothermic nature of the steam reforming reaction requires the continuous addition of heat into the catalyst bed. Otherwise, the catalyst bed temperature will rapidly drop as reaction occurs, which slows reaction kinetics and decreases equilibrium conversion. This condition is commonly referred to as a "heat transfer limited reaction." The catalytic reactor must be designed such that the rate of heat transfer into the catalyst bed is similar to the rate of heat consumed in the endothermic reaction. In a traditional steam reformer, this is accomplished by keeping the diameter of the catalyst bed as small as practical to decrease the heat transfer path.

It is common for the reactor to be composed of a series of small diameter (2–5 in. (5–11 cm) in diameter) metal alloy tubes filled with particulate catalysts.

Since heat is transmitted from the outer diameter inward and the rate is proportional to the temperature, a higher rate exists at the outer radial portion of the bed (nearest the heat source) than in the middle of the catalyst bed. This results in an ineffective use of the inner-radial bed catalyst (lower effectiveness factor). To improve the effectiveness factor, hundreds of small diameter metal tubes, packed with catalyst, are used to minimize the radial conduction path toward the radial center of the bed. Small diameter packed beds translate to higher linear velocities and higher pressure drop. A balance of enhanced heat transfer and allowable pressure drop is factored into the process economics. Space velocities of about $3000\,h^{-1}$ are typically used and are dictated by the extent of heat transfer resistance. The system is designed to deliver equilibrium product concentrations. Figure 6.2 shows a cartoon depicting a severely heat transfer limited reaction for CH_4 steam reforming. The temperature and the rate are highest nearest the heat source.

The catalyst-filled tubes are located within a natural gas-fueled furnace that provides the heat of reaction. The furnace is arranged such that there are alternating rows of tubes and burners. The burners are located on the top of the potentially 30 m high furnace and burn natural gas in the downward direction. At the top of the steam reformer, where the reaction rate is highest and thus greater heat transfer rates are needed, the catalyst tubes are directly adjacent to the gas combustion allowing heat transfer due to both convection and a significant radiation component. The combustion products flow downward in the furnace ultimately leaving the furnace through ceramic-lined tunnels. In order to recover the heat of combustion, the hot gases exiting the furnace are drawn into a very large heat exchange section commonly referred to as the "convection section" of the steam reformer. Within this large ceramic box, the hot combustion gases pass through a maze of piping recovering the combustion enthalpy to preheat reformer reactants, generate steam to be used elsewhere in the plant, heat process gases, and so on.

The Ni-based catalyst is prepared by multiple impregnations of α-Al_2O_3 with nickel salts to achieve the proper loading. Holes are often present in the carrier, ~0.2 in. diameter, in order to reduce pressure drop and reduce any pore diffusion that

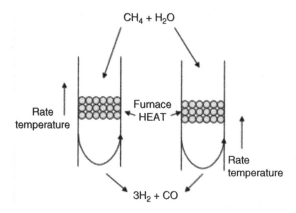

Figure 6.2 A series of metallic tubes filled with particulate catalysts bathed in a furnace of burning natural gas providing the required heat of reaction. The rate of reaction and temperature are highest near the heat source.

could further limit the reaction rate. The catalyst is typically 30–35% Ni/α-Al$_2$O$_3$ (tablets approximately 0.75–1 in. in cylinder diameter). About 14% CaO is added to the α-Al$_2$O$_3$ to improve strength, by formation of cement component CaAl$_2$O$_4$. About 1% K$_2$O is also added to decrease acidity minimizing coke formation and to catalyze gasification of the coke with steam.

To be active for steam reforming, the nickel oxide must be reduced to its metallic state. However, once activated, the nickel metal will oxidize spontaneously when exposed to air generating a large and potentially unsafe exotherm. Given the relatively high loading of Ni on the catalyst (~30 wt%) and high combustion enthalpy associated with Ni oxidation (−244 kJ/mol), spontaneous oxidation of the catalyst has the potential to release enough energy to raise the catalyst from room temperature to over 1000 °C. Certainly sufficient to destroy the catalyst, however more significantly, the potential energy release is sufficient to create serious safety concerns in the plant. As a result, nickel-based catalysts are always shipped from the catalyst manufacturer and loaded into the reactor in an oxidized state (NiO) or passivated in some way. In the case of shipping in its oxidized state, the catalyst is loaded as NiO, and then slowly reduced in the reactor to Ni metal prior to use (Equation 6.6).

$$NiO + H_2 \rightarrow Ni + H_2O, \quad \Delta H° = -500 \, kJ/mol \tag{6.6}$$

The reduction reaction is highly exothermic, so care must be taken to avoid damage to the catalyst and metallurgy of the reactor due to uncontrolled high temperatures. The bed temperature is monitored during the activation procedure, so when it rises to unacceptable levels, the H$_2$ flow and its concentration are decreased to safe bed temperatures (<800 °C). To buffer the exotherm, steam is added to the H$_2$ (H$_2$O/H$_2$ ~ 7). But since steam is a strong oxidizing agent, the balance of H$_2$ and H$_2$O must be respected to maintain a reducing environment. During the activation procedure, the higher the reduction temperature, the greater the oxidizing power of steam and therefore the H$_2$ must be increased (e.g., lower H$_2$O/H$_2$) to achieve satisfactory activation. This process may take up to 7 days to avoid reactor and catalyst damage (see Figure 6.3). Once the catalyst is suitably activated, the steam reforming reaction is initiated.

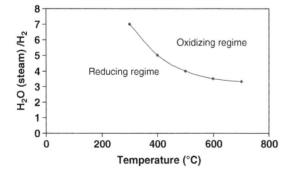

Figure 6.3 Reduction or activation of Ni SR catalyst: H$_2$O (steam)/H$_2$ as a function of temperature for redox of NiO/Ni.

6.2.3.2 *Deactivation of Steam Reforming Catalyst*   For the Ni/Al_2O_3 catalyst, nickel sintering and mechanical failure are common deactivation modes. Both
the processes are favored at high reaction temperatures and steam partial pressures.
These conditions are also unfavorable due to metallurgy limitations of the reactor
materials. Yet, it is also these conditions that favor both reaction kinetics and
thermodynamics. Thus, an optimal balance must be achieved among the reactor
material failure rate, the rate of catalyst deactivation, steam reforming kinetics, and
thermodynamics. To reduce the issues of attrition and breakage of the tablet catalyst,
high-strength α-Al_2O_3 is used as a carrier. Another issue associated with nickel is its
tendency to facilitate the accumulation of coke on the catalyst surface. First, nickel
catalyzes carbon-forming reactions, which most notably include

$$\text{Cracking and dehydrogenation:} \quad CH_4 \rightarrow C + 2H_2 \tag{6.7}$$

$$\text{Reverse gasification:} \quad CO + H_2 \rightarrow C + H_2O \tag{6.8}$$

$$\text{Boudourd:} \quad 2CO \rightarrow C + CO_2 \tag{6.9}$$

The Boudouard reaction is thermodynamically favored between roughly 450
and 550 °C, so the reaction kinetics associated with different catalytic components and
reforming conditions can be controlled to minimize this reaction. The other two
reactions can be limited by the catalyst composition, metal dispersion, acidity of the
carrier, and the H_2O/C ratio. A second issue is the accumulation of coke on the
catalyst surface. In the case of nickel, carbon formed on the metal surface can migrate
through the nickel crystallite to the carrier–metal interface, where the carbon
accumulates. More carbon forms on the exposed nickel surface, and migrates to
the carrier interface where it adds to the accumulation. This process repeats continuously and the carbon accumulation at the metal–carrier reaches a point where the
nickel crystallite lifts from the carrier surface and a carbon "filament" forms. This
accumulation definitely masks catalytic sites and occludes pores, but also has the
potential to be so severe that it fractures catalyst particles and fills the interstitial space
between catalyst particles, which increases pressure drop. To control coke accumulation, the H_2O/CH_4 ratio is increased with operational temperature for coke-free
operation as described in Figure 6.4.

Ultimately, coke will accumulate to significant levels especially as a lower H_2O/CH_4 ratio is used to decrease the energy needed to vaporize excess water. Coke can be

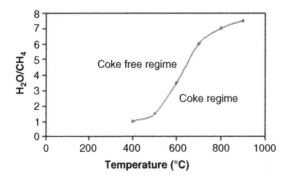

Figure 6.4 H_2O/C versus temperature: a high H_2O/CH_4 ratio
allows higher temperatures for
coke-free operation. To the right
of the line is the coke forming
regime.

removed and the catalyst regenerated at 400–500 °C (673–773 K) by oxidation or gasification. Steam gasification is generally used for several reasons. First, since steam is a reactant, it is readily available. Second, gasification is an endothermic reaction; thus, temperature control during catalyst regeneration is not an issue. Gasification, the reverse of Equation 6.8, is generally conducted at 673 K in steam over a period of roughly 6 h. The catalyst formulation often includes K as a gasification catalyst. Not only would such a promoter facilitate a purposeful regeneration during a maintenance period, but it also helps slow coke accumulation during the normal operation by promoting gasification on the catalyst surface. Any residual sulfur on the catalyst is desorbed as H_2S. Sometimes, a mild air treatment is needed to remove all the graphitic coke that is not easily removed during gasification. After the regeneration sequence, the catalyst must again be activated (reduced) according to the procedures described for start-up.

6.2.3.3 Pre-reforming

Methane, the major component in natural gas, is relatively unreactive and requires higher temperatures to achieve significant steam reforming rates. Hydrocarbons such as ethane, propane, and butane and their olefin counterparts are often present in natural gas and lead to coking at methane steam reforming temperatures. They can be catalytically pre-reformed to CH_4 at <450 °C preventing excessive coking during CH_4 steam reforming. As a result, heavy hydrocarbons are often called "coke precursors." Olefins are generally more reactive than paraffins and have a greater tendency to result in coke accumulation. The pre-reforming step is shown in Figure 6.1 upstream of the high-temperature steam reforming. The products are CH_4, CO, and H_2. Methane is formed (methanation) as the reverse of steam reforming shown in Equation 6.2. Pre-reforming is performed using Ni-based catalysts (3 mm × 3 mm) at about 400 °C (673 K). Once the heavier hydrocarbons are converted to syngas and methane, the higher temperature methane steam reforming can be performed with less coke formation.

6.2.3.4 Partial Oxidation and Autothermal Reforming

In endothermic catalytic steam reforming, the carbon atoms in the feed are partially oxidized to $H_2 + CO$ using steam. Alternatively, it is also possible to exothermically partially oxidize the carbon to CO using a substoichiometric amount of oxygen yielding synthesis gas (Equation 6.10).

$$CH_4 + \tfrac{1}{2}O_2 \text{ (in air)} \rightarrow 2H_2 + CO \tag{6.10}$$

Partial oxidation (PO) eliminates the heat transfer issue associated with steam reforming simplifying the reactor design. Actually, the thermal issue now becomes heat removal in order to prevent thermal damage to the metallurgy of the reactor materials and excessive deactivation of the catalyst. However, lower inlet temperatures can be used to offset the reaction enthalpy. Thus, the issue of heat removal is far less than that of heat transfer to maintain reaction temperature during steam reforming. Another difference from steam reforming, which may be an advantage or disadvantage, is the formation of less hydrogen per mole of CO. Depending on the ultimate use of the synthesis gas, this lower H_2/CO ratio could be an advantage for methanol synthesis that requires $2H_2/CO$ syngas or Fischer–Tropsch (F-T) synthesis for production of higher molecular weight hydrocarbons such as diesel fuel. If air is used as a source of O_2, then the product synthesis gas is diluted with N_2. So these factors must be factored into the process design that depends on the end use.

Another alternative is the combination of steam reforming and partial oxidation, which when combined together are referred to as autothermal reforming (ATR) (net reaction shown in Equation 6.11). In autothermal reforming, steam and oxygen (air) are mixed with the fuel. The catalyst used is one that has either two active components for oxidation and steam reforming reactions or a single active component for both reactions. The sequence described in Equation 6.11 suggests that the reaction proceeds net isothermally along the reactor length due to equalizing reaction enthalpy of the two reactions. However, in reality, a significant temperature gradient exists along the reactor length. The oxidation reaction is kinetically much faster than the steam reforming reaction. Thus, as the feed gas mixture enters the catalyst bed, the oxidation reaction occurs rapidly until all oxygen is depleted, which results in a rapid increase in reaction temperature. Then, as the slower steam reforming reaction occurs over the remaining reactor length, the sensible heat gained from the oxidation reaction is lost to the steam reforming reaction and the reaction temperature decreases. In theory, if we neglect heat loss with the environment, the outlet temperature could be the same as the inlet temperature. Recognizing that there will be this sharp increase in reaction temperature due to the oxidation reaction, the inlet temperature must be set to prevent exceeding the maximum allowable temperature for the reactor materials and the catalyst that otherwise could lead to thermal deactivation.

$$\text{ATR:} \quad \text{(unbalanced)} \, CH_4 + [O_2] \, \text{air} + H_2O \rightarrow H_2 \, (H_2O) + CO \, (CO_2) \quad (6.11)$$

$$\text{CPO:} \quad CH_4 + \tfrac{1}{2}O_2 \, \text{(in air)} \rightarrow 2H_2 + CO \quad (6.12)$$

$$\text{SR:} \quad CH_4 + H_2O \rightarrow 3H_2 + CO \quad (6.13)$$

In summary, three reaction scenarios are available for hydrogen generation: steam reforming, partial oxidation, and autothermal reforming. The difference between each option involves process design requirements (required heat exchange into or out of the reactor) and, perhaps more significant for industrial processes, the ultimate syngas composition in terms of the H_2/CO ratio. As will be discussed later in this chapter, different downstream processes require different optimal ratios and thus will make different reaction scenarios more or less ideal. Also in this chapter, small-scale hydrogen generation for fuel cells will be discussed. In small-scale applications, process intensification or miniaturization creates new heat transfer challenges. In these cases, the implications of each reaction scenario on process design will become more significant.

6.2.4 Water Gas Shift

The effluent from the steam reformer typically contains 14% dry CO, which for various reasons must be removed. The steam reformer effluent also contains a high concentration of unreacted steam, which provides the opportunity for it to catalytically react with CO to reduce the CO concentration and generate additional H_2.

For historical reasons, this catalytic reaction has been named the "water gas shift" reaction. At the beginning of the 20th century, the Haber–Bosch ammonia synthesis process obtained hydrogen from blowing steam over red hot coal (carbon). The resulting mixture of H_2 and CO was named "water gas." It was also discovered that cooling the water gas somewhat and contacting it with an appropriate catalyst shifted the CO to CO_2. Thus, this reaction was appropriately named "water gas shift."

$$CO + H_2O \rightleftharpoons H_2 + CO_2, \quad \Delta H° = -44\,kJ/mol \qquad (6.14)$$

For the generation of pure H_2, the water gas shift is carried out with two different catalysts at two different process conditions. The high-temperature shift (HTS) requires a Fe-containing catalyst and a low-temperature shift (LTS) requires a Cu-containing catalyst.

Regardless of the temperature regime, the reaction sequence occurring on the catalyst surface is basically the same. For the water gas shift reaction to occur, the following sequence of events must occur: (1) CO and H_2O adsorb onto the catalyst surface, (2) H_2O dissociates to form OH and H species, (3) adsorbed CO and OH species react and recombine to form CO_2 and H_2, (4) and finally the CO_2 and H_2 products desorb. Despite the apparent simplicity of the water gas shift reaction, the sequence of surface reactions comprising step 3 has been the topic of debate for decades. Many have suggested that the reaction of adsorbed CO and OH follows a redox sequence where the OH oxidizes the catalyst surface ultimately liberating H_2 as a product, and the oxidized catalyst surface is subsequently reduced by CO to form CO_2. The other side of this mechanistic debate has proposed an associative pathway where CO and OH species react directly to form an intermediate that then decomposes to form CO_2 and H_2. However, even the identity of the intermediate product is a debated topic where possible intermediates include carbonate (COO^{2-}), carboxyl ($HCOO^-$), and formate ($HOOC^-$) species. In the end, both reaction pathways are plausible and it is reasonable that either pathway could be followed depending on the catalyst used and the operating conditions. Given the right catalyst and conditions, it is even reasonable that the reaction could proceed through both pathways simultaneously at different locations on the catalyst surface. The mechanism is also a function of the catalytic elements used.

Regardless of the actual mechanism occurring on the surface, the catalyst must be designed to perform two fundamental processes. First, there must be sites for water to adsorb and activate to form OH and H species. In the case of the Cu-based catalyst, this likely occurs directly on the Cu metal site. In the case of the Fe-based catalyst, it is suspected that some iron exists as an oxide in close proximity to the Fe metal, and it is the iron oxide that activates water through a redox cycle. Second, the catalyst must also provide sites for CO to adsorb strongly enough to maintain it on the surface for the subsequent reaction with the OH species, but not too strong as to inhibit the subsequent reaction. Numerous metals will adsorb CO, including Cu, Fe, Au, Pt, Ni, and Ru. The binding strength of the CO to the surface is a function of the properties of the active metal. Studies have been performed correlating the adsorption strength of metal–CO to the rate of the WGSR. Of all the metals tested, this work found Cu–CO binding energy to be optimal for WGS where

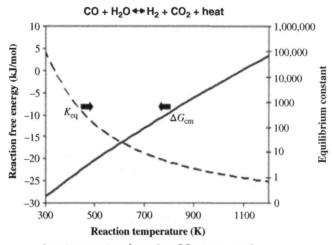

Figure 6.5   WGS equilibrium: free energy and equilibrium constant for WGS as a function of temperature.

other metals with lower binding energies (e.g., Au–CO) and higher binding energies (e.g., Fe–CO and Pt–CO) yielded lower WGSR rates relative to Cu–CO. The binding energy for Fe–CO is considerably stronger than Cu–CO. Thus, it becomes clear why Cu-based catalysts are good low-temperature catalysts and Fe-based catalysts require higher temperatures to achieve equivalent reaction rates. This "volcano-shaped" relationship between binding energy of the adsorbed reactant and reaction rate is not exclusive to the WGSR and this trend is found repeatedly throughout heterogeneous catalysis.

From a thermodynamic perspective, the slightly exothermic WGS reaction is favored at low temperatures (increasing equilibrium constant or decreasing free energy) as shown in Figure 6.5.

The equilibrium concentration of CO is shown as the dotted line in Figure 6.6. The solid line shows the exit concentration of CO that approaches equilibrium about 340 °C for the HTS process. The product gas from the steam reformer is typically 14% dry CO with 28–36% steam and with mostly H_2 and 3–5% CO_2. The HTS catalyst is composed of 90% Fe_2O_3 and 10% Cr_2O_3 and is thermally stable up to about 420 °C and can be safely used to reduce the CO to about 2% CO at an inlet temperature of about 320 °C. The higher temperature allows a space velocity of about 5000–7000 h^{-1}. The FeCr is the least expensive of the two WGS catalysts. The catalyst is produced by coprecipitation of the precursor salts of Fe and Cr. The precipitate composition is $Fe_2O_3–Cr_2O_3$. A paste is prepared and extruded to give 5–8 mm diameter particulates. The active catalyst is produced by a prereduction with H_2, where FeO with small amounts of Fe is produced and believed to be the active state (Equation 6.15).

$$Fe_2O_3 + H_2 \rightarrow 2FeO\,(Fe) + H_2O \tag{6.15}$$

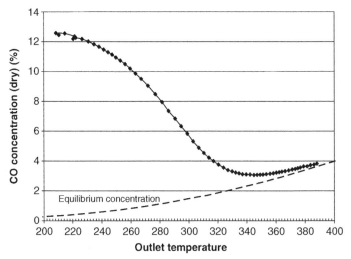

Figure 6.6   Typical performance of a HTS WGS catalyst with respect to exit CO.

Cr_2O_3 is added as an activity promoter and to minimize sintering. WGS catalysts are unsupported and are subject to sintering and therefore great care must be taken to avoid excessive temperature increases due to the exothermic activation procedures. Equation 6.15 shows only what is believed to be the active components (Fe). The promoter Cr_2O_3 is not reduced in the activated catalyst and is therefore not shown. Activation is accomplished at about 400 °C (673 K) in H_2 with steam present to dampen the rise in temperature due to the exotherm. It is also possible to activate the catalyst with mixtures of H_2 and CO depending on the individual plant start-up procedure. The temperature is monitored with thermocouples in the bed and the flow of reducing agent is carefully controlled.

The LTS CuZnAl catalyst is more active but more expensive of the two WGS catalysts and thus can be operated at inlet temperature of about 180 °C favoring equilibrium. It is used to reduce the CO from about 2% CO to <0.5% while generating an equivalent amount of H_2. It is typically composed of 30% CuO, 65% ZnO, and 5% Al_2O_3. It is coprecipitated and extruded to 2–4 mm particles. Like the HTS catalyst, it is unsupported and therefore prone to sintering (>280 °C) and mechanical failure. ZnO and Al_2O_3 are added as stabilizers and promoters. It is active in the reduced state, so it too must be carefully reduced with H_2. Only the active Cu phase is shown below since neither the ZnO nor the Al_2O_3 is catalytically active.

$$CuO + H_2 \rightarrow Cu + H_2O, \quad \Delta H° = -85 \, kJ/mol \qquad (6.16)$$

A consequence of lower temperature operation is a large catalyst bed to compensate for poorer kinetics (lower space velocities $\sim$2000–3000 h^{-1}).

The LTS catalyst can be avoided by using pressure swing adsorption during the production of pure H_2 to remove the remaining 2% CO. The exit composition from the HTS process contains about 2% CO, 65% H_2, 13–15% CO_2, and about 18–20%

steam. Essentially, pure H_2 can be separated from the residual gases by the PSA system that is described below.

There are now mid-temperature range WGS catalysts that operate at around 300 °C inlet temperature composed of both HT and LT shift components, for example, FeO with CuO and Cr_2O_3, that are finding use when producing pure H_2 followed by a PSA unit (Section 6.2.6.1).

6.2.4.1 Deactivation of Water Gas Shift Catalyst

Both high- and low-temperature WGS catalysts are deactivated by sulfur present in the feed but the main source of deactivation is mechanical failure such as attrition or particle breakage. Furthermore, these catalysts can be easily oxidized and mechanically weakened by exposure to liquid water contained in the synthesis gas, so shutdown must be performed in a dry and inert environment to preserve the integrity of the catalyst and for safety of the workers. Catalyst lifetimes of up to 3 years are typical.

6.2.5 Safety Considerations During Catalyst Removal

Both the steam reforming and water gas shift catalysts are active in their reduced state. Thus, when the process is to be shut down for maintenance, great care must be taken to avoid exposing all the catalysts to air. In the case of Cu and Ni, the oxidation enthalpy (ΔH°) is relatively high as shown in Equations 6.17 and 6.18. If these catalyst were allowed to oxidize rapidly, as would be the case if the reactor was opened and exposed to ambient air, significant heating would occur creating fire and thermal hazards that could damage equipment or injure workers. This occurs during routine maintenance of shutdown.

$$Cu + \tfrac{1}{2}O_2 \rightarrow CuO, \quad \Delta H^\circ = -150\,kJ/mol \qquad (6.17)$$

$$Ni + \tfrac{1}{2}O_2 \rightarrow NiO, \quad \Delta H^\circ = -250\,kJ/mol \qquad (6.18)$$

These hazards are prevented by performing a controlled air oxidation to generate a dense surface oxide or shell around the core of the metal that is impervious to air. The process is called passivation. Controlled oxidation is conducted at 50 °C with small amounts of dilute air (usually 1% O_2 in N_2) while monitoring the bed temperature to avoid a runaway reaction. When the exotherm is small, the catalyst can safely be discharged.

6.2.6 Other CO Removal Methods

Depending on how the hydrogen will be used, it is sometimes required to reduce CO concentrations to very low levels. Beyond catalytic processes, pressure swing adsorption is used for the production of high-purity hydrogen. As an alternative to water gas shift, one can use methanation (reverse of Section 6.2.6.2) or preferential oxidation of CO (Section 6.2.6.3).

6.2.6.1 Pressure Swing Absorption

The exit composition from the HTS process contains about 2% CO, 65% H_2, 13–15% CO_2, and about 18–20% steam.

In some plants, low-temperature water gas shift is replaced with pressure swing adsorption, which produces high-purity H_2 and a tail gas composed primarily of residual CO, CO_2, H_2O, and up to about 10% H_2. The pressurized process operates by adsorbing all components except 90–95% of the H_2, in a bed of high surface area zeolite. The gas-phase H_2 is removed in a relatively pure state at process pressure. The pressure in the PSA unit is then reduced and the tail gas desorbed in a batch process. The tail gas can be combusted providing some of the heat necessary for the endothermic SR reaction. There are usually banks of PSA units working in tandem to allow continuous operation.

6.2.6.2 *Methanation*

The methanation catalyst is typically about ~25% Ni on high surface area Al_2O_3 (5 mm × 5 mm tablets) but may also contain about 7% rare earth oxides for enhanced stability against sintering of both the Ni and Al_2O_3 carrier. Here, the CO is hydrogenated to CH_4 and H_2O (Equation 6.19). An undesirable side reaction that consumes 4 mol of H_2 per mole of CO_2 is the methanation of CO_2 (Equation 6.20). For this reason, the CO_2 must be removed prior to the CO methanation reaction. This is mostly carried out by scrubbing the CO_2 in mono-ethanolamine (MEA). The CO_2 is then recovered and used for methanol synthesis (see the next chapter). CO removal by methanation is shown in Figure 6.7.

Methanation: $CO + 3H_2 \rightarrow CH_4 + H_2O, \quad \Delta H° = -320 \, kJ/mol$ (6.19)

Undesired: $4H_2 + CO_2 \rightarrow CH_4 + 2H_2O, \quad \Delta H° = -1000 \, kJ/mol$ (6.20)

The methanation reaction is conducted at around 200–300 °C at a space velocity of ~7000 h^{-1}. It is highly exothermic ($K \sim 4 \times 10^3$) at 300 °C, but at 200 °C $K \sim 10^9$ and thus some cooling is required depending on the CO levels. The CO is reduced to <100 ppm. A final drying step is used to remove the H_2O prior to entering downstream such as the NH_3 synthesis reactor.

6.2.6.3 *Preferential Oxidation of CO*

An alternative technology to methanation is called preferential oxidation (PROX). A small amount of air ($O_2/CO \sim 1$) is injected into the effluent from the LTS process over 65% H_2, 15% CO_2, and 15–20% H_2O with about 0.5% CO. The highly selective PROX catalyst promotes the oxidation of 0.5% CO to less than 10 ppm without oxidizing appreciable amounts of the H_2 present. One commercial catalyst contains a low level of Pt deposited on Al_2O_3 spheres (3–5 mm in diameter) promoted with a trace of Fe and about 5% Cu. It operates at a SV = ~20,000 h^{-1}. The catalyst with Pt only requires almost 200 °C for CO oxidation with selectivity toward CO over H_2 of ~30%. The presence of Cu and Fe enhances the reaction kinetics, so it occurs at about 100 °C with a selectivity for CO of greater than 60%.

Desired: $CO + \frac{1}{2}O_2 \rightarrow CO_2$ (6.21)

Undesired: $H_2 + \frac{1}{2}O_2 \rightarrow H_2O$ (6.22)

This technology reduced the CO to <10 ppm. One advantage for the PROX system is that it can function in the presence of CO_2 while with methanation the CO_2 reacts with

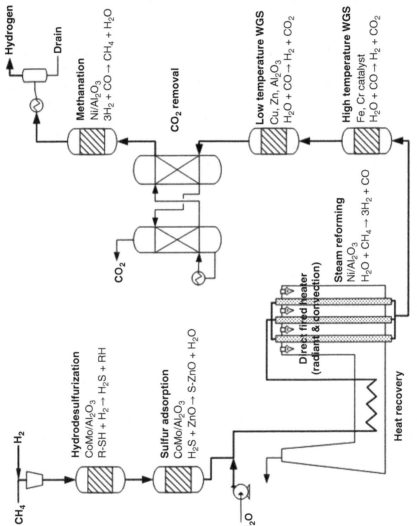

Figure 6.7 Reformer schematic for pure H_2.

valuable H_2 liberating a large exotherm. A disadvantage of PROX is that it adds N_2 to the product stream when air is used as the source of O_2.

6.2.7 Hydrogen Generation for Ammonia Synthesis

Ammonia is produced by the catalytic hydrogenation of nitrogen described in Equation 6.23. This reaction will be discussed in detail in the next chapter. However, the hydrogen required for ammonia synthesis is produced using the processes discussed here.

$$N_2 + 3H_2 \rightleftharpoons 2\,NH_3 \qquad (6.23)$$

One unique feature of hydrogen generation for ammonia synthesis is the use of two reforming processes: steam reforming followed by autothermal reforming (Figure 6.8). This combination of steps is performed to yield the required stoichiometry of hydrogen and nitrogen, where the latter is introduced with the air for autothermal reforming.

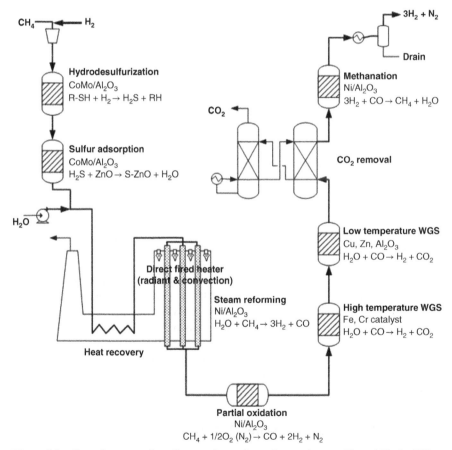

Figure 6.8   Overall process flow diagram for preformed natural gas to H_2 and N_2 for NH_3 production.

The process begins with steam reforming of desulfurized natural gas (30 atm) in the first reformer unit. Then, before entering the secondary reformer, a precise amount of air is added resulting in both partial oxidation and steam reforming occurring in the second reformer (ATR). In addition to heat generation, the addition of air brings the nitrogen into the synthesis gas that will be needed for downstream ammonia synthesis.

The upstream primary reformer reaction is limited to conversions of only about 50–60% of the CH_4 by adjusting heat input and the space velocity. The remaining CH_4 is then combined with a predetermined amount of air for the secondary reforming reaction over the Ni/α-Al_2O_3 particulate catalyst.

The ATR product gas containing about 5% CO is cooled to about 350 °C, where it undergoes HTS followed by additional cooling to about 180 °C where the CO (and H_2O) is further converted to CO_2 (and H_2) in the LTS catalyst bed. The effluent gas contains ~0.5% CO, CO_2, H_2O, and N_2 and is rich in H_2. All oxygenates, for example, CO, CO_2, and H_2O, must be removed since they are poisons to the NH_3 synthesis Fe catalyst. The product gas is passed through a reactor containing liquid MEA that selectivity removes the CO_2. This is necessary to avoid its hydrogenation in the final methanation step used to remove the last traces of CO. The methanation catalyst typically used consists of ~25% Ni on high surface area Al_2O_3.

The process for $H_2 + N_2$ generation is shown in Figure 6.8.

6.2.8 Hydrogen Generation for Methanol Synthesis

The synthesis of methanol can be written stoichiometrically as

$$2H_2 + CO \rightleftharpoons CH_3OH \tag{6.24}$$

It is believed to involve a water gas shift reaction followed by hydrogenation of CO_2. This appears reasonable since both methanol synthesis and water gas shift are catalyzed by Cu-containing catalysts.

$$CO + H_2O \rightleftharpoons H_2 + CO_2 \tag{6.25}$$

$$CO_2 + 3H_2 \rightleftharpoons CH_3OH + H_2O \tag{6.26}$$

The methanol synthesis reaction will be discussed in greater detail in the next chapter. $3H_2 + CO$ is the synthesis gas produced during steam reforming. The reaction stoichiometry for methanol is only 2 mol of H_2 for every mole of CO. Thus, one can use partial oxidation of methane (Equation 6.10) using pure O_2 to generate the proper 2:1 ratio needed. It is also possible to use reverse WGS (reverse of Equation 6.12) following steam reforming to bring the H_2/CO ratio to 2:1. CO_2 plays a key role in the synthesis of methanol and for this reason the source of CO_2 is from the MEA scrubber before methanation as shown in Figure 6.8. In general, a H_2/CO ratio of 3:1 with up to 10% CO_2 is obtained. Therefore, some processes use regular steam reforming with WGS to obtain the proper ratio.

6.2.9 Synthesis Gas for Fischer–Tropsch Synthesis

Synthesis gas is appropriately named since many different hydrocarbons and oxygenated hydrocarbons can be produced depending on the catalyst, process conditions, and the proper H_2/CO ratio. This important hydrocarbon synthesis process will be discussed in the next chapter.

Noncatalytic partial oxidation in which CH_4 is combined with O_2 (separated from N_2) to generate a H_2/CO ratio of ~2 is used by Shell for their F-T process. This is an exothermic reaction (>1300 °C) carried out at 70 atm for the feed gas to the F-T process.

$$CH_4 + \tfrac{1}{2}O_2 \rightarrow 2H_2 + CO, \quad \Delta H^\circ = -33\,\text{kJ/mol} \qquad (6.27)$$

Since natural gas often contains light hydrocarbons that can complicate the partial oxidation reaction by coking, they are removed in a separation pre-reforming step along with desulfurization. Pure O_2 is then mixed with pure CH_4 and delivered to the PO reactor.

Catalytic autothermal reforming (Equation 6.11) combines catalytic partial oxidation with catalytic steam reforming and generates approximately a H_2/CO ratio of 2:1 also suitable for the synthesis of a variety of hydrocarbons including diesel, gasoline, olefins, and alcohols. It is less complicated than SR since heat is generated *in situ*. Furthermore, with steam present endothermic SR occurs buffering the maximum temperature in the catalytic bed compared with PO or CPO causing less thermal stress on the materials of construction and the catalyst. ATR was discussed in the section on H_2 generation for ammonia production but pure O_2 is used since N_2 plays no role in F-T hydrocarbon production. Primary reforming gives a H_2/CO ratio of 3:1, while ATR gives 2:1. The catalyst for both ATR and CPO is Ni/α-Al_2O_3 particulates. If necessary, some WGS can be used to adjust the H_2/CO ratio to that required for the desired hydrocarbon or oxygenated hydrocarbon product.

6.3 HYDROGEN GENERATION FOR FUEL CELLS

Low-, medium-, and high-temperature fuel cells are now commonly found in the marketplace. Applications for central and distributed power generation, residential combined heat and power, and portable power offer advantages over more conventional power generation. These will be discussed more thoroughly in a later chapter. The common denominator for fuel cell applications is the utilization of a H_2-rich fuel for the anode electrochemical reaction. Fuel cell vehicles, operating with H_2 (stored on-board) with O_2 from air for the cathode reaction, are now available throughout the world but on a limited scale. Introduction to the commercial vehicle market occurred in 2015 depending on the availability of hydrogen service stations. Stationary fuel cell applications are now a reality for a number of applications. It is understood that the ideal hydrogen economy will utilize solar power and wind as natural sources of energy to generate H_2 by water electrolysis, but these technologies are localized and would require transportation of H_2 to its application site. Therefore, the transitional solution is to use the existing infrastructures of natural gas and LPG as a source of hydrogen.

For low-temperature fuel cells (proton-exchange membrane and phosphoric acid fuel cells), the unit operations for H_2 generation from natural gas or LPG are similar to those already been described for large-scale H_2 production for chemical and petroleum applications. Initially, most fuel cell integrators and manufacturers adapted the same technologies since a large database existed. They used traditional particulate structures (spheres, extrudates, tablets, etc.) usually of Group VIII metals such as Ni or Ru for steam reforming, Fe or Cu for WGS, and Ru for the preferential oxidation reaction of CO (for fuel cell quality hydrogen). As indicated above, desulfurization of the fuel is necessary since sulfur poisons catalysts and deteriorates downstream equipment and disturbs downstream processing steps. This is accomplished in the chemical industry with the HDS process that requires high-pressure H_2 (up to 30 atm) and temperatures up to 400 °C not suitable for small-scale applications. Therefore, solid adsorbents (e.g., carbons, zeolites, and metal-impregnated Al_2O_3 particulates) have been especially developed for selectively adsorbing organic (mercaptans and thiophenes) and inorganic sulfur compounds (COS and H_2S) commonly found (or added) in natural gas (1–5 ppm) and LPG (5–20 ppm). Removal is accomplished at atmospheric pressure without the need for H_2. Organic sulfur compounds are removed by physical adsorption on activated carbon or zeolites favored at low temperatures, for example, room temperature. Some inorganic compounds such as COS and H_2S are absorbed chemically (chemisorption or activated adsorption) onto metal-exchanged carriers and therefore reaction is favored at higher temperatures. Therefore, the feed (natural gas or LPG) is first delivered at room temperature where the organic sulfur compounds are adsorbed followed by a slight elevation in temperature for removal of the inorganic sulfur compounds. Outlet sulfur levels close to 1–2 ppb are achieved with a capacity approaching 2 wt%. Replacement of the adsorbents is part of the routine maintenance of the unit.

For low-temperature fuel cells (<200 °C), the steam reforming processes utilize Ni and sometimes Ru as particulates packed into metallic tubes and the traditional WGS catalysts (both high and low temperature) similar to those used in the chemical industry deliver approximately 0.5% CO to the PROX reactor. The PROX catalyst is either Ru on Al_2O_3 or Pt with traces of Fe (particulates) that, with the addition of O_2 (air) slightly above CO oxidation stoichiometry, oxidizes the small amount of CO present to less than 10 ppm without significant oxidation of the large excess of H_2 present. Typical operating conditions are 100–150 °C with an O_2/CO ratio of 1–1.5 with a SV ~ 10,000 h^{-1}.

All of the traditional catalysts when used for small-scale fuel cell applications require careful control in start-up and shutdown. For example, all catalysts must be in the reduced state; therefore, an activation procedure (H_2 reduction) must be used to initiate reforming. This is done immediately after the unit is installed as part of start-up. Shutdown must be carefully controlled to avoid the reduced catalyst reacting with air and liquid water, which causes structural damage to particulate catalysts. Nevertheless, they were and are still being used with additional controls installed.

6.3.1 New Catalyst and Reactor Designs for the Hydrogen Economy

In many cases, large-scale hydrogen plants cannot simply be reduced in size to meet the economic, safety, and frequent duty cycle requirements for some fuel cell applications with demands for varying H_2 levels for transient power requirements. In order to meet the new challenges, engineers have drawn on the successful

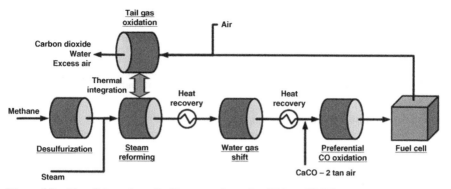

Figure 6.9   Monolith catalysts for H_2 generation using PSA or PROX.

experience of automotive pollution control where precious metal-containing wash-coated catalysts are deposited onto structured reactors (i.e., monoliths or heat exchangers). Structured reactors offer many significant advantages over traditional packed bed catalysts and reactors such as reductions in reactor size, lower weight, and pressure drop, rapid response to transient operations, and greater chemical and mechanical structural stability and robustness. Consequently, they are viable materials for small-scale distributed hydrogen generation. The high activity density of precious metal catalysts permits thin washcoats to be used on structured reactors, a situation not feasible for the less active base metal catalysts. Furthermore, no activation procedure is necessary for start-up nor any elaborate controls for shutdown given their resistance toward air and/or water exposure. Their unique properties provided the rationale for use for fuel processing of natural gas to generate H_2 integrated to low-temperature fuel cells and stand-alone hydrogen generators. The high initial cost of precious metals is compensated for by reactor capital saving, fewer controls needed to protect and activate catalysts, and an existing infrastructure for precious metal recycling. Figure 6.9 illustrates the hypothetical arrangement of unit operations involved in the small-scale hydrogen facility that would be used for a fuel cell application. In the figure, the unit operations are described as discrete units; however, in reality the system would be highly integrated thermally to maximize energy efficiency and highly integrated mechanically to minimize footprint.

The washcoat thickness on the walls of the monolith, typically about 100–200 μm, allows greater structural stability since temperature gradients do not exist appreciably relative to 4–6 mm diameter particulates. Thin catalyst washcoat layers allow rapid response to transient operation, with turndown ratios of up to 10:1 for fuel cells with varying power demands. Therefore, precious metal washcoated catalysts allow new compact reformer designs with high activity and durability for both stand-alone fuel processors and those integrated to stationary fuel cells.

6.3.2 Steam Reforming

Washcoats of supported precious metals can be deposited on metal surfaces with excellent adhesion. Thus, it is possible to coat the metal surfaces of a heat exchanger

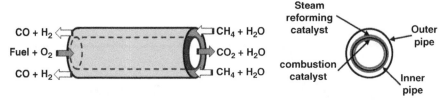

Figure 6.10   Illustration of a highly simplified catalyzed double pipe heat exchanger where a combustion catalyst is applied to the inside surface and a steam reforming catalyst is applied to the outside surface of the inner tube.

with a catalyst. This allows highly efficient thermal integration between highly exothermic and endothermic processes such as combustion and steam reforming. This concept is illustrated in Figure 6.10 for a very simple example of a catalyzed double pipe heat exchanger. Commercialized designs, which can vary from one designer to another, are often significantly more complex and designed to maximize available area for heat transfer, which in turn decreases the thickness of the required catalyst coating per unit area.

Natural gas and steam flow through the process side of a precious metal-catalyzed heat exchanger. The endothermic heat is provided by either a thermal burner or a catalytic oxidation reaction on the inner tube side. With the proper system design, heat transfer no longer limits the rate of reaction allowing for higher throughputs (higher space velocities) and smaller reactors.

Catalyzed heat exchanger designs allow a 5–10-fold reduction in reactor size for CH_4 steam reforming relative to the process utilizing catalyst particulates in a packed bed. On the inner side of the exchanger, catalytic or thermal combustion of the tail gas (CO and smaller amounts of H_2) from a PSA provides the heat for the endothermic steam reforming reaction. Equilibrium concentrations of CH_4, H_2, CO, H_2O, and CO_2 are achieved at the exit temperature of the process. The goal is to maximize the throughput of process gas while maintaining as high a temperature as the reactor metallurgy will permit. A Rh-containing washcoat is sufficiently active to achieve equilibrium product gas. Space velocities greater than $30,000\,h^{-1}$ have been achieved with precious metal-containing catalysts deposited on highly efficient heat exchangers.

Rh-containing catalysts do not promote coke formation, especially from the higher hydrocarbons present in natural gas, and thus no pre-reformer is necessary. In contrast, catalysts based on many other metals such as Ni, Pd, and Ru are very prone to catalyzing coke formation and thus a pre-reformer is usually added at the front end of those processes.

6.3.3   Water Gas Shift

The water gas shift reaction (Equation 6.14) occurs immediately downstream from the steam reformer. The inlet to the WGS reactor is cooled to about 250–300 °C. The reaction is slightly exothermic and thus is thermodynamically favored at low temperature at the sacrifice of slower kinetics.

A ceramic (parallel channel) monolith with washcoated (Pt, Re)/Al_2O_3 offers the same advantages as for SR such as increased activity (i.e., smaller reactor volume), low pressure drop, rapid response to temperature transients, and robust performance characteristics as experienced in complicated duty cycles. No activation procedures are necessary for start-up nor control to protect catalysts from air exposure during shutdown.

A family of precious metal-containing WGS catalysts supported on monoliths are now commercially available that address many of the limitations of the traditional base metal particulates. These catalysts were designed to have high activity, good hydrothermal stability, and low methanation activity. Furthermore, precious metal washcoated formulations can be exposed to air and liquid water during shutdown with no loss in performance or mechanical instability. This is an importantly distinction from base metal WGS catalysts such as those containing Cu.

6.3.4   Preferential Oxidation

As described above, PROX catalysts have also been adapted for CO selective oxidation and are capable of producing fuel cell quality H_2 with a CO level of <5 ppm. The Pt/Fe-based catalysts with Cu additions have been reformulated as washcoats on ceramic monoliths and can function adiabatically with inlet temperatures between 90 and 120°C and an outlet temperature of ~150 °C with up to 40% selectivity at an O_2/CO ratio of 1.5:1 and a CO level of 0.5% (5000 ppm). As a monolith, this catalyst operates at $30,000 \, h^{-1}$. Figure 6.11 demonstrates that CO levels close to 0 can be obtained over a broad range of inlet temperature operating adiabatically.

6.3.5   Combustion

During start-up, when H_2 or reformate are not available, a hydrocarbon fuel such as natural gas is catalytically combusted to heat the unit and to generate steam.

$$CH_4 + 2O_2 \rightarrow CO_2 + 2H_2O \tag{6.28}$$

Thus, the light-off temperature of natural gas/air mixtures is an important criterion for catalyst selection. Washcoats composed of precious metal (Pt and/or Pd) on stabilized Al_2O_3 on monoliths or heat exchanger surfaces will light off CH_4, CO, and H_2 tail gas from the fuel cell anode compartment or PSA unit once the system is operating. The catalyst operates at space velocities approaching $100,000 \, h^{-1}$

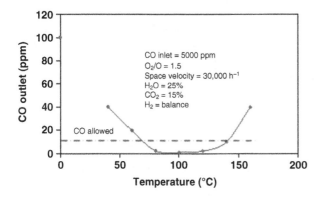

Figure 6.11   Preferential oxidation of 0.5% CO using a (Pt, Fe, Cu)/Al_2O_3 monolith catalyst.

depending on the inlet temperature and degree of combustion required. There are a number of engineering requirements to take into consideration when working with catalytic combustion of H_2, such as fuel/air ratios and mixing, water content, flow rate, gas linear velocities, combustion dynamics, heat transfer rates, channel geometry, and the possibility of hydrogen flashback.

6.3.6   Autothermal Reforming for Complicated Fuels

The ATR (combined catalytic partial oxidation and steam reforming) process is being utilized for LPG fuels that often contain coke precursors such as olefins. It is commonly believed that LPG is only propane, but it does contain a variety of other hydrocarbons, many of which are coke precursors. ATR is more resistant to coking since the O_2 (air) combusts the more reactive coke precursors allowing the remaining fuel to be reformed coke-free. A combination of Pt and Rh species in washcoats deposited on ceramic monoliths is used. Inlet temperatures are roughly 300–400 °C with a H_2O/C ratio of about 0.4 and a H_2O/C ratio of 2–3 depending on the presence of coke precursors in the LPG. Knowledge of the LPG composition is critical for successful design of the proper catalyst composition, and operating and feed conditions. The CPO reaction is much faster than SR, so the inlet section of the monolith first experiences elevated temperatures, which then cool as SR occurs downstream. Catalysts must be designed to resist deactivation due to these high inlet temperatures.

6.3.7   Steam Reforming of Methanol: Portable Power Applications

Methanol steam reforming is also becoming an important low-temperature source of H_2. Traditional CuO, ZnO, and Al_2O_3 (similar to LTS) catalysts as particulates are used to generate a $H_2/CO/CO_2$ gas mixture. The catalyst must be activated (reduced) in the unit. It must then be protected from air or liquid reformate to avoid deactivation and loss of mechanical strength. A Pd membrane downstream, permeable to H_2, is separated from the tail gases (e.g., CO, CO_2, and excess H_2O) that are directly fed to the fuel cell. The tail gases are combusted to provide some of the heat for the endothermic methanol reforming reaction. A new supported PdZn catalyst, operated above 300 °C, is deposited on a heat exchanger. This design eliminates attrition and enhances heat transfer permitting smaller and more mechanically durable units relative to particulate designs. The tail gases are combusted on the tube side of the heat exchanger while the endothermic SR catalyst generates H_2 on the shell side. The higher the steam/methanol ratio, the less likely deactivation due to coking occurs but at a cost of energy consumption for vaporizing excess water. A $PdZnO/CeO_2$-containing washcoated steam reforming catalyst can operate close to stoichiometric H_2O/methanol ratio with long life. It is reduced in process gas (methanol and H_2O) at about 450 °C to form a PdZn alloy, which is the active catalytic phase.

6.4   SUMMARY

An overall summary of the catalytic steps involved in producing H_2 and/or synthesis gas is shown in Figure 6.12.

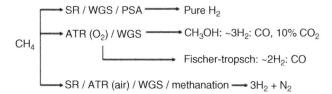

Figure 6.12   Various catalytic processes for generating H_2 and synthesis gas from desulfurized natural gas (methane) and methanol.

QUESTIONS

1. The extraction of H_2 from water requires a temperature of over 3000 °C. How this is accomplished in chemical and petroleum plants?

2. **a.** Why are sulfur compounds added to pipeline natural gas?

   b. Why must they be removed when natural gas is converted to H_2?

   c. How are they removed in large chemical plants?

   d. Why is it preferred not to add steam (H_2O) until all the H_2S is removed?

   e. Why is the HDS catalyst presulfided?

3. Steam reforming of natural gas is endothermic and the equilibrium constant is more favored as the temperature is increased ($K_{571 K} = 9 \times 10^{-10}$ and $K_{1173 K} = 1.4 \times 10^3$). What limits the upper temperature of the reaction for the catalytic reaction?

4. Methane steam reforming generates an expansion of the products relative to the reactants.

   a. What process parameter would enhance equilibrium H_2 production?

   b. Why is this not used?

   c. Why are the advantages and disadvantages of using high steam concentrations?

5. Describe the rationale for the catalyst components in combination with the active Ni steam reforming catalyst.

6. Why is it necessary to use small diameter metal tubes to house the catalyst in the SR reaction?

7. How is the Ni catalyst activated?

8. Why are pre-reformers sometimes used?

9. What is the rationale for using high- and low-temperature shift reactors and two different catalysts?

10. **a.** How are the steam reforming and WGS catalysts activated?

    b. Why is it necessary to passivate each catalyst?

11. How will pressure affect the WGS reactions?

12. Explain pressure swing adsorption and how it functions in the production of pure H_2.

13. What is ATR and how and why is it used for the production of NH_3?

14. **a.** Why is it necessary to remove the CO from the feed gas for NH_3 production and how is this accomplished?

    b. Why is it necessary to remove CO_2 and how is this accomplished?

15. **a.** Why is it necessary to remove the CO present in H_2 for low-temperature PEM fuel cell use?

    b. How is this accomplished?

16. What are the advantages of washcoating a heat exchanger for SR?

17. What is the advantage of using ATR for fuels containing coke precursors?

18. **a.** Why can methanol be steam reformed at a much lower temperature than methane?

    b. Why is it preferred to carry out methanol SR on a catalyzed heat exchanger rather than a packed bed for portable power fuel cells?

19. What characterization tools are useful in giving direction to plant personnel when activating the catalysts in a H_2 generation process?

20. What laboratory tools are helpful in understanding coke burn-off from the Ni SR catalyst using air?

21. Why is it important to understand the reaction kinetics (i.e., activation energies and reaction orders for the reactants and products) for various reactions?

BIBLIOGRAPHY

Armor, J. (2005) Catalysis for the hydrogen economy. *Catalysis Letters* 101 (3–4), 131–135.

Bartholomew, C. and Farrauto, R.J. (2006) Chapter 6, in *Fundamentals of Industrial Catalytic Processes*, 2nd edn, John Wiley & Sons, Inc., Hoboken, NJ, pp. 339–370.

Farrauto, R.J. (2014) New catalysts and reactor designs for the hydrogen economy. *Chemical Engineering Journal* 238, 172–177.

Farrauto, R., Liu, Y., Ruettinger, W., Ilinich, O., Shore, L., and Giroux, T. (2007) Precious metal catalysts supported on ceramic and metal monolithic structures for the hydrogen economy. *Catalysis Reviews* 49, 141–196.

Imperial Chemical Industries (1970) *Catalyst Handbook: Ammonia and Hydrogen Manufacture*, Wolfe Scientific Books.

Rostrup-Nielson, J.R. (1984) Catalytic steam reforming, in *Catalysis, Science and Technology* (eds J. Anderson and M. Boudart), Springer.

Zhang, Q. and Farrauto, R.J. (2012) Selective oxidation of CO in H_2 with a Cu, Fe, Pt monolith catalyst. *International Journal of Hydrogen* 37 (14), 10874–10880.

AMMONIA, METHANOL, FISCHER–TROPSCH PRODUCTION

7.1 AMMONIA SYNTHESIS

N_2 is relatively inert and consequently does not react easily with other chemicals to form new compounds. Nitrogen fixation is the process by which relatively inert atmospheric N_2 is converted into a more reactive species, such as the nitrogen atom in NH_3, which can occur through natural or synthetic processes. Fixed nitrogen is essential for plant growth. Despite being surrounded by an abundance of diatomic nitrogen in the atmosphere, only fixed nitrogen is useful for plant growth. Before 1910, fixed nitrogen was obtained from natural deposits of saltpeter (KNO_3) and Chilean saltpeter ($NaNO_3$). Natural sources are limited, and by the early 1900s it was becoming increasingly more clear that natural sources of fixed nitrogen would be insufficient to keep up with increased fertilizer demands of a growing global population. Thus, a synthetic route was needed. In 1914, researchers from BASF in Ludwigshafen, Germany developed the first commercial route to ammonia, which can then be processed further to form fertilizer components such as ammonium nitrate (NH_4 (NO_3)) and urea (NH_2CONH_2). NH_3 is also needed for production of explosives.

7.1.1 Thermodynamics

To form NH_3, one must combine dissociated N_2 and H_2 in an exothermic ($\Delta H° = -109$ kJ/mol of N_2) reaction thermodynamically favorable at low temperatures and high pressures (Figure 7.1). Thermally dissociating N_2 requires $1000\,°C$ given the slow kinetics of breaking triple-bonded N_2. This temperature is too far removed from equilibrium for any suitable process. Catalysts for low-temperature dissociation of H_2 are well known (Ni, Pd, Pt, etc.); however, no suitable catalyst existed for N_2 dissociation in the early twentieth century.

$$N_2 + 3H_2 \rightleftharpoons 2NH_3, \quad \Delta H° = -109 \text{ kJ/mol} \tag{7.1}$$

Introduction to Catalysis and Industrial Catalytic Processes, First Edition. Robert J. Farrauto, Lucas Dorazio, and C.H. Bartholomew.
© 2016 John Wiley & Sons, Inc. Published 2016 by John Wiley & Sons, Inc.

$$K_{eq} = \frac{[NH_3]^2}{[N_2][H_2]^3}$$

$$K_{573\,K} = 4 \times 10^{-2}$$
$$K_{673\,K} = 9 \times 10^{-3}$$
$$K_{773\,K} = 3 \times 10^{-3}$$

7.1.2 Reaction Chemistry and Catalyst Design

The function of the ammonia synthesis catalyst is to adsorb and dissociate hydrogen and nitrogen, and then provide the required environment for the hydrogenation of the dissociated nitrogen species. While hydrogen adsorption and dissociation occurs rapidly on a relatively wide variety of materials, the dissociation of nitrogen at mild reaction conditions is considerably more challenging due to the large dissociation energy of N_2. This is the distinguishing characteristic of an ammonia synthesis catalyst: its ability to adsorb and dissociate nitrogen. The optimal catalytic material must adsorb N_2 strongly enough to weaken its triple bond, but not adsorb the N species too strongly as to inhibit its hydrogenation to ammonia. The dissociation of N_2 to nitrogen atoms is the slowest rate of the sequence of surface reactions and is the rate-limiting step (RLS). It is interesting to note that the dissociation of nitrogen is very endothermic ($944\,kJ/mol\,N_2$), yet the overall reaction to form ammonia is moderately exothermic ($-109\,kJ/mol\,N_2$). This is a good example that demonstrates the overall heat of reaction observed macroscopically in the reactor is the net sum of all the microscopic sequence of surface reactions that take place on the catalyst surface as reactants are transformed to products.

BASF developed the first ammonia synthesis Fe catalyst derived from a very special iron oxide extracted from magnetite (Fe_3O_4) found in special locations in Sweden. It contained a number of what were believed to be minor impurities, for example, K_2O, CaO, MgO, SiO_2, and Al_2O_3, that surprisingly was necessary for an

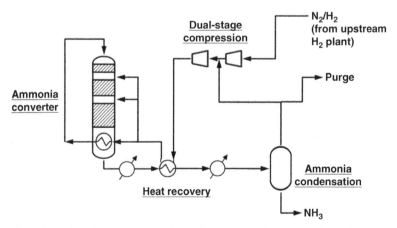

Figure 7.1   Simplified flow sheet for NH_3 synthesis illustrating a "quench"-type ammonia converter and two-stage feed gas compression.

active catalyst. This is an excellent example of promoters that themselves are not active but for a variety of reasons promote the performance of the catalyst. The catalyst is activated when the promoted magnetite is carefully reduced to Fe metal. It is the Fe metal crystallites that form during the reduction process that are the active centers that adsorb and dissociate nitrogen. Without the promoters present during the reduction process, metallic iron is a relatively poor ammonia catalyst. The highly active catalyst forms only due to the presence of each of the promoters that participate in the reduction process and interact with the metallic Fe in different ways. For example, K is a key "activity" promoter that is believed to interact with the metallic Fe, altering the chemistry of the metallic Fe crystallite to create the suitable site to adsorb and dissociate nitrogen. As another example, Al_2O_3 and CaO are believed to be "structural" promoters that inhibit sintering of the Fe crystallites during the reduction of the magnetite.

The active catalyst is produced by melting the iron oxide along with its promoters (K_2O may also be added), which is then poured into cold water shattering the combined oxides to tiny particles more suitable for reduction and activation:

$$Fe_3O_4 + \text{promoters (melted)} \rightarrow \text{pour into cold water} \rightarrow \text{shattered particles}$$

The catalyst particles, approximately $1\ m^2/g$ in BET surface area, are charged to the reactor in a fixed bed arrangement and slowly reduced with H_2 at 450–500 °C. This is an exothermic reaction ($\Delta H° = -600\ kJ/mol$). The reduction reaction is carried out very slowly to avoid contacting the product H_2O with reduced iron given the potential for Fe metal to be reoxidized.

$$Fe_3O_4 + 2H_2 \longrightarrow 3Fe + 2H_2O + O_2 \tag{7.2}$$

If the H_2O/H_2 is maintained below ~0.16 during the reduction, the catalyst BET surface area is increased to about $20\ m^2/g$ (pore size = 20–40 nm). This increase is due to the loss of oxygen through the formation of H_2O. If the reduction is not carefully controlled, the H_2O produced will enhance sintering and oxidation of the Fe. In a typical NH_3 process, reduction can require as much as 7 days for catalyst activation.

Ruthenium is another metal found to be active for adsorption and dissociation of nitrogen; hence, it is also an active catalyst for ammonia synthesis. Similar to metallic iron, ruthenium by itself is a poor ammonia catalyst. However, when combined with the right "structural" and "activity" promoters, such as K, Cs, or Ba, ruthenium-based catalysts are even more active than iron-based catalysts. At least one reason for the higher relative activity of the ruthenium catalyst is the order of magnitude greater dispersion (i.e., more catalytic sites) of the active metallic ruthenium crystallites compared to that observed with iron catalysts. Ru/C (graphite) has been used as a polishing reactor in tandem with processes that operate with the Fe catalyst. Although the Ru catalyst, promoted with K, Cs, or BaO, is more active than Fe, it is the product inhibited by NH_3. When exposed to large concentrations of NH_3, the catalyst activity falls. For this reason, it is primarily used as a final polishing bed where lower NH_3 concentrations are experienced due to its removal upstream. It allows operation at low temperatures and pressures than the Fe-catalyzed process and increases the final yield of NH_3. It is sensitive to O_2, but not as sensitive to other impurities as Fe. With time on stream, the graphite carrier slowly undergoes Ru-catalyzed methanation with H_2 causing deactivation.

7.1.3 Process Design

Thermodynamic equilibrium indicates higher pressures are required when the process is operated at higher temperatures to favor kinetics. The ideal catalyst would be highly active for N_2 dissociation at low temperatures, thus allowing the reaction to proceed at low pressures. In reality, the catalyst is operated above 300 °C; thus, high reaction pressures are necessary, which in turn significantly impacts process design and manufacturing cost. The thermodynamics and kinetics of this reaction require a balance to be established between compression costs, equipment cost, and the amount of catalyst required. Generally, this balance results in accepting a low single-pass conversion, with recycle of unreacted N_2 and H_2 between each pass required. This suggests the process operate at high feed pressures of N_2 and H_2 with periodic removal of the NH_3. Thus, a recycle reactor is operated at 300 atm and at about 400 °C (673 K).

The N_2 and $3H_2$ from the upstream steam reforming process, with about 0.1% methane produced in the synthesis gas reformer where CO is methanated (see chapter on H_2 generation), is fed to the ammonia synthesis unit. The operating pressure is a function of the commercial process technology and type of catalyst used. In the traditional process, the reactants are compressed in two stages: first to 30 atm and then to roughly 300 atm. Recent processes using more advanced technologies are able to operate at lower pressure in the 150–250 atm range. Although thermodynamics favors higher conversions at lower temperatures, kinetics will dictate the lowest practical inlet temperature into the reactor. Since the reaction is moderately exothermic, the reaction gas temperature will rise along the reactor length as reaction enthalpy is converted into sensible heat and results in an unfavorable shift in chemical equilibrium. Modern ammonia synthesis reactors are designed to incorporate heat removal from the reaction gas to minimize the temperature rise along the reactor length. The simplest example is the quench reactor design where cold feed gas is injected between reaction stages that quench the temperature and increases reactant concentration, which work for both shift equilibrium and product formation.

In the quench-type reactor design, the cold feed is introduced between catalyst bed stages that act to both cool the reaction gas and dilute reaction products, both of which favorably shift the chemical equilibrium. Typically, the feed gas enters the first catalyst bed at 400 °C and the catalyst bed sized to yield a 100 °C temperature increase. After the first bed, the reaction gas is quenched with cold feed (125–200 °C) and fed into subsequent beds where this process repeats. There exists an optimal temperature where kinetics and thermodynamics are in optimal balance. The nature of the quench design results in the actual temperature fluctuating between higher and lower values, which yields only a small fraction of residence time spent at this optimal temperature. Indirect cooling by integrating heat exchangers into the catalyst beds yielded a more effective design. However, while integrating heat exchange into the reactor design reduces the amount of catalyst required, it comes at the cost of larger physical size of the reactor. The capital cost of the high-pressure vessels increases significantly with the increase in size. Additionally, the mechanics of pressure vessels generally requires a cylindrical geometry to be used. These issues must be considered in the overall design of the process.

Equilibrium limits the one-pass conversion to roughly 15%. The effluent from the reactor is chilled to $-17\,°C$ where a large fraction of NH_3 is condensed and removed. The noncondensable fraction containing unconverted reactants and a small amount of ammonia is then reheated and recycled to the inlet of the reactor. Figure 7.1 illustrates a simplified flow sheet for ammonia synthesis. In practice, the flow sheet for ammonia synthesis is considerably more complex when including heat integration, product recovery, and hydrogen recovery from the purge gas.

As the demand for ammonia grew during the latter half of the twentieth century, the size of the ammonia converter increased. As described above, the synthesis of the ammonia catalyst involved water cooling of molten materials resulting in the formation of small particulate ($\sim$1–2 mm). To increase the capacity of the ammonia converter, larger catalyst beds were required. A design using longer catalyst beds resulted in unacceptable large pressure drop. To compensate for pressure drop, larger catalyst particles were needed ($\sim$6–10 mm). However, larger catalyst particles were found to have significantly lower activity attributed partly to greater pore diffusion associated with larger particles (lower effectiveness factors), but also due to differences in the activation of the iron-based catalyst. Together, these two issues forced the use of small particulate materials. To lower pressure drop of a larger bed without increasing the size of the catalyst particle, the cross-sectional area for flow must be increased to decrease the linear velocity. For a traditional tubular reactor, this requires increasing the diameter. However, the cost of the pressure vessel becomes unacceptably high as the reactor diameter is increased. A novel solution to this problem was the development of the radial flow reactor. In its simplest form, the radial flow reactor consists of two concentric screens or perforated materials with the annulus filled with particular catalyst, illustrated in Figure 7.2. The annular catalyst bed is contained within a cylindrical pressure vessel. Reactants enter the pressure vessel and are directed to the outer periphery of the vessel. The screen or perforated material is designed to allow gas to pass through while still holding the catalyst particles in place. Instead of flow direction proceeding axially through the tubular reactor, the reactor is designed to force flow radially through the bed, which significantly increases the flow cross-sectional area, decreasing the linear velocity and hence the pressure drop. The flow is being directed from the outside radius inward. The illustration in Figure 7.2 is of a simple single-stage radial flow bed. In practice, other more complex designs incorporating thermal integration and multiple radial flow beds are used. Since the development of the radial flow reactor, other novel designs have been developed to maximize the area for flow allowing the use of small catalyst particles without the issue of pressure drop. The common constraint in all designs is the requirement that they must be incorporated within a pressure vessel, which creates significant mechanical design challenges.

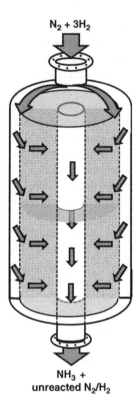

N$_2$ + 3H$_2$

**NH$_3$ +
unreacted N$_2$/H$_2$**

Figure 7.2 Simplified illustration of a single-stage radial flow ammonia converter.

7.1.4 Catalyst Deactivation

Some sintering of the Fe occurs during the reaction, but life is typically 10 years. Poisons such as oxygenates CO, CO_2, H_2O are removed to <200 ppm, preferably <100 ppm, in the synthesis gas process upstream.

7.2 METHANOL SYNTHESIS

Methanol is a common laboratory solvent and is made in large volumes worldwide. The reaction is written in two steps, the second of which is the familiar water gas shift reaction (WGSR):

$$CO_2 + 3H_2 \rightleftharpoons CH_3OH + H_2O, \quad \Delta H^\circ = -49 \text{ kJ/mol} \tag{7.3}$$

$$CO + H_2O \rightleftharpoons H_2 + CO_2, \quad \Delta H^\circ = -41 \text{ kJ/mol} \tag{7.4}$$

The addition of the two reactions yields the general equation:

$$CO + 2H_2 \rightleftharpoons CH_3OH, \quad \begin{aligned} \Delta H^\circ &= -90 \text{ kJ/mol} \\ \Delta G^\circ &= -39 \text{ kJ/mol} \end{aligned} \tag{7.5}$$

It should be noted that the ΔS will decrease due to volume contraction of the products, and therefore the free energy is less negative than the reaction enthalpy. Because the reactions are exothermic with a net decrease in molar volume, the reaction will be favored at low temperatures and high pressures. This is the challenge for the catalyst development. The ideal catalyst should be highly active at low temperatures to avoid the requirement of high pressure to achieve the desired one-pass conversion.

The process is carried out with added CO_2; thus, Equation 7.3 is important for synthesis. The early BASF process utilized a Cr_2O_3–ZnO catalyst requiring about 300 atm at 350 °C. Today, due to the developments by ICI, the preferred catalyst and process, referred to as the low-pressure process, operate at 200 °C and 50 atm pressure (similar to what is delivered in the synthesis gas generation of H_2 and CO). This significantly reduces compression costs. The Cu, ZnO, and Al_2O_3 particulate catalysts are similar to that used for WGS and thus it is clear that WGS also plays a role in the overall synthesis. The methanol catalyst produced is a tablet 6 mm × 4 mm and has an approximate composition of 60%CuO, 30%ZnO, and 10%Al_2O_3. The active phase of the catalyst is produced when the mixed Cu–Zn–Al metal oxide is reduced yielding copper metal in intimate contact with Zn and Al oxides. The Zn and Al oxides are structural promoters in that they stabilize the dispersion of the active Cu metal and resist sintering. However, there is also speculation that the ZnO also serves as an activity promoter. As described in Equations 7.3 and 7.4, the reaction sequence on the catalyst surface is believed to proceed through two major steps. First, CO and H_2O adsorb onto the Cu metal surface, where water activates to form OH and H, and CO_2 is produced via the water gas shift reaction. Evidence has been found that CO can be hydrogenated directly to methanol, but at a considerably slower rate than the hydrogenation of CO_2 to methanol. Thus, CO is more likely to react via the WGSR than hydrogenation. However, the hydrogenation of CO_2 will require more H species than generated in the activation of water. Thus, hydrogen is also fed into the reactor, which adsorbs onto the catalyst surface and dissociates to form additional H species that proceed to hydrogenate CO_2. The hydrogen may adsorb and dissociate on the copper metal. However, it is also speculated that in addition to stabilizing the dispersion of Cu, ZnO also acts to store dissociated H species feeding them to CO_2 adsorbed on the Cu site, which in turn would accelerate the hydrogenation of CO_2.

This reaction also demonstrates the importance of the proper catalyst directing the reactants to the more desired product (i.e., selectivity). Such is the case here since H_2 and CO can combine to form methane with a $\Delta G° = -144$ kJ/mol, while the Cu-containing catalyst directs them to CH_3OH ($\Delta G° = -39$ kJ/mol). Thermodynamics favors methane, but in the presence of the correct catalyst methanol is formed. Thus, the rate of reaction to methanol is accelerated to a much greater extent than toward methane with the Cu catalyst. In contrast, Ni favors the formation of methane. Therefore, provided both products are favorable, a catalyst can accelerate a reaction to specific products that are less favorable thermodynamically than others. Thus, catalysts are of great importance in selectivity as well as overall rate.

The contraction of product gas volume relative to reactants indicates the equilibrium concentration of methanol is favored at high pressures. The reaction is exothermic indicating the reaction is more thermodynamically favorable at low temperatures. But low temperatures lead to poor kinetics. Therefore, by raising the reaction temperature, the kinetics are improved, but at the sacrifice of less favorable equilibrium. It is a compromise necessary to optimize the process economics.

7.2.1 Process Design

The methanol reactor consists of a fixed bed catalyst with some type of cooling incorporated into the design. As discussed earlier, the reactions involved in the production of methanol are exothermic. Thus, as the reaction proceeds, the reaction temperature will increase unless heat exchange is incorporated into the reactor design to remove reaction enthalpy. While higher temperatures increase kinetics, the reaction becomes less thermodynamically favorable and the maximum possible conversion decreases. Four general reactor types have been used for methanol synthesis to integrate cooling into the design: quench, staged cooling, tube cooled, and shell cooled.

7.2.1.1 Quench Reactor This is the same design discussed previously for ammonia synthesis. Multiple catalyst beds are used with cold feed injected between. The reaction gas proceeds through a catalyst bed adiabatically and temperature increases. The injection of cold feed reduces the reaction temperature before entering the next catalyst bed, as well as increases the reactant concentration, both work to favorably influence chemical equilibrium. The resulting temperature profile along the length of the reactor follows a sawtooth pattern of increases during adiabatic reaction

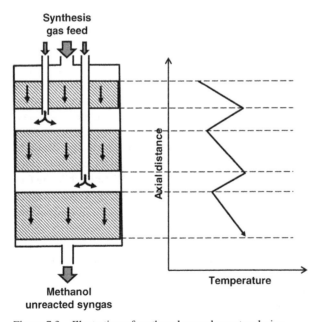

Figure 7.3   Illustration of methanol quench reactor design.

followed by decreases as cold feed is introduced. The advantage has been a simple reliable design that has been used widely in industrial processes. The disadvantage is the limited time spent at the optimal reaction temperature. Due to the balance between thermodynamics and kinetics, there exists an optimal temperature where the two are balanced. The quench reactor design results in the temperature fluctuating above and below this optimal temperature, but not much time spent at the optimal temperature. An illustration of this design and the associated "sawtooth" temperature profile is provided in Figure 7.3.

7.2.1.2 Staged Cooling Reactor

7.2.1.2 Staged Cooling Reactor   This reactor design integrates heat exchange between the catalyst beds. As with the quench design, the reaction proceeds adiabatically within the catalyst bed, then the reaction gas is cooled to extract reaction enthalpy and lower the temperature before entering the next catalyst bed. This concept is illustrated in Figure 7.4. The temperature follows the same sawtooth pattern discussed with the quench reactor and thus subject to the same disadvantage. The primary advantage of this design is that more feed gas contacts all of the catalyst volume. However, an added disadvantage is the added complexity and cost associated with the heat exchangers required between the catalyst beds.

7.2.1.3 Tube-Cooled Reactor   The tube-cooled reactor design consists of a shell and tube style vessel where the catalyst is contained on the shell side and the reaction feed gas on the tube side as the cooling fluid, as illustrated in Figure 7.5. As the feed gas flows through the tubes, it is heated by the reaction occurring on the shell side of the heat exchanger. The advantage of this design is that the temperature profile along the length more closely follows optimal temperature that maximizes reaction rate (i.e., balance between thermodynamics and kinetics).

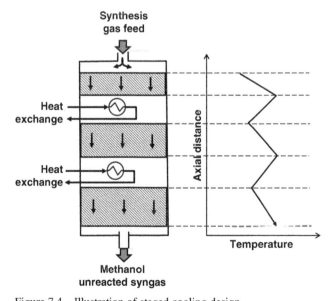

Figure 7.4   Illustration of staged cooling design.

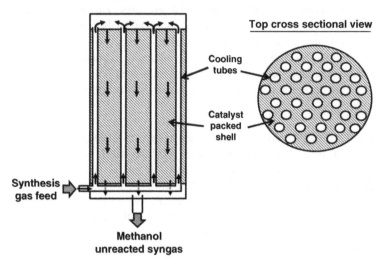

Figure 7.5   Illustration of cooled tube reactor design.

7.2.1.4   *Shell-Cooled Reactor*

In the shell tube design, the tubes of the shell and tube heat exchanger are packed with catalyst (Figure 7.6). On the shell side, steam is raised that is used elsewhere in the chemical plant. This is just one possible

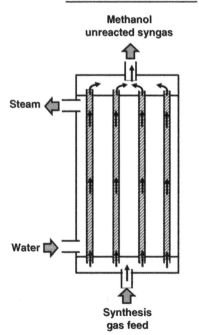

Figure 7.6   Illustration of shell-cooled reactor design.

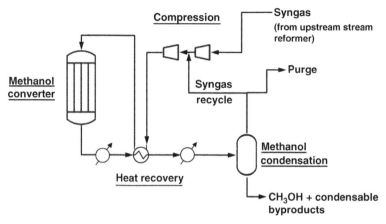

Figure 7.7   Flow sheet for methanol synthesis.

configuration of cooling via steam generation and a variety of other configurations have been used, including raising steam on the tube side. The advantage of steam generation is the near isothermal temperature profile that provides the optimal balance between thermodynamics and kinetics, and thus the lowest possible catalyst requirement.

The one-pass conversion dictated by the balance between thermodynamics and kinetics requires the recycle stream be included into the process illustrated in the flow sheet shown in Figure 7.7. The catalyst is carefully prereduced using H_2 to form the active species in a manner similar to what was used for the Cu-containing WGS reaction described in the previous chapter.

The feed gas is adjusted to give H_2/CO ratio of about 2:1 with about 10% CO_2 (from WGS). The feed gas enters the reactor filled with the Cu-containing particulate catalyst. The space velocity is about $30,000\,h^{-1}$ with about 15% conversion per pass with an exit temperature of 250 °C. The product gas is cooled to below the condensation point for methanol. The condensed phase is distilled separating the pure methanol for the heavy hydrocarbon by-products. Unreacted H_2, CO, and CO_2 are preheated and compressed if necessary during recycle to the reactor inlet. Periodically, an inert purge is needed to remove the built-up impurities.

7.2.2   Catalyst Deactivation

Deactivation occurs due to sintering, breakage of the particulates as evident by pressure drop increases, and selective poisoning by traces of impurities, mainly sulfur that may not have been adequately removed in the reforming process. Recent evidence indicates coking to be a major source of deactivation. Evidence that coke combustion leading to regeneration credence to coking is at least one cause of deactivation. However, during reaction, water is produced at the surface of the metal that likely oxidizes it to an inactive state. Hydrogen reduction, after coke combustion, converts the oxide surface to the active metallic state that may now be redispersed contributing to regeneration. Regardless of the main mechanism of deactivation, commercial processes do employ regeneration indicating some degree of reversibility.

7.3   FISCHER–TROPSCH SYNTHESIS

Synthesis gas (H_2 and CO) are the starting reactants for production of pure H_2 for the chemical and petroleum industries, fuel cells, alcohols, and a large distribution of low and high molecular weight hydrocarbons. Synthesis gas can be generated from sources such as coal, wood (biomass), and municipal wastes and fossil fuels currently found in nature such as natural gas. These sources are to be discussed later in the chapters related to alternative energy. Synthesis gas can also be converted to useful fuels and chemicals using the proper catalyst and process. The first Fischer–Tropsch (F-T) synthesis was developed in Germany and patented in the United States in 1930. The first plant was built in 1936 utilizing Germany's large reserves of coal to generate the synthesis gas. Later, South Africa advanced the technology as an alternative to petroleum utilizing their abundance of coal. In its simplest, but least desirable reaction, H_2 and CO can be converted to CH_4 by the reverse of steam reforming provided the proper catalyst is used. According to the principle of microscopic reversibility, a catalyst accelerates the rate of both the forward and reverse reactions, ultimately bringing the reactants and products into equilibrium. Therefore, just as supported Ni was the preferred steam reforming catalyst, it is also active for the reverse reaction. Supported Ru and Rh also catalyze the reactions in both directions; however, for large H_2 plants, Ni is preferred.

$$3H_2 + CO \rightleftharpoons CH_4 + H_2O, \quad \begin{aligned} \Delta H° &= -204\,\text{kJ/mol} \\ \Delta G° &= -136\,\text{kJ/mol} \end{aligned} \tag{7.6}$$

For the production of olefins such as n-hexene using an Fe catalyst, we write

$$12H_2 + 6CO \rightleftharpoons C_6H_{12} + 6H_2O, \quad \begin{aligned} \Delta H° &= -900\,\text{kJ/mol} \\ \Delta G° &= -500\,\text{kJ/mol} \end{aligned} \tag{7.7}$$

General equation for olefins:

$$2xH_2 + xCO \rightleftharpoons C_xH_{2x} + xH_2O \tag{7.8}$$

For larger paraffins (C_{12} + molecules—the lower boiling and molecular range for diesel fuel), the active catalyst is Co.

$$12CO + 26H_2 \rightleftharpoons C_{12}H_{26} + 13H_2O, \quad \begin{aligned} \Delta H° &= -1600\,\text{kJ/mol} \\ \Delta G° &= -960\,\text{kJ/mol} \end{aligned} \tag{7.9}$$

General equation for paraffins:

$$(2x + 2)H_2 + xCO \rightleftharpoons C_xH_{2x+2} + (x + 1)H_2O \tag{7.10}$$

When the Fischer–Tropsch process is used to make hydrocarbon liquids, it is referred to as gas-to-liquid (GTL) technology.

Clearly, the heat of reaction (enthalpy) increases with the carbon number and, therefore, the reaction is favored at lower temperatures. The contraction in volume indicates pressure favors the products. Therefore, to produce a large array of large molecular weight hydrocarbons for either chemicals (olefins) or fuels (gasoline or diesel), the process temperature and pressure must be controlled. These reactions demonstrate the importance of catalysis in product selectivity. When a number of reactions are all favorable thermodynamically, the product is determined mainly by the composition of the catalyst and the process conditions. This is a key reason catalysts are used because they can generate selective products that are thermodynamically favorable, but less so than others, based on kinetics. This is clearly demonstrated in F-T technology where methane, ethane, and propane have more favorable free energies $(\Delta G^{\circ}_{CH_4} = -150\,\text{kJ/mol}, \quad \Delta G^{\circ}_{C_2H_6} = -115\,\text{kJ/mol},$ $\Delta G^{\circ}_{C_3H_8} = -105\,\text{kJ/mol})$, respectively, than methanol $\Delta G^{\circ}_{CH_3OH} = -40\,\text{kJ/mol}$, but with a Cu-containing catalyst methanol can be produced in high yields with small amounts of undesirable products. Thus, catalyst kinetics dominates the product distribution through its ability to accelerate the reactants to desired (selective) products. In contrast, Ni catalyzes the production of methane in large yields.

We can make some general statements regarding Fe and Co, the two most commonly used commercial catalysts in F-T processes. Fe is preferred for olefin and gasoline production and for low H containing feeds (i.e., coal) that require some water gas shift to adjust H_2 to CO to the proper ratio of ~2. It is used for high-temperature F-T processes. Supported Co is used for producing higher molecular weight paraffins (i.e., waxes > C_{22}+) that are subsequently hydrocracked to C_{12}–C_{18} diesel fuels. It is referred to as a low-temperature F-T catalyst and process. Since biomass can be a source of synthesis gas, it is not surprising that attention is being given to these technologies for generating chemicals and fuels as alternatives to fossil fuel. An added advantage is the absence of sulfur and low production of aromatics.

F-T is a polymerization process in which a chain propagation step competes with a chain termination step. The ratio of the rate of propagation (R_p) and rate of termination (R_t) is expressed as in (7.11):

$$\alpha = R_p/(R_p + R_t) \tag{7.11}$$

This equation can be generalized statistically in the Anderson–Schultz–Flory expression that shows a distribution of hydrocarbons based on probability of chain growth and termination. It is assumed that each individual step is independent of the previous. This provides guidance statistically, but the catalyst and process conditions allow a wide variety of products to be produced. High α-values are preferred when larger molecular products are desired such as diesel fuel. Low α-values reflect lower molecular weight products with the extreme being methanation CH_4. It is believed that the CO dissociates on the catalytic site forming an active carbon species (C^*) while the dissociated H atoms add to the carbon with increasing chain lengths.

One commercial Fe-containing catalyst is a combination of 70% Fe, 5%K_2O, 5% Cu, and 25% SiO_2. The components are fused (melted) with magnetite (Fe_2O_3) at 1000 °C, cooled, and milled to a powder. The powder is then admixed with water and extrusion aids (lubricants such as dialcohols) to make a paste that is extruded as a particulate for fixed bed processes. Before use, it is reduced in H_2 to a lower and activated state (Fe_3O_4 or Fe). It is speculated to form some defect Fe_5C_2 structure during reaction that is suggested to be the active phase for catalysis.

The Al_2O_3 supported cobalt catalyst ($\sim$10–15%Co/Al_2O_3) is combined with a trace of promoter ($\sim$0.1%) such as Pt or Ru (which dissociates H_2 into nascent H atoms), which facilitates reduction of Co oxide to the metallic and active state. The Al_2O_3 pore size is optimized to minimize size restrictions from large molecular products produced to minimize pore diffusion and coking due to pore blockage.

7.3.1 Process Design

7.3.1.1 Bubble/Slurry-Phase Process
In the SASOL process, operating in Qatar, catalytic (Ni) autothermal reforming (ATR) of natural gas is used to generate synthesis gas. A 2:1 H_2/CO ratio at 30 atm is sparged into small bubbles, through a distributor plate, into the reaction solvent (i.e., diesel range liquid) to enhance dissolution. The reactor is operated in the upflow mode (Figure 7.8). A powdered 10% Co/Al_2O_3 with promoters is slurried into the reaction solvent. Catalysis occurs with the generation of large quantities of heat that is removed by cooling coils containing flowing steam to maintain the temperature between 210 and 260 °C. The process operates under conditions of high α-values to maximize production of wax that is removed through filters from the high-density liquid phase. The filters retain the catalyst in the slurry for continuous use. The wax (> C_{22}, BP > 370 °C) represents about 70% of the product and is hydrocracked at roughly 400 °C, using a Pt/zeolite

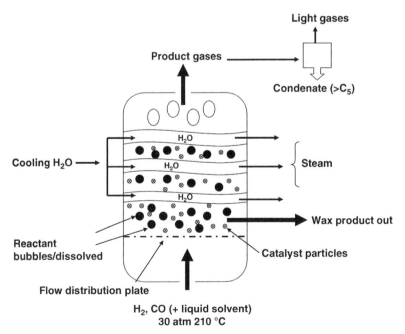

Figure 7.8   Bubble slurry reactor for Fischer–Tropsch.

catalyst, into the diesel fuel range (C_{12}–C_{18}). Product gases are removed from the top of the reactor and condensed. The condensed phase (representing about 30% of the product) is typically 44% (%v) C_5 at 160 °C (or gasoline range), 43% at 160–270 °C (diesel range), and about 13% at 270–370 °C (wax). The lower molecular weight hydrocarbons ($<C_5$) are separated and used for chemical applications. Liquid solvent and catalyst are continuously added as required. The process generates 34,000 barrels per day (B/D) of diesel fuel.

SASOL is also working closely with Chevron for another plant in Nigeria utilizing natural gas as a source of synthesis gas.

7.3.1.2 Packed Bed Process Shell oil uses a series of packed bed tubular reactors. The Shell middle distillate synthesis (SMDS) process is designed for 14,700 barrels per day (B/D) of middle distillate liquid product. The plant is located in Bintulu, Malaysia. Synthesis gas is generated by noncatalytic partial oxidation (PO) of processed "pure" CH_4 from local natural gas. Multitubular reactors, packed with a promoted Co catalyst, are used at 70 atm and 230 °C and designed for long chain wax production. The wax is subsequently hydrocracked and hydroisomerized to kerosene, gasoline, and diesel using precious metal supported on zeolites. Some of the C_{10}–C_{18} paraffins are used for detergent production. The products contain virtually no sulfur, nitrogen, or aromatics making them suitable environmentally for fuel and chemical use. The tubular reactors are specifically designed with 2 in. diameters to enhance heat extraction into cooling water. The unreacted CO and H_2 are recycled into the inlet of the reactor, while the liquid products are separated and further processed.

In 2011, Shell, in concert with Pearl GTL, began operating a 140,000 BBD process in Qatar using technology scaled up from their experiences in Malaysia.

7.3.1.3 Slurry/Loop Reactor (Synthol Process) This process uses a combination of Fe, Cu, and K_2O supported on SiO_2 extrudates. The catalyst is prereduced with H_2 and CO. The process is designed mainly for gasoline and olefin production. It operates at 30 atm pressure and 340 °C, and thus is considered a high-temperature F-T process. The catalyst and reactants are slurried in a suitable solvent and continuously looped into a cooling zone to remove the reaction heat (Figure 7.9). After cooling, the reaction mix passes through a separator where the desired products are removed. At this point, more feed and, if necessary, more catalyst can be added as the loop reaction proceeds further.

7.3.2 Catalyst Deactivation

A loss in activity is believed related to oxidation of the Co by the steam produced at its surface. The catalyst is designed for maximum attrition resistance, but sintering of the active components does occur. For catalysts with poorly designed pore structures, coking and deposition of waxy products causing pore blockage can occur to varying degrees. This is usually reversible by air calcination to remove the coke-like products. The Fe catalyst is subject to the same deactivation modes as Co, but undesirable carbide formation is common for Fe. Poisoning by sulfur and nitrogen compounds is a problem for both catalysts.

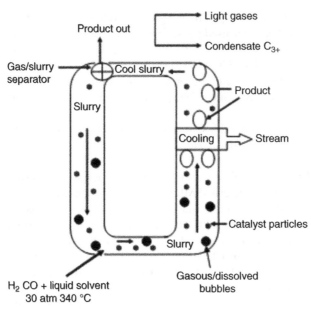

Figure 7.9 Loop reactor for Fischer–Tropsch.

QUESTIONS

1. Why is ammonia synthesis carried out at such high pressures and moderate temperatures?

2. Why is recycle used and what are the key elements of the process?

3. **a.** How is the traditional ammonia synthesis catalyst prepared?

   b. What characterization tools would be useful to use in the laboratory to assist in the catalyst preparation of the NH_3 catalyst?

   c. What simple laboratory tools would be useful in determining causes of deactivation?

4. **a.** Ru/carbon with promoters has a higher activity than the Fe catalyst for ammonia synthesis. It is used in some plants. What is its advantage regarding process conditions and hence economics?

   b. Why is Ru not used in place of Fe and what is the role of kinetics?

5. What are the major uses of NH_3?

6. **a.** What are the two reactions believed to lead to methanol synthesis?

   b. Why is conversion limited for each pass?

7. **a.** Why must the reaction products for methanol synthesis be shifted after each pass?

   b. How is this done for at least two reactor designs?

8. What are the deactivation modes for the Cu-containing methanol synthesis catalysts and what laboratory tools are used to determine this?

9. What parameter is used to understand hydrocarbon product distribution in Fischer–Tropsch synthesis?

10. What are the key parameters that determine product distribution?

11. Why is it difficult to predict extent of heat removal necessary in F-T?

12. What are the major deactivation mechanism for both Fe and Co Fischer–Tropsch catalysts and explain the methods used to determine?

13. Consider the following rate equations developed by A. Egbebi and J. Spivey (2008), Effect of H_2/CO ratio and temperature on methane selectivity in the synthesis of ethanol on a Rh-based catalyst, *Catalysis Communications* 9, 2308–2311.

$$\text{Rate (ethanol)} = 6.3 \times 10^{17} \exp(-128\,\text{kJ/mol/RT})(P_{H2})^{0.9} \text{ and } (P_{CO})^{-0.76}$$
$$\text{Rate (methane)} = 9.0 \times 10^{15} \exp(-157\,\text{kJ/mol/RT})(P_{H2})^{0.79} \text{ and } (P_{CO})^{-0.6}$$

What reaction conditions would favor ethanol production over methane?

14. If there was a new ammonia synthesis catalyst that was significantly more active below 300 °C compared to existing catalysts, how might this change the design ammonia synthesis reactor?

15. Given equal feed conditions and total amount of catalyst, why is a higher reactant conversion possible using a "quench reactor" design compared to a traditional axial flow packed bed reactor.

BIBLIOGRAPHY

Bartholomew, C. and Farrauto, R.J. (2006) Chapter 6, in *Fundamentals of Industrial Catalytic Processes*, 2nd edn, John Wiley & Sons, Inc., Hoboken, NJ, pp. 339–370.

Dry, M. (1981) The Fischer–Trospch synthesis, in *Catalyst Science and Technology*, Vol. 1, (eds J. Anderson, and M. Boudart), Springer, New York, pp. 160–255.

Ertl, G. (2003) Ammonia synthesis, in *Encyclopedia of Catalysis*, Vol. 6 (ed I. Horvath,), John Wiley & Sons, Inc., pp. 329–352.

Fischer, F. and Tropsch, H. (1930) *Process for the production of paraffin-hydrocarbons with more than one carbon atom*. U.S Patent 1,746,464.

Hoek, A. (2006) The shell GTL process. *DGMK International Conference: Synthesis Gas Chemistry*, October 4–6, Dresden, Germany.

Imperial Chemical Industries Limited, Agricultural Division (1970) *Catalyst Handbook: With Special Reference to Unit Processes in Ammonia and Hydrogen Manufacture*, Springer, New York.

Jager, B., Steynberg, A.P., Kelfkens, J.R., Smith, R.C., Michael A., and Malherbe, F.E.J. (1984) *Process for producing liquid and, optionally, gaseous products from gaseous reactants*. U.S Patent 5,599,849

McGrath, H., Rubin, E., and Rubin, L. (1951) *Preparation of reduced magnetite synthesis catalyst*. U.S. Patent 2,543,327.

Methanol. (1982) *Kirk–Othmer Encyclopedia of Chemical Technology*, 3rd ed., Vol. 16, John Wiley & Sons, Inc., New York, pp. 299–315.

Stynberg, A.P. and Dry, M. (2004) *Fischer–Trospch Technology*, Studies in Surface Science and Catalysis, Vol. 152, Elsevier, Amsterdam, The Netherlands.

Tsakounis, N. E., Ronning, M., Borg, Q., Rytter E., and Holmen, A. (2010) Deactivation of cobalt based Fischer–Tropsch catalysts: a review. *Catalysis Today* 154, 162.

Zhang, Q., Kang, J., and Wang Y. (2010) Development of novel catalysts for Fischer–Tropsch synthesis: tuning product selectivity. *ChemCatChem* 2, 1030–1058.

CHAPTER 8

SELECTIVE OXIDATIONS

8.1 NITRIC ACID

We have previously discussed the importance of nitrogen for plant and human life. Ammonium-containing fertilizers represent one of the most important chemicals necessary for providing nutrients to plants that yield substantial amounts of food to feed the world's growing population. The development of the ammonia synthesis process by catalytically combining N_2 and H_2 remains one of the most important processes and achievements of catalytic technology. How ammonia is utilized as a fertilizer is the subject of the ammonia oxidation process. Ironically, nitrogen sustains life via amino acids but can also be destructive in the formation of explosives.

8.1.1 Reaction Chemistry and Catalyst Design

Nitric acid is formed through sequential oxidation of NH_3 to NO (reaction 8.1) and then NO to NO_2 (reaction 8.2). The NO produced is absorbed/reacted with water to form nitric acid (reaction 8.3). Several undesirable reactions can also occur. Non-catalytic reactions 8.5 and 8.6 are used for the production of fertilizers.

Ammonia oxidation (AMOX):

$$4NH_3 + 5O_2 \rightarrow 4NO + 6H_2O,$$
$$\Delta H_{900\,°C} = -850\,kJ/mol, \quad K_e(900\,°C) = 10^{10} \tag{8.1}$$

$$NO + \tfrac{1}{2}O_2 \rightarrow NO_2 \tag{8.2}$$

$$3NO_2 + H_2O \rightarrow 2HNO_3 + NO \tag{8.3}$$

Undesired reaction:

$$4NH_3 + 3O_2 \rightarrow 2N_2 + 6H_2O,$$
$$\Delta H_{900\,°C} = -1200\,kJ/mol, \quad K_e(900\,°C) = 10^{15} \tag{8.4}$$

Ammonium nitrate:

$$NH_3 + HNO_3 \rightarrow NH_4NO_3 \tag{8.5}$$

Introduction to Catalysis and Industrial Catalytic Processes, First Edition. Robert J. Farrauto, Lucas Dorazio, and C.H. Bartholomew.
© 2016 John Wiley & Sons, Inc. Published 2016 by John Wiley & Sons, Inc.

Urea:

$$2NH_3 + CO_2 \rightarrow NH_2COONH_4 \rightarrow NH_2CONH_2 + H_2O \qquad (8.6)$$

Reaction 8.1 is conducted by the catalytic oxidation of 10–11% NH_3 in air between 1 and 13 atm pressure depending on the plant. The NH_3 concentration depends on the flammability and pressure of operation. High-pressure plants (~13 atm) must operate further from the 14% lower flammability limit for safety reasons than those operated at lower pressures. The high-pressure reaction is initiated at about 250 °C over stacks of woven wires composed of 90% Pt and 10% Rh alloy gauze (an alloy containing 90% Pt, 5% Rh, and 5% Pd is also used) with a mesh size of 80 (wire diameter is about 0.003 in.), which provides suitable geometric surface area to maximize mass transfer. The active component is Pt, but Rh is added mainly to maintain mechanical strength during the wire drawing and weaving operations. The gauze sheets (15–30) are stacked on top of each other on a support structure of Inconel to give turbulent flow and high mass transfer conversion rates. The highly exothermic reaction 8.1 generates an adiabatic temperature rise of about 900–950 °C well into the mass transfer control regime. Nitric oxide (NO) is the thermodynamically stable form at these temperatures, which must be cooled to <100 °C and reacted with O_2 to favor NO_2 (reaction 8.2). The NO_2 is further combined in a H_2O spray tower producing HNO_3 (reaction 8.3) and more NO. HNO_3 is continuously removed shifting the equilibrium to more NO_2 and HNO_3. The effluent typically contains about 0.3% NO and some NO_2 ($NO + NO_2$ is referred to as NO_x) and about 3% O_2 that is then catalytically abated in a NO_x reduction system (SCR to be discussed in the stationary environmental chapter).

8.1.1.1 *The Importance of Catalyst Selectivity*

The competing reaction is the oxidation of NH_3 to N_2 shown in Equation 8.4. Its K_e value at reaction conditions is 10^5 times more favorable than the desired reaction 8.1 based on a comparison of their respective K_e values. Thus, the role of the catalyst is to direct the reactants selectively to the desired product even though the undesired reaction 8.4 is more favorable thermodynamically. This illustrates the power of the selective catalyst to enhance the desired reaction over the less desired one by providing the proper chemical path favoring NO over N_2.

8.1.1.2 *The PtRh Alloy Catalyst*

During catalysis, the Pt evolves from the surface of the gauze creating a roughened or "sprouted" morphology, resembling cauliflower (see Figure 8.1). The evolution of Pt is equivalent to about 1–2 g/t of HNO_3 and is largest for the high-pressure processes. The gauze surface area increases from 12 to 200 cm^2/g in the first few days of operation. The activity gradually increases as the geometric surface area increases to a final steady state. This increase in activity and geometric surface area (GSA) is consistent with the mass transfer control of the reaction; increased GSA enhances mass transfer rate. The composition of the Pt species evolving is not clearly understood but it occurs only in the presence of NH_3 and air and we will write it as PtO_xN_y. The Pt depleted surface becomes enriched with Rh (approximately 50% of each metal) over its life and typically after 90–120 days the activity declines to an unacceptable level and the gauze must be

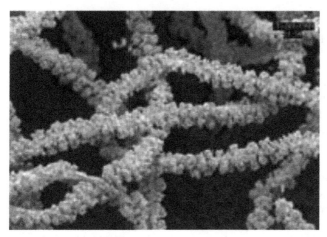

Figure 8.1 Surface roughening (sprouting of PtRh gauze). (Reproduced from Chapter 8 of Bartholomew, C. and Farrauto, R.J. (2006) *Fundamentals of Industrial Catalytic Processes*, 2nd edn, John Wiley & Sons, Inc., Hoboken, NJ.)

replaced. This is the main cause for deactivation. The spent gauze is returned to the supplier where the precious metals are recovered for future use.

The kinetic mechanism is still in dispute; however, a popular version is that NH_3 is dissociatively adsorbed on the Pt forming NH radicals that react on the surface with adsorbed O atoms in what resembles a Langmuir–Hinshelwood mechanism. What is not in dispute is the fact that the reaction is mass transfer controlled. For this reason, about 30–50 gauzes are stacked in series in a high-pressure plant producing turbulent flow that enhances mass transfer. For low-pressure plants, lifetimes of 6 months to 1 year are typical since the rate of Pt depletion is lower.

8.1.2 Nitric Acid Production Process

A 13 atm (high-pressure) plant operates with a gauze diameter of 3–4 ft while one operating at atmospheric pressure is about 12 ft in diameter. The high-pressure plant has a selectivity of 94% versus about 98% for the plant operating at atmospheric pressure. Furthermore, the Pt volatilized is higher for the high-pressure process. It operates at a gas hourly space velocity (GHSV) of over $500,000 \, h^{-1}$. Figure 8.2 shows a schematic of a high-pressure plant. Positioned immediately below the AMOX gauze is a stack of Pd-rich woven getter gauzes that decompose and capture the volatile Pt species evolved from the AMOX gauze due to the extreme temperatures and oxidizing conditions (Figure 8.3), where the Pt metal forms an alloy with Pd. Interestingly, the capture rate is controlled by diffusion of Pt into the bulk of the Pd. The effluent from the reactor is primarily nitric oxide and H_2O. The formation of nitrogen dioxide (reaction 8.2) is a slow reaction and favored at low temperatures. After exiting the reactor, the product gases are cooled in a heat exchanger train designed to recover heat for steam generation and integration back into the process. As the reaction gas cools, the oxidation of nitric oxide to nitrogen dioxide occurs noncatalytically in the gas phase. Some nitric acid is produced during

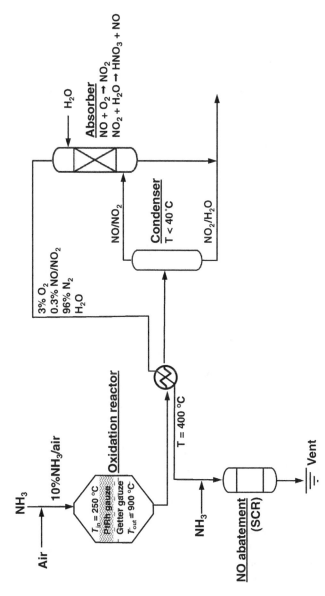

Figure 8.2 High-pressure NH_3 oxidation/HNO_3 plant with Pd getter gauze.

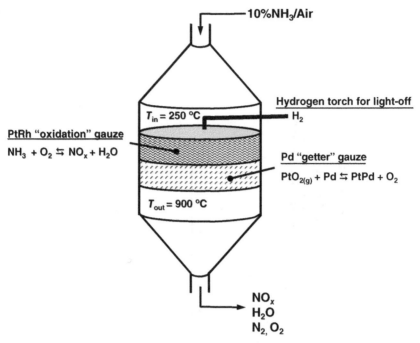

Figure 8.3 An expanded view of the reactor containing the stacks of oxidation and getter gauzes.

cooling since NO_2 combines with the H_2O produced in reaction 8.1. Due to the corrosive nature of hot liquid nitric acid, the extent of cooling after the reactor is limited by the dew point of the nitric acid. After cooling, the reactor effluent is further cooled in a condenser, which is constructed of more expensive exotic corrosion-resistant materials. The liquid nitric acid is separated from the gaseous reaction products, which are subsequently fed to a downstream absorber. In the absorber, nitric acid and nitrogen dioxide are absorbed into water forming nitric acid, which is combined with the condensate from the upstream condenser. The gaseous effluent from the absorber still contains small quantities of nitrogen oxides that must be removed for environmental reasons before being discharged to the atmosphere. This gas stream is heated and brought to a catalytic abatement unit where NH_3 is injected into a highly selective V_2O_5 catalyst deposited on a monolithic structure where the NO_x is selectively converted to N_2. This is discussed in the environmental stationary abatement chapter.

The "getter" gauze, of a similar geometry as the AMOX catalyst, is positioned immediately downstream from the Pt gauzes. It is also recovered and refined for its metal value. Typically, 60–80% of the Pt is recovered in the getter gauze.

8.1.3 Catalyst Deactivation

The AMOX gauze is brought to reaction temperature by the use of a H_2 torch that a skilled technician directs over the surface watching for the red glow indicating initiation of the reaction. Once the glow is uniform across the diameter of the reactor, the torch is extinguished and the enthalpy generates sufficient heat for the reaction to proceed without addition of external heat.

The position of the torch during start-up is critical since overexposure at any given spot can literally burn a hole in the gauze. Thus, a successful start-up dictates the life of the process but presenting a uniform mixture of the NH_3, O_2, and product NO is critical. If there are pockets of unreacted NH_3, it can react with NO according to the highly exothermic reaction 8.7 causing melting of the AMOX and getter gauze.

$$4NH_3 + 6NO \rightarrow 5N_2 + 6H_2O, \quad \Delta H = -1700\,\text{kJ/mol}, \quad K_e = 10^{14} \quad (8.7)$$

Deactivation can also occur due to Fe-containing compounds depositing on the gauze from upstream reactor corrosion. The Fe not only masks the gauze, but also catalyzes NH_3 decomposition to N_2.

8.2   HYDROGEN CYANIDE

Nylon is produced by the polymerization of adiponitrile (CN—$(CH_2)_4$—CN) by the hydrogen cyanide (HCN) reaction with CH_4 (Equation 8.8). Even in the presence of a catalyst, the required reaction temperature to achieve acceptable conversion rate is 1000–1200 °C. This relatively high reaction temperature, coupled with the highly endothermic nature of the reaction, results in heat transfer effectiveness dominating the design of the catalytic reactor. There are two commercially acceptable processes where the primary difference is the means of providing the enthalpy required to maintain the required reaction temperature: the BMA process and the Andrussow process.

The BMA process was developed by Degussa, where BMA stands for German words Blausaure-Methan-Ammoniak, which mean hydrogen cyanide–methane–ammonia. As the name suggests, this process involves only the primary reactants methane and ammonia to form hydrogen cyanide (Equation 8.8). The advantage of this approach is higher ammonia conversion. The disadvantage is the high heat demand from the surroundings required to maintain the desired reaction temperature. In order to achieve the needed heat transfer rate, the reaction is conducted in many small diameter tubes where the platinum catalyst is coated directly onto the interior tube wall. The small diameter tubes provide high external tube area for heat transfer as well as high internal tube area to support the required quantity of catalyst. By coating the catalyst directly onto the tube wall, the resistance to heat transfer is decreased. Thus, heat is conducted through the tube wall directly to the catalyst coating.

The most widely used manufacturing process is the Andrussow process (Equation 8.9), which introduces air (oxygen) in addition to methane and ammonia. A 90% Pt and 10% Rh woven gauze, similar to that used for AMOX, catalyzes the oxidation of a portion of methane and the combustion enthalpy is used to supply the heat required to maintain the reaction temperature effectively supplying the heat from within the reactor. Using this approach, once the reaction is initiated, the reaction can thermally sustain itself.

BMA:

$$CH_4 + NH_3 \rightarrow HCN + 3H_2, \quad \Delta H^\circ = 252 \, kJ/mol \tag{8.8}$$

Andrussow:

$$2CH_4 + 3\frac{1}{2}O_2 + NH_3 \rightarrow HCN + CO_2 + 5H_2O, \\ \Delta H^\circ = -474 \, kJ/mol \tag{8.9}$$

8.2.1 HCN Production Process

The Andrussow process will be discussed here. The heat required is generated adiabatically by the oxidation of a portion of the methane. The NH_3 feed concentration is 12%, CH_4 is 13% (desulfurized), and the balance is air at about 2 atm pressure. The linear velocity is about 3 m/s to avoid precombustion of the feed gases (i.e., flashback). This mixture is above the flammability limits permitting safer operation. An electric igniter initiates the reaction and the adiabatic temperature rise is then sufficient for sustaining the kinetics. The PtRh gauze sheets ($\sim$2 m in diameter) are stacked in series in the reactor similar to the reactor design for nitric acid production. The process flow diagram is shown in Figure 8.4.

The net reaction is reducing, which eliminates Pt loss, so the morphological changes accompanying nitric acid production are much less severe for HCN. Recrystallization of the Pt does occur, mainly due to the high temperatures experienced, and results in an increase in geometric surface area during the first 100 h of operation. The rate of reaction increases during this time consistent with the increase in mass transfer area. The yield of HCN is about 70% based on NH_3.

The product HCN is rapidly chilled to <350 °C to avoid its decomposition followed by an acidification with H_3PO_4 in the absorber. This must be done to avoid polymerization of the HCN. Excess NH_3 is separated in the stripper from the diammonium phosphate. In some processes, the ammonium phosphate is decomposed and the NH_3 recycled as shown exiting the fractionator. The final acidification of the HCN (the H_2SO_4 added to the final fractionator) prevents polymerization of the HCN generating a 30% solution. The HCN (boiling point 26 °C) is steam distilled to 99% purity.

8.2.2 Deactivation

A main catalyst purity issue, to ensure reasonable lifetimes of up to 1 year, is the absence of Pd contamination that decomposes the NH_3. Thus, the purification of the starting precious metals is critical. Upstream reactor corrosion can deposit iron oxides onto the gauze decomposing the NH_3 to N_2 resulting in a loss in selectivity. The morphological changes can lead to mechanical failure shortening catalyst life.

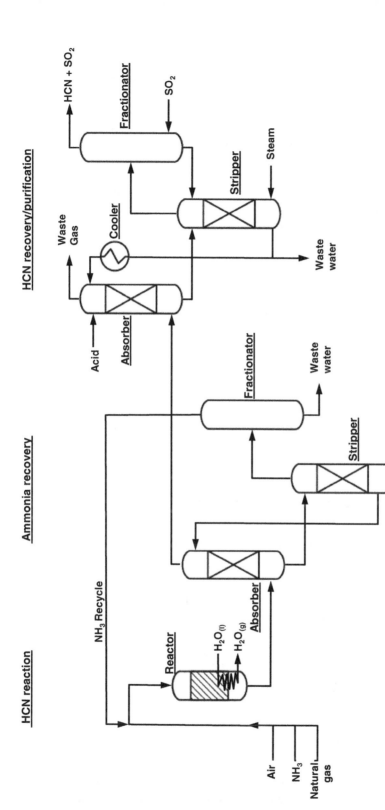

Figure 8.4   HCN process flow diagram.

8.3 THE CLAUS PROCESS: OXIDATION OF H$_2$S

Increasing amounts of sulfur are available in the marketplace due to deep desulfurization of both fossil and shale petroleum. Environmental issues are mandating reduced levels especially in transportation fuels such as gasoline and diesel. Lower sulfur values enhance the effectiveness and life of the catalysts used for pollution abatement as well as minimizing corrosive effects. The sulfur is removed from these feeds mainly though catalytic hydrodesulfurization (HDS) producing H$_2$S generating streams greater than 10%. At such high levels of H$_2$S, sulfur recovery is viable and the Claus process can be applied.

$$2H_2S + SO_2 \rightleftharpoons 2H_2O + \frac{3}{x}S_x \tag{8.10}$$

8.3.1 Clause Process Description

The reaction has some unusual thermodynamic properties mainly due to the nature of the sulfur compounds produced as a function of temperature (see Figure 8.5). The catalytic step yields an eight-member polymer S$_8$, which is highly exothermic favored at lower temperatures. Above 575 °C, an endothermic thermal reaction (no catalyst) is favored for producing the S$_2$ (dimer). The required heat and the SO$_2$, needed for subsequent reactions, are provided by the addition of O$_2$ according to the oxidation reaction shown in Equation 8.11.

An excess of H$_2$S thermally reacts with added O$_2$ generating temperatures as high as 1300 °C. In addition to heat, SO$_2$ and S$_2$ are formed. Upon cooling, 70% of the total sulfur initially present becomes the dimer.

$$10H_2S \text{ (excess)} + 5O_2 \rightarrow 2H_2S + SO_2 + \tfrac{7}{2}S_2 + 8H_2O,$$
$$\Delta H° = -4150\,kJ/mol \text{ of } H_2S \tag{8.11}$$

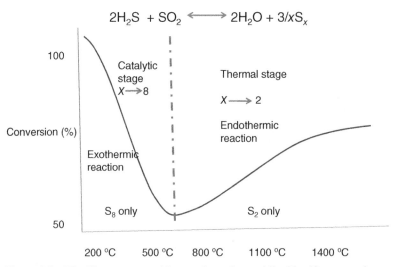

Figure 8.5 The Claus process with staged reaction and liquid sulfur removal.

The S_2 dimer and the residual H_2S and SO_2 are cooled to 370 °C and passed through a fixed bed of catalyst (high surface area γ-Al_2O_3 or TiO_2 or combinations of γ-Al_2O_3 and TiO_2, in reactors up to 2 in. in diameter), where the equilibrium reaction for S_8 formation occurs. At this point, 70% of the sulfur has been removed.

$$4S_2 \rightleftharpoons S_8 + \text{heat} \tag{8.12}$$

The remaining 30% of sulfur existing as H_2S and SO_2 is reheated to 300 °C and passed through a second catalyst bed. The effluent is again cooled where the sulfur is condensed yielding 87% of the remaining sulfur. A succession of catalytic steps, with intermittent effluent condensation steps, proceed for liquid sulfur recovery to 200 °C bringing the sulfur total recovery to 99.8%. The process of removing the product and progressively cooling the reactants is a nice example of chemical thermodynamics in practice. No technology currently exists to remove the H_2O with the liquid sulfur, which would shift the equilibrium to an even greater extent.

8.3.2 Catalyst Deactivation

A common cause of catalyst deactivation is condensation of liquid sulfur in the catalyst pores blocking reactant accessibility. This can be avoided by not allowing the mixture to reach the dew point of liquid sulfur in the catalyst bed. Traces of aromatics can also react with the sulfur forming carbon–sulfur polymers called "carsuls" that accumulate in the pores occluding the catalytic sites.

8.4 SULFURIC ACID

Sulfuric acid is a high-demand chemical that is one of the top manufactured products in the chemical industry. It has a wide range of applications including use in manufacturing other chemicals, as a homogeneous catalyst for production of chemicals, solubilizing minerals, electrolyte, and even use in a variety of domestic products. It is expected that more and more elemental sulfur will be available in the future as higher sulfur content crudes (e.g., oil sands and sour crudes) are processed to fuels. Also the environmental regulations are now requiring lower and lower sulfur contents in fuels (gasoline ~20 ppm, diesel ~10 ppm) that will also generate more elemental sulfur. Using the Claus process discussed in the previous section, this fuel-bound sulfur can be processed to elemental sulfur. Using the processes discussed in this section, the elemental sulfur can be processed to produce sulfuric acid.

8.4.1 Sulfuric Acid Production Process

The precursor for sulfuric acid may include a wide variety of sulfur-containing chemicals. However, the primary sulfur-containing precursor is either elemental sulfur or hydrogen sulfide. For these two materials, the production of sulfuric acid involves three basic steps: noncatalytic oxidation of elemental sulfur or hydrogen sulfide to sulfur dioxide (Equations 8.13 and 8.14), catalytic oxidation of sulfur

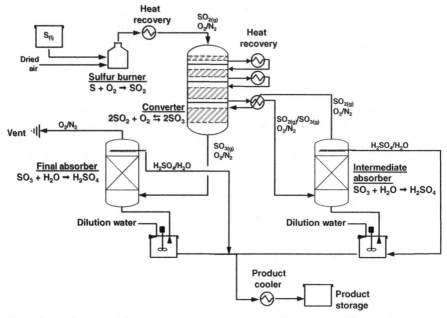

Figure 8.6   Elemental sulfur is reacted with dry air at 900 °C producing SO$_2$. Staged air injection into the second and third stages for cooling is shown in Figure 8.8.

dioxide to sulfur trioxide (Equation 8.15), and absorption of sulfur trioxide into water forming sulfuric acid (Equation 8.16). The overall process is shown in Figure 8.6. The catalytic oxidation of sulfur dioxide to sulfur trioxide is reversible. Given its exothermic nature, the equilibrium conversion becomes increasingly more unfavorable at temperatures greater than 450 °C. The SO$_2$/SO$_3$ equilibrium is shown in Figure 8.7. Unfortunately, this temperature is also the minimum required to achieve acceptable kinetics over known catalysts. Thus, the reactor must be designed to remove reaction enthalpy to keep the reaction temperature low enough to minimize thermodynamic limitations, yet high enough to achieve acceptable kinetics. The solution used in the traditional manufacturing process is a series of catalyst beds with staged cooling. Figure 8.7 shows the adiabatic temperature rise with subsequent staged cooling as equilibrium is approached.

$$S_{(s)} + O_{2\,(g)} \rightarrow SO_{2\,(g)} \tag{8.13}$$

$$2H_2S_{(g)} + 3O_{2(g)} \rightarrow 2H_2O_{(g)} + 2SO_{2(g)}, \quad \Delta H^\circ_{rxn} = -518 \text{ kJ/mol} \tag{8.14}$$

$$2SO_{2(g)} + O_{2(g)} \rightleftharpoons 2SO_{3(g)}, \quad \Delta H^\circ_{rxn} = -99 \text{ kJ/mol} \tag{8.15}$$

$$SO_3 + H_2O \rightarrow H_2SO_4 \tag{8.16}$$

As described in Figure 8.6, elemental sulfur is oxidized to form SO$_2$, which is then cooled to about 400 °C, where 10% SO$_2$ in air enters a catalytic reactor. The

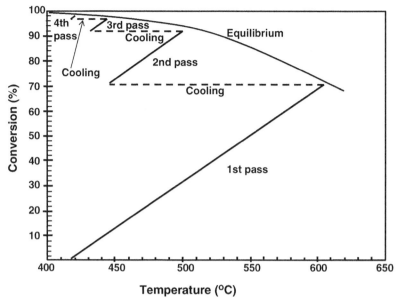

Figure 8.7 SO_2/SO_3 equilibrium as a function of temperature. (Reproduced from Chapter 8 of Bartholomew, C. and Farrauto, R.J. (2006) *Fundamentals of Industrial Catalytic Processes*, 2nd edn, John Wiley & Sons, Inc., Hoboken, NJ.)

traditional catalyst is V_2O_5–K_2SO_4 dispersed on SiO_2. The reaction is conducted at atmospheric pressure in a reactor approximately 10 ft in diameter. The materials of construction are cast iron and steel. Maintaining the SO_2/SO_3/air dry is essential to minimize corrosion. The air is dried by passing it through hydroscopic H_2SO_4 before it is mixed with the SO_2.

For an adiabatic bed design, the first catalyst bed is sized such that the temperature rise is from 400 to about 600 °C with 60% of the SO_2 converted based on equilibrium constraints. This is shown in Figure 8.6 and can be visualized in Figure 8.7. The hot product gas is diverted to an external heat exchanger, where it is cooled to about 450 °C and reintroduced into a second bed of catalyst. The conversion now reaches about 88% with the temperature increasing to 500 °C. The cycle of reaction followed by cooling can be repeated several times. In most plants, small amounts of air are injected into the second and third stages to cool the mixture. As the concentration of SO_3 increases, equilibrium is approached and no further significant conversion is possible without removing some of the SO_3. Additionally, the large amount of SO_3 present adsorbs on the catalyst kinetically inhibiting the forward reactions 8.15 and 8.16, so its removal accomplishes both kinetic and thermodynamic functions.

$$r \sim \frac{P_{SO_2} P_{O_2}}{P_{SO_3}} \tag{8.17}$$

Depending on the catalysts used, which are more active at lower temperatures, slightly higher conversions are possible. Prior to air pollution regulations imposed

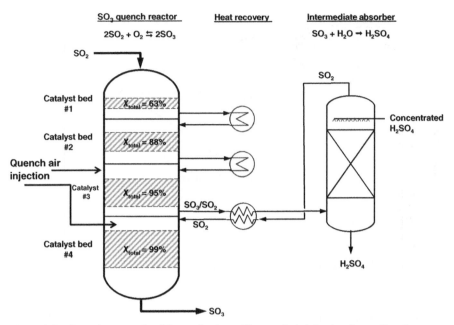

Figure 8.8 Quench reactor for SO_3 production with staged air injection for cooling for stages 2 and 3.

during the 1970s, this conversion was acceptable and unreacted SO_2 was vented to the atmosphere. This process was referred to as the "single contact process" as only one postreaction absorption step was used. To meet environmental standards by converting all SO_2 to SO_3 in the oxidation reactor, a conversion greater than 99.7% is required. This is accomplished in a quench reactor (Figure 8.8) by diverting the product to a chilled SO_3 adsorption tower reducing the temperature to 70 °C where SO_3 is contacted with a H_2SO_4/H_2O mixture, adsorbing SO_3 from the gas phase, and sending $SO_2/O_2/N_2$ mixture back into the oxidation reactor and the final reaction state. The cooled gas is then mixed with a portion of the feed gas (10% SO_2 in air) at 400 °C elevating the feed temperature delivered to the third reactor to 400 °C where essentially the remaining SO_2 is converted to SO_3.

After the final catalyst bed, the product gas is cooled and contacted with a H_2SO_4/H_2O mixture in an absorption tower. The reaction of SO_3 with H_2O is also exothermic ($\Delta H° = -140\,kJ/mol$ of SO_3), so cooling is also necessary for this operation. When two absorption steps are used, intermediate absorption and final absorption, this process is referred to as the "double contact process" for manufacturing sulfuric acid.

8.4.2 Catalyst Deactivation

The original catalyst used for the contact process was Pt on SiO_2. It has been replaced with ring-shaped particulates or pellets of $V_2O_5 + K_2O/SiO_2$. During reaction, K_2SO_4 forms producing a low melting temperature eutectic mixture. The catalyst therefore exists as a melt, coating the SiO_2 carrier, during the reaction. Because of this, the major deactivation mode is volatilization, dusting, and sintering, especially for the first and second beds. The ring shape minimizes pressure buildup from dust accumulation. The typical catalyst life is 5–10 years.

8.5 ETHYLENE OXIDE

This specialty chemical is used in the production of polyesters, bottles, polyurethane, surfactants, and detergents, but is probably best known as the precursor to ethylene glycol used as antifreeze for car radiators and polyesters in the production of carpets. Ethylene oxide (EO) is produced by the catalytic partial oxidation of ethylene with a highly selective partial oxidation catalyst composed of Ag and Re supported on α-Al_2O_3 (Equation 8.18). The use of low surface area α-Al_2O_3 is unusual in that we most often use high surface area carriers to disperse the catalytic components. The rationale for this choice will be discussed later but it does influence selectivity. For the manufacture of ethylene glycol, the epoxide is subsequently reacted with water to produce the glycol ($C_2O_2H_2$).

Ethylene oxide:
$$\underset{O}{CH_2\!-\!CH_2}$$

$$CH_2CH_2 + \tfrac{1}{2}O_2 \rightarrow CH_2CH_2O, \quad \Delta H^{\circ}_{rxn} = -105\,kJ/mol \qquad (8.18)$$

The undesirable side reaction is the complete oxidation of ethylene to CO_2 and H_2O, which is highly exothermic (Equation 8.19). The enthalpy from complete oxidation would raise the reaction temperature and further decrease selectivity toward the desired product. Successful implementation of the ethylene oxide process requires a highly selective catalyst as well as an optimized process design.

$$CH_2CH_2 + 3O_2 \rightarrow 2CO_2 + 2H_2O, \quad \Delta H^{\circ}_{rxn} = -1323\,kJ/mol \qquad (8.19)$$

8.5.1 Catalyst

The spherical or cylindrical catalyst (1/8–3/8 in. in diameter) is composed of 15–20% Ag with Re and BaO promoters, deposited on low surface area ($3\,m^2/g$) α-Al_2O_3. The low surface area Al_2O_3 is used to enhance selectivity ($\sim$80%, life $\sim$3 years) to the epoxide and minimize complete oxidation to CO_2 and H_2O. Because both the desired and mostly the undesired reactions are exothermic, removing heat from the catalyst bed is necessary to avoid complete oxidation at elevated temperatures. This suggests that the complete oxidation has higher activation energy (greater temperature sensitivity) than the desired reaction. Al_2O_3, in the α-structure, is sufficiently dense to be thermally conductive and enhances heat removal. Also the high concentration of Ag in multilayers on the carrier also enhances thermal conductivity assisting in heat removal. Interestingly, a few ppm of a chloride compound such as dichloroethane, in the process gas, greatly enhances selectivity by decreasing the activity for the undesired reaction. It is another example of how intentional poisoning can be used to enhance selectivity.

8.5.2 Catalyst Deactivation

Sintering of the Ag is a common source of deactivation given its low dispersion on low surface area α-Al_2O_3. The inclusion of Re and BaO in the catalyst minimizes sintering. Sulfur and alkali (from the CO_2 scrubbers present in the recycle loop) must be scrubbed from the process gas to avoid undesired poisoning. Coking is also possible since traces of acetylene, strong coke precursors, may form. Regeneration by burn-off in controlled air flow is commonly employed.

8.5.3 Ethylene Oxide Production Process

Historically, there have been two processes used for EO production: one using air for the oxygen source, and the other that uses pure oxygen. The air-based system dominated industrial design up until the 1990s. To maximize selectivity to ethylene oxide, the single-pass conversion is kept at 10–20% and a product recycle is employed to yield a high overall conversion. One problem created by using air as the oxygen source is the large purge required in the recycle to remove nitrogen and other impurities that build up in the recycle loop. In the 1990s, new EO plants were designed to use pure oxygen. The absence of nitrogen eliminated the need for a purge reactor, which simplified the overall process. The generation of CO_2 still requires a small purge unit in the recycle stream.

In the O_2 process, the clean feed consisting of a 3:1 ratio of ethylene and O_2, with a few ppm of dichloroethane (to control selectivity), is combined with recycle gas, and is compressed to 10–20 atm and preheated through the outlet gas heat exchanger (Figure 8.9). The reactors are oil-cooled tubular reactors containing the particulate catalyst maintained around 200–250 °C. Great care must be taken in controlling the mixing of the feed gases since the ethylene concentrations are within the flammable range (13–32%). Conversion is limited to less than 15% to control temperature and selectivity. The space velocity is about 15,000 h^{-1}. The product gases are further cooled where ethylene oxide (BP = 10.7 °C) can be separated by condensation from the unreacted ethylene (BP = −104 °C). The ethylene oxide is distilled and separated from acetylene and acetaldehyde impurities. The process gas, containing ethylene, CO_2, and traces of acetylene, is passed through an alkali scrubber to remove the CO_2 and the more concentrated stream of ethylene is recycled to the front end of the process where it is mixed with fresh feed. The recycled gas is periodically purged to remove the buildup of acetylene and other impurities.

8.6 FORMALDEHYDE

Formaldehyde is used in the manufacture of urea–formaldehyde polymers and resins, adhesives, wood products, and moldings. There are two major processes used for its production, both utilizing the partial oxidation of methanol with varied concentrations.

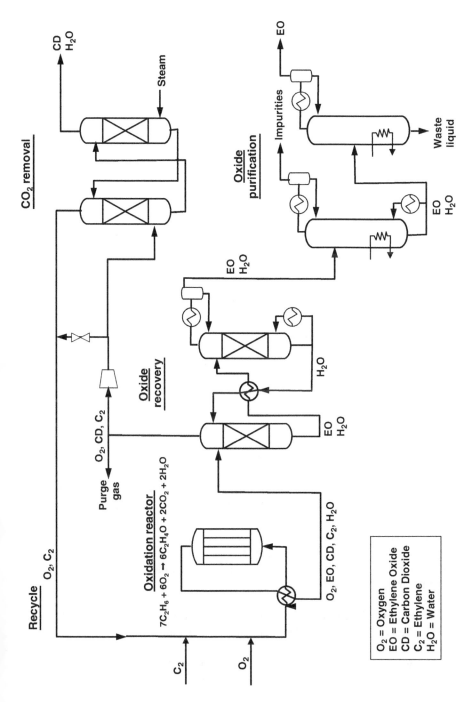

Figure 8.9   The O₂ process for ethylene oxide production.

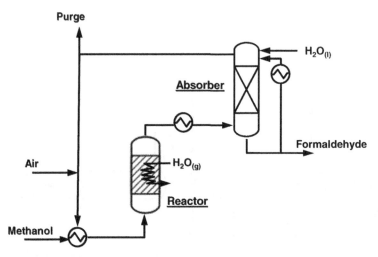

Figure 8.10 Process for low methanol concentration process to formaldehyde over a (Fe, Mo)/SiO$_2$ catalyst.

The oldest process uses Ag needles or a woven gauze catalyst with highly concentrated methanol. This has been mostly replaced by a lean or low methanol concentration process utilizing a SiO$_2$-supported Fe- and Mo-containing catalyst.

8.6.1 Low-Methanol Production Process

The most commonly used process uses 9% methanol in air and steam over a Fe- and Mo-containing catalyst at atmospheric pressure (Figure 8.10). This process is now preferred over the high-concentration system because of its higher selectivity (98% versus 90%) and lower catalyst bed temperatures (275 °C versus 500–700 °C) relative to the Ag process. The presence of a large excess of steam moves the mixture outside of the flammable range. The 9% methanol–air–steam mixture is preheated from the reactor exit gas heat exchanger, and is fed to tubular water (steam)-cooled reactors containing the particulate catalyst. The bed temperature is maintained at about 275 °C. The product gas is further cooled through the exit heat exchanger. A water spray tower cools the product to 100 °C and the product is distilled for purification from formic acid and acetone by-products. Any unreacted methanol is recycled to the front of the process. The tail gas, containing traces of the oxygenated impurities, is properly abated consistent with local emission regulations. Periodically, a purge gas is used to remove the buildup of impurities and N$_2$ (from the air) in the recycle loop.

8.6.1.1 *Fe + Mo Catalyst* The catalytic components are composed of 17% Fe$_2$O$_3$ and 81% MoO$_3$ dispersed on SiO$_2$ particulates 2–3 mm in diameter. The space velocity is 10,000 h^{-1}. Because of the high surface area, it is less subject to poisoning than the low surface area Ag catalyst and therefore has a lifetime approaching 2 years compared with 2–6 months for the Ag-catalyzed process described below.

8.6.2 High-Methanol Production Process

The concentrated methanol process feeds a mixture of 45% methanol and about 25% air (balance steam) to the Ag catalyst. The concentration of methanol is above the flammability range (6.7–36.5%) allowing relatively safe operating conditions.

Desired reaction:

$$CH_3OH + \tfrac{1}{2}O_2 \rightarrow CH_2O + H_2O, \quad \Delta H_{600\ °C} = -155\ kJ/mol \quad (8.20)$$

Undesired reactions: unbalanced:

$$CH_3OH + \tfrac{1}{2}O_2 \rightarrow CO_2 + H_2O \quad (8.21a)$$

$$CH_3OH + \tfrac{1}{2}O_2 \rightarrow HCOOH\ (formic\ acid) \quad (8.21b)$$

$$CH_3OH + \tfrac{1}{2}O_2 \rightarrow CH_3{-}O{-}CH_3\ (methyl\ formate) \quad (8.21c)$$

The mechanism proposed is that chemisorbed O on Ag reacts with gas-phase CH_3OH forming chemisorbed CH_3O on Ag with chemisorbed OH. The methoxy species then decomposes to formaldehyde and a H atom chemisorbed on Ag. The chemisorbed H and OH on adjacent Ag atoms react to produce H_2O.

$$CH_3OH + Ag\cdots O \rightarrow CH_3O\cdots Ag + OH\cdots Ag \quad (8.22)$$

$$CH_3O\cdots Ag \rightarrow CH_2O + H\cdots Ag \quad (8.23)$$

$$OH\cdots Ag + H\cdots Ag \rightarrow H_2O \quad (8.24)$$

The atmospheric process is shown schematically in Figure 8.11.

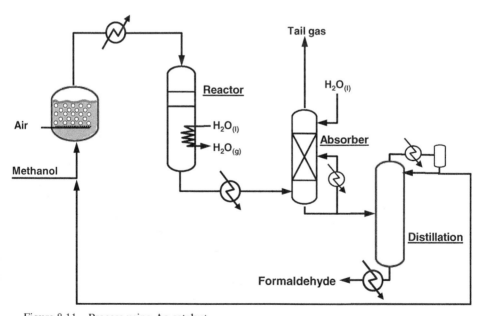

Figure 8.11   Process using Ag catalyst.

The feed gases, composed of 45% CH_3OH and 25% air, are preheated by the injection of steam to raise the temperature to 500 °C and passed through a reactor containing either Ag needles or Ag gauze (designed from reduced pressure drop) where adiabatic reaction occurs raising the temperature to 700 °C. The product gas is cooled to 300 °C and passed into two sequential water spray towers producing an aqueous solution of formaldehyde (also referred to as methylene glycol). The tail gases of CO_2 and H_2O are vented. If a sufficiently high concentration of methanol is present, it can be separated and recycled to the front of the process. The desired products are distilled separating the CH_2O solution from the higher boiling formic acid and acetone impurities.

8.6.2.1 Ag Catalyst The catalyst is essentially pure Ag mostly woven into gauze for reduced pressure drop. The space velocity is 10^5 with a selectivity of 90% and life of 2–6 months. The catalyst is regenerated by washing to remove S, N, and halogens that selectively adsorb and poison the catalyst. Another major source of deactivation is the Ag fusion of the catalyst mass due to its low melting temperature leading to large pressure drops. Any coke can be easily removed by physical separation.

8.7 ACRYLIC ACID

Acrylates, derived from acrylic acid, have super adsorption properties and are extensively used in baby diapers. They also have use in water-based paints, polishes, and adhesives.

8.7.1 Acrylic Acid Production Process

The process is carried out in three stages starting with propane being dehydrogenated to propylene with a Cr_2O_3-based catalyst (Equation 8.25). Propylene is reacted with a limited amount of O_2 using a Bi- and Mo-containing catalyst producing acrolein (Equation 8.26). The latter is reacted with additional O_2 using a Mo-, V-, W-, and Fe-containing catalyst to acrylic acid (Equation 8.27).

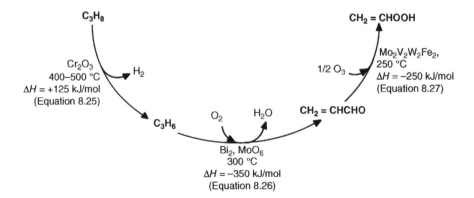

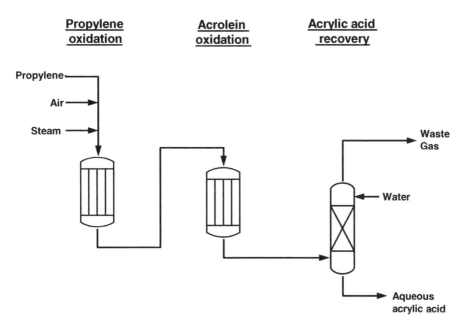

Figure 8.12   Propylene to acrolein to acrylic acid process flow diagram. Tubular reactor with a diameter of about 2.5 cm and a length of about 4 m cooled by a molten carbonate.

The two-step process flow diagram in Figure 8.12 reflects the catalytic conversion of propylene to acrolein, which in turn is further selectively oxidized to acrylic acid. Five to seven percent propylene, in 30–50% steam, with the balance air, is preheated to 300 °C and passed into a series of tubular bed reactors (about 2–3 cm in diameter and 4 m long) containing oxides of Bi_2Mo particulate catalyst where the selective oxidation to acrolein takes place. The series of tubular reactors are cooled with a molten carbonate eutectic. The selectivity is 85% at 95% conversion at a space velocity of $3000 \, h^{-1}$. Catalyst life is 3–4 years.

The acrolein is then fed at 250 °C to another series of tubular reactors filled with catalyst tablets composed of oxides of Mo and V, dispersed on SiO_2. Low catalyst porosity minimizes loss of selectivity. Conversion is 90% and selectivity to acrylic acid is 92–98% depending on the catalyst and process at a space velocity varying from 1000 to $3000 \, h^{-1}$. The product is sprayed with water at 70 °C in a tower where the crude acrylic acid is produced. It is later distilled to remove impurities such as acetic acid, acetaldehyde, CO, and CO_2. Very little coke forms due to the presence of a high steam content. The reaction is approximated to be first order in acrolein and pseudo-zero order in O_2 provided the O_2 is present in a large excess.

8.7.2   Acrylic Acid Catalyst

Most proprietary catalysts contain a mixture of the oxides of V, Mo, and W supported on either SiO_2 or Al_2O_3. Other promoters (Fe, Ce, Zn, and MgO) are added to enhance selectivity.

8.7.3   Catalyst Deactivation

Moderate sintering and occasional coke formation, minimized by the presence of steam, are the major sources of deactivation. When deactivation is observed, the temperature is slowly increased to maintain at least 90% conversion. Coke that does form can be regenerated with air.

8.8   MALEIC ANHYDRIDE

It is used as a monomer for polyesters when reacted with propylene glycol. One process uses benzene, which is partially oxidized over a 10–20% $2V_2O_5$–MoO_3/ α-Al_2O_3 particulate low surface area carrier at 2–5 atm and 400 °C. Some catalysts also contain P_2O_5 as a promoter. The oxygen content in the feed gas is critical to retain the catalyst components in the proper oxidation states. The catalyst is housed in a series of tubular reactors cooled with KNO_3–$NaNO_3$ eutectic molten salts. The process achieves 90% conversion with a selectivity of 65%.

$$\text{benzene } (C_6H_6) + 4\tfrac{1}{2}O_2 \rightarrow \text{maleic anhydride } (C_4H_6O_3) + 2CO_2 + 2H_2O, \tag{8.28}$$
$$\Delta H° = -1800 \text{ kJ/mol}$$

Selectivity toward H_2O varies, which is reflected in the unbalanced equation for hydrogen.

The size of the exotherm is also influenced by undesired side products. Selectivity is about 70%. Some processes operate with a fluidized bed.

Butane can also be used in tubular reactors using V_2O_5 and P_2O_5 catalyst at 500 °C. Conversions of 85% are achieved with a selectivity of 50%. The extent of the exotherm also depends on undesired side products.

$$C_4H_{10} + 3\tfrac{1}{2}O_2 \rightarrow \text{maleic anhydride } (C_4H_2O_3) + 4H_2O, \tag{8.29}$$
$$\Delta H° = -1200 \text{ kJ/mol}$$

8.8.1   Catalyst Deactivation

If temperatures in the partial oxidation reaction are not properly controlled, there can be a loss of MoO_3 due to volatilization for the benzene process. Poisoning by sulfur compounds and aromatics can also lead to deactivation. Loss of P_2O_5 by high-temperature volatilization is commonly believed to be the cause of deactivation in the butane process.

8.9   ACRYLONITRILE

It is a monomer for the production of acrylonitrile–butadiene–styrene (ABS) polymers for synthetic rubber fibers, carpets, and furniture.

8.9.1 Acrylonitrile Production Process

$$CH_2{=}CHCH_3 + NH_3 + 1\tfrac{1}{2}O_2 \rightarrow CH_2{=}CHCN,$$
$$\Delta H^\circ = -500\,\text{kJ/mol} \tag{8.30}$$

The reaction is carried out at 400 °C and 1–3 atm in a fluidized bed reactor equipped with cooling coils to control the exotherm (Figure 8.13). The relative concentrations of reactants vary widely depending on the process but on average the oxygen to propylene ratio is ~2:1, the ammonia to propylene ratio is ~3 or 4 to 1, and the water to propylene ratio is about 2 or 3 to 1. The reactants may be fed close to the flammability limits, so safety is of utmost importance in operation. The residence time in the fluidized bed is around 10 s. The selectivity is 70–80% based on propylene converted to acrylonitrile. Because acrylonitrile may polymerize during the reaction by free radical reactions, often some hydroquinone is added to the purification train to quench the radicals. The reaction is conducted essentially free from mass transfer effects and is therefore limited by chemical kinetics. A unique feature of the reaction is its first-order dependence on propylene, while the undesired reactions producing HCN, CO, and CO_2 have a higher reaction order. Therefore, the propylene content is maintained relatively low to minimize impurity production and maximize yield of the desired product. This is a nice example of why knowing the kinetics is important in designing the process. The reaction is close to zero order in both NH_3 and O_2. Product purity is achieved by a series of distillations where the unreacted ammonia and undesired HCN are removed by distillation and water scrubbing.

The mechanism of the reaction is still being studied, but there is no universally accepted theory.

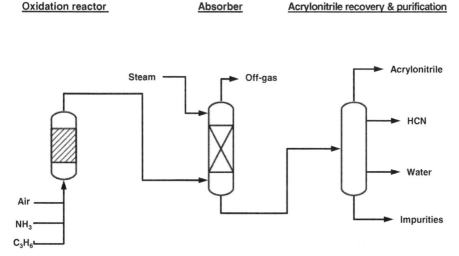

Figure 8.13   Process for converting propylene to acrylonitrile.

8.9.2 Catalyst

The actual nature of the catalyst has been extensively studied, but no specific active site structure has been reliably reported. Oxides of Bi, Fe, Mo, and Cr, and other promoters used are maintained highly proprietary by producers, so little information is available. The catalyst is about 200 μm in size for ease of fluidization. At this size, there is little or no pore diffusion. The mixing is sufficient to eliminate bulk mass transfer effects.

8.9.3 Deactivation

Loss of performance is mainly due to volatilization of the oxides of Mo and Cr. Mo is often replenished during the process to maintain activity and selectivity.

QUESTIONS

1. Refer U.S. Patent 4,863,893 "Low temperature light-off ammonia oxidation" for the production of NO for nitric acid.

   a. Explain the usual procedure for starting up the PtRh gauze. What are the possible problems encountered in such a procedure?

   b. The desired reaction for NH_3 oxidation to NO has $\Delta G_{900\,°C} = -360\,kJ/mol$, while the undesired reaction has $\Delta G_{900\,°C} = -560\,kJ/mol$. Why does the desired reaction occur?

   c. What is the function and mechanism of the getter gauze?

   d. What is the proposed mechanism for deactivation of the PtRh gauze?

   e. Compare Claims 1, 12, and 16.

   f. Compare Claims 16 and 17.

   g. Compare Claims 7 and 18.

   h. What are the proposed advantages of the Pt coating on the gauze relative to the uncoated gauze regarding light-off, life, and product (HNO_3) yield?

   i. For what reason is H_2 oxidation considered in this patent?

2. **a.** Why is the "poison" Cl^- added to the process feed gas for the selective oxidation of ethylene to ethylene oxide?

   b. Give two other examples in heterogeneous catalysis where poisons are intentionally added to the feed and explain the effect.

3. Explain the safety issues associated with the two different processes for the formation of formaldehyde.

4. Explain the multiple process steps in the production of H_2SO_4.

5. Explain the process steps in the Claus process for generating elemental sulfur.

6. Explain the sequential process steps in the conversion of propane to acrylic acid.

7. In the production of maleic acid from benzene or butane, why is the heat of reaction not exactly equal to the products in the desired reaction?

8. In the production of NO from ammonia, explain why the decomposition of NH_3 to N_2 does not occur when it is thermodynamically more favorable (by about 10^5) than NH_3 to NO?

9. If a new SO_2 catalyst is developed that is active at 200 °C, how might the SO_2 reactor design change? Is it possible that the overall process scheme might change?

10. What is meant by the name "double contact process" for SO_2 oxidation?

11. PtRh is used for both ammonia oxidation (AMOX) and HCN production and both gauzes undergo morphological changes during each process. Why does the Pt in the AMOX process volatilize and must be captured downstream in a getter gauze while the same catalyst with NH_3 and CH_4 does not lose Pt in the production of HCN?

BIBLIOGRAPHY

General Selective Oxidations

Centi, G. and Perathoner, S. (2003) Selective oxidations—industrial, in *Encyclopedia of Catalysis* (ed. I. Horvath), John Wiley & Sons, Inc., Hoboken, NJ, pp. 239–298.

Nitric Acid

Bartholomew, C. and Farrauto, R.J. (2006) Chapter 8, in *Fundamentals of Industrial Catalytic Processes*, 2nd edn, John Wiley & Sons, Inc., Hoboken, NJ, pp. 570–575.
Farrauto, R.J. and Lee, H. (1990) Ammonia oxidation catalysts with enhanced activity. *Industrial & Engineering Chemical Research* 29, 1125.
Lee, H. and Farrauto, R.J. (1989) Catalyst deactivation due to transient behavior in nitric acid production. *Industrial & Engineering Chemical Research* 28, 1.
Maxwell, G. (2012) Synthetic nitrogen products, in *Handbook of Industrial Chemistry and Biotechnology*, Vol. 2, 12th edn (ed J. Kent), Springer, New York, pp. 903–910.
McCabe, R., Smith, G., and Pratt, A. (1986) The mechanism of reconstruction of rhodium–platinum catalyst gauzes. *Platinum Metals Review* 30 (2), 54.
Schmidt, L. and Luss, D. (1971) Physical and chemical characterization of platinum–rhodium gauze catalysts. *Journal of Catalysis* 22, 269.

Sulfuric Acid

Bartholomew, C. and Farrauto, R.J. (2006) Chapter 8, in *Fundamentals of Industrial Catalytic Processes*, 2nd edn, John Wiley & Sons, Inc., Hoboken, NJ, pp. 562–570.
Catalyst suppliers. Available at http://www.sulphuric-acid.com/techmanual/contact/contact_catalysts.htm.
Clark, P.D. (2006) Sulfur and hydrogen sulfide recovery, in *Kirk-Othmer Encyclopedia of Chemical Technology*, 5th edn, John Wiley & Sons, Inc., pp. 597–620.
d'Aquin, G. and Fell, R. (2012) Sulfur and sulfuric acid, in *Handbook of Industrial Chemistry and Biotechnology*, Vol. 2, 12th edn (ed. J. Kent), Springer, New York, pp. 997–1016.
King, M., Davenport, W., and Moats, M. (2013) *Sulfuric Acid Manufacture, Control and Optimization*, 2nd edn, Elsevier, Amsterdam, The Netherlands.

Hydrogen Cyanide

Anonymous (n.d.) Cyanides, in *Kirk-Othmer Encyclopedia of Chemical Technology*, Vol. 8, 4th edn, John Wiley & Sons, Inc., New York, pp. 171–199.

Bartholomew, C. and Farrauto, R.J. (2006) Chapter 8, in *Fundamentals of Industrial Catalytic Processes*, 2nd edn, John Wiley & Sons, Inc., Hoboken, NJ, pp. 575–578.

Maxwell, G. (2012) Synthetic nitrogen products, in *Handbook of Industrial Chemistry and Biotechnology*, Vol. 2, 12th edn (ed J. Kent,), Springer, New York, pp. 925–929.

Ethylene Oxide

Anonymous (2000) Ethylene oxide, in *Kirk-Othmer Encyclopedia of Chemical Technology*, Vol. 10, John Wiley & Sons, Inc., New York, pp. 632–673.

Bartholomew, C. and Farrauto, R.J. (2006) Chapter 8, in *Fundamentals of Industrial Catalytic Processes*, 2nd edn, John Wiley & Sons, Inc., Hoboken, NJ, pp. 597–604.

Porcelli, J. (1981) Ethylene oxide: exploratory research. *Catalysis Reviews—Science and Engineering*, 23 (1–2), 151–162.

Formaldehyde

Anonymous (n.d.) Formaldehyde, in *Kirk-Othmer Encyclopedia of Chemical Technology*, Vol. 12, 5th edn, John Wiley & Sons, Inc., New York.

Barnicki, S. (2012) Synthetic organic chemicals, in *Handbook of Industrial Chemistry and Biotechnology*, Vol. 1, 12th edn (ed. J. Kent), Springer, New York, pp. 346–347.

Bartholomew, C. and Farrauto, R.J. (2006) Chapter 8, in *Fundamentals of Industrial Catalytic Processes*, 2nd edn, John Wiley & Sons, Inc., Hoboken, NJ, pp. 584–597.

Stiles, A. and Koch, T. (1995) *Catalyst Manufacture*, 2nd edn, Marcel Dekker, New York.

Acrylonitrile

Anonymous (n.d.) Acrylonitrile, in *Kirk-Othmer Encyclopedia of Chemical Technology*, Vol. 1, 5th edn, John Wiley & Sons, Inc., New York, pp. 397–414.

Bartholomew, C. and Farrauto, R.J. (2006) Chapter 8, in *Fundamentals of Industrial Catalytic Processes*, 2nd edn, John Wiley & Sons, Inc., Hoboken, NJ, pp. 604–609.

Callahan, J., Grasselli, R., Millberger, E., and Strecker, A. (1970) Oxidation and ammoxidation of propylene over bismuth–molybdenum catalysts. *Industrial & Engineering Chemistry Product Research and Development* 9 (2), 134–142.

Centi, G. and Perathoner, S. (2003) Selective oxidations—industrial, in *Encyclopedia of Catalysis* (ed. I. Horvath), John Wiley & Sons, Inc., Hoboken, NJ, pp. 239–298.

Grasselli, R. (2011) New insights in heterogeneous oxidation catalysis. *Topics in Catalysis* 54 (10), 587.

Maleic Anhydride

Barnicki, S. (2012) Synthetic organic chemicals, in *Handbook of Industrial Chemistry and Biotechnology*, Vol. 1, 12th edn (ed J. Kent), Springer, New York, pp. 345–346.

Bartholomew, C. and Farrauto, R.J. (2006) Chapter 8, in *Fundamentals of Industrial Catalytic Processes*, 2nd edn, John Wiley & Sons, Inc., Hoboken, NJ, pp. 610–618.

Acrylic Acid

Anonymous (n.d.) Acrylic acid and derivatives, in *Kirk-Othmer Encyclopedia of Chemical Technology*, Vol. 1, John Wiley & Sons, Inc., New York, pp. 341–369.

Naohiro, N., Sakai, Y., and Watanabe, Y. (1995) Two catalytic technologies of much influence on progress in chemical process development in Japan. *Catalysis Reviews—Science and Engineering* 37 (1), 145.

CHAPTER 9

HYDROGENATION, DEHYDROGENATION, AND ALKYLATION

9.1 INTRODUCTION

Selective catalytic hydrogenation of functional groups contained in organic molecules is one of the most useful, versatile, and environment-acceptable reaction routes available for organic synthesis. This important area of catalytic chemistry has been and continues to be the foundation for the development of numerous, diverse, large- and small-scale commercial hydrogenation processes, including (i) fine chemicals, (ii) intermediates for the pharmaceutical industry, (iii) monomers for the production of various polymers, and (iv) fats and oils for edible and nonedible products.

Dehydrogenation reactions find a wide application in production of hydrogen, alkenes, polymers, and oxygenates. In recent years, the demand for light alkenes has grown dramatically due to increased demand for polypropylene, acrylonitrile, oxo alcohols, and propylene oxide. As a result, dehydrogenation of lower alkanes to alkenes is a rapidly expanding business.

Alkylation allows smaller molecules to be coupled to for larger molecules mostly for petroleum applications.

9.2 HYDROGENATION

9.2.1 Hydrogenation in Stirred Tank Reactors

With exception of a few continuous hydrogenation processes in petroleum refining, hydrogenation processes are often conducted in stirred tank reactors. This chapter focuses on hydrogenation occurring within stirred tank reactors, which are ideally suited and extensively used for liquid-phase hydrogenation reaction. For reactions where the hydrocarbon to be hydrogenated is in the liquid phase, stirred tank reactors are ideal. For hydrogenation, stirred tank reactors can be designed in two configurations, semibatch and continuous, which are illustrated in Figure 9.1. In a semibatch stirred tank reactor, the "batch" of liquid hydrocarbon to be hydrogenated is stirred

Introduction to Catalysis and Industrial Catalytic Processes, First Edition. Robert J. Farrauto, Lucas Dorazio, and C.H. Bartholomew.
© 2016 John Wiley & Sons, Inc. Published 2016 by John Wiley & Sons, Inc.

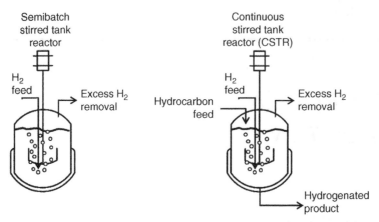

Figure 9.1 Illustration comparing difference between a semibatch stirred tank reactor and a continuous stirred tank reactor.

inside the reactor and gaseous hydrogen is continuously introduced into the reactor and withdrawn from the head space until the hydrogenation reaction is complete. In a continuous stirred tank reactor (CSTR), both liquid hydrocarbon and hydrogen are continuously added and withdrawn from the stirred tank. The volume of the CSTR and the liquid hydrocarbon feed rate determine the mean residence time of the liquid hydrocarbon inside the reactor. Which design is used is a function of the production scale and reaction dynamics.

The stirred tank reactor in its most simplistic form is analogous to a household blender. The liquid reactants, sometimes in an inert solvent, are combined with solid catalysts that are vigorously stirred while hydrogen is sparged into the slurry. Mixing dynamics inside of the tank is a function of its design (height/diameter, baffle design) and the type of impeller selected. A variety of mixing impellers are available, each capable of producing different mixing dynamics inside the reactor. On one end of the mixing spectrum is the "high-efficiency" hydrofoil impeller that is designed to maximize axial fluid circulation inside the reactor and minimize turbulence in the region around the impeller. On the opposite end of the mixing spectrum is the radial flow impeller that produces radial circulation patterns characterized by highly turbulent flow in the impeller region. For hydrogenation reactions, the turbulent flow produced by the radial turbine is effective to disperse the sparged hydrogen gas into tiny bubbles. As will be discussed later in the chapter, good dispersion of the sparged hydrogen maximizes the gas–liquid interface area for hydrogen diffusion into the liquid phase, which in turn increases the rate of reaction. In addition to dispersing hydrogen, the agitator and overall stirred tank design must sufficiently suspend the solid catalyst in the liquid. The hydrogen is introduced into the reactor through a "dip tube" that typically injects the gas very near to the impeller outflow, where shear forces required to disperse the hydrogen is greatest. The greater the gas–liquid surface area, the greater the rate of hydrogen transport into the liquid phase. The temperature of the reactor is maintained using either heat transfer coils immersed in the circulating liquid or heat transfer surface incorporated into the walls of the vessel. In one design

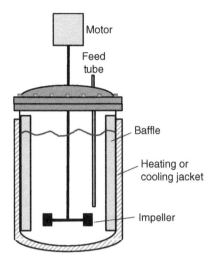

Motor

Feed
tube

Baffle

Heating or
cooling jacket

Impeller

Figure 9.2   Illustration of a semibatch stirred tank reactor (STR). The sparger (or also called dip tube) is used for continuous addition of a reactant, which is hydrogen for hydrogenation reactions. Not shown is the removal of unreacted hydrogen from the headspace, which must occur to maintain the desired reactor pressure.

option, the hydrogen feed is adjusted to maintain the headspace above the mixed slurry at a constant pressure. The consumption of H_2 is measured by the decrease in H_2 pressure in the reservoir. The rate is measured by determining the consumption of H_2 as a function of time. This rate is linear until the reactant concentration is small after which time it will approach zero (Figure 9.2).

When operating in semibatch configuration, the consumption of hydrogen over time will change as the fixed charge of hydrocarbon is hydrogenated. Initially, the concentration of unreacted hydrocarbon is high, which results in a large consumption of hydrogen. However, as time progresses, the concentration of unreacted hydrocarbon decreases as reaction occurs, which decreases the amount of hydrogen being consumed. Ultimately, the concentration of unreacted hydrocarbon approaches zero, which results in the consumption of hydrogen approaching zero. This is illustrated in Figure 9.3.

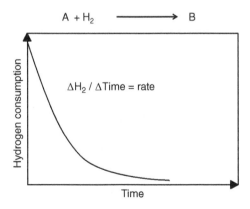

$A + H_2 \longrightarrow B$

Hydrogen consumption

$\Delta H_2 / \Delta \text{Time} = \text{rate}$

Time

H_2 consumption = disappearance of A

Figure 9.3   Illustration showing hydrogen consumption versus time during typical hydrogenation reaction.

9.2.2 Kinetics of a Slurry-Phase Hydrogenation Reaction

Consider the reaction of liquid-phase hydrogenation of nitrobenzene to aniline with a powdered catalyst:

$$3H_2 + C_6H_5NO_2 \rightarrow C_6H_5NH_2 + 2H_2O \qquad (9.1)$$

Up to this point, we have limited the discussion of diffusion to single-phase systems (i.e., gas diffusion from bulk to catalyst surface). However, hydrogenation reactions involve a multiphase system. Gaseous hydrogen dispersed into the liquid must diffuse into the liquid phase. Liquid-phase hydrocarbon and dissolved hydrogen must diffuse through the liquid to the catalyst surface. In addition to bulk diffusion, the diffusion resistance at each interface, gas–liquid and liquid–solid, affects the overall rate of hydrogen diffusion to the catalyst surface. Hydrogenation reactions are typically controlled by rate of diffusion of H_2 to the active sites. The diffusion of H_2 involves the following steps, which is illustrated in Figure 9.4:

1. Be transported through the gaseous bubble
2. Diffusion through the gas–liquid interface
3. Transport through the liquid phase

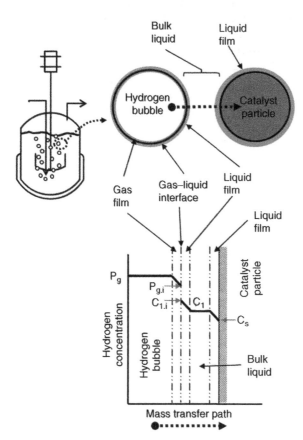

Figure 9.4 Illustration of the mass transfer path taken by hydrogen as it diffuses from the gas bubble, through the bulk liquid, and ultimately to the catalyst particle. In most hydrogenation reactions, the rate of this diffusion process limits the overall rate of reaction.

4. Diffusion through the liquid film surrounding the catalyst particle
5. Diffuse through the pore structure
6. Reaction at the catalyst surface

For most systems, the rate of mass transfer from the gas into the liquid is rate limiting. Thus, we can assume transport of the H_2 through the liquid interface, through the liquid, the liquid film surrounding the catalyst, and transport through the small catalyst particle size (pore diffusion) are fast and thus are seldom rate-limiting steps. Thus, we can simplify the overall kinetics with the following treatment.

Since the chemical reaction can occur no more rapidly than the rate at which hydrogen is supplied, we can equate rate of chemical reaction to the rate of mass transfer into the liquid:

$$r_{MT} = k_{MT} \frac{\left([H_2]_g - [H_2]_s\right)}{[H_2]_g}$$

(9.2)

$$r_{MT} = k_{MT} \left(1 - \frac{[H_2]_s}{[H_2]_g}\right) = k_{MT}(1 - h')$$

$h' = [H_2]_s/[H_2]_g$, which is effectively similar, but not the same, as the Henry's law constant. k_{MT} in this instance includes the geometric surface area of the gas bubble. $[H_2] = H_2$ in the gas (g) and in solution (s).

The chemical reaction rate at the catalyst site is

$$r = k_f(W_{cat})h'$$

(9.3)

The mass transfer and chemical rates are in series and thus their rates are dependent on each other and can be equated to give r_{net}.

Here, $H = r_{net}/k_f W_{cat}$, which when substituted into (9.3) gives the general expression for the net rate of reaction:

$$r_{net} = \frac{k_f k_{MT}}{k_f W_{cat} + k_{MT}}$$

(9.4)

When the amount of catalyst (W_{cat}) is small, (9.4) reduces to the chemical rate being limited:

$$r_{net} = k_f W_{cat}$$

(9.5)

When Q is large, the rate is limited by mass transfer and (9.4) reduces to the mass transfer coefficient k_{MT}:

$$r_{net} = k_{MT}$$

(9.6)

In searching for the best catalyst, one makes a plot such as in Figure 9.5, at the expected reaction temperature and compares the initial slope (low catalyst loadings W_{cat}). This reflects the kinetic rate constant for each catalyst. After the catalyst is chosen, one must select the proper process parameter to operate efficiently. At high

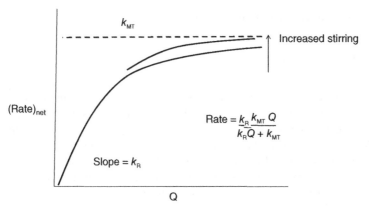

Figure 9.5 Kinetic rate for a catalytic slurry-phase batch reaction.

catalyst loadings, mass transfer of H_2 into the reaction mixture becomes rate limiting. For this condition, an increase in stirring rate (or pressure) will increase the net rate of reaction until which time H_2 mass transfer is no longer rate limiting. One must be careful to use too high a stirring rate since catalyst attrition due to high shear rates may occur, making separation by filtration and catalyst reuse difficult.

9.2.3 Design Equation for the Continuous Stirred Tank Reactor

For large-volume chemicals, one can scale from a batch process to a continuous process with the same catalyst in a CSTR. Here, both the hydrogen and hydrocarbon liquid are continuously fed and product and excess hydrogen are continuously withdrawn while the catalyst is retained in the reactor (refer to Figure 9.1). A common simplifying assumption of the CSTR is that it is "well mixed," which means the concentration of reactants and products is uniform and no concentration gradient exists. In reality, a concentration gradient will exist within the reactor, the extent and significance being a function of the reactor design, feed rate, and reaction rate.

One can derive the reactor design equation by performing a mole balance around the stirred tank. Consider the generic hydrogenation reaction of hydrocarbon "A" being hydrogenated to "B".

$$A + H_2 \rightarrow B$$

The steady-state mole balance around the stirred tank is

$$F_{Ai} - F_{Ao} + \int_0^V r_A dV = 0 \tag{9.7}$$

For a well-mixed reactor, the rate of reaction is constant throughout the reactor volume (V); thus, the third term on the left-hand side simplifies to

$$\int_0^V r_A dV = r_A V \tag{9.8}$$

Rearranging the mole balance provides an equation for the design of the stirred tank:

$$V = \frac{F_{Ao} - F_{Ai}}{-r_A} = \frac{F_{Ao}X}{-r_A} \tag{9.9}$$

r_A can also be equated to kC_{Ao}, where C_{Ao} is the known concentration of unreacted A at the exit. The rate constant k is determined from the stirred tank reaction as shown in Figure 9.3.

Using this design equation, we can calculate the reactor volume required for a desired conversion and feed rate. Conversely, by solving for reaction rate, we can use this equation to calculate reaction rate for laboratory-scale kinetic experiments when measuring conversion at a known reactor volume and feed rate.

9.3   HYDROGENATION REACTIONS AND CATALYSTS

9.3.1   Hydrogenation of Vegetable Oils for Edible Food Products

Plant-derived oils such as soy, cottonseed, peanut, canola, and corn, are natural sources of edible products such as baking dough for cakes, cooking oils, salad dressing, chocolates, and margarine. Nonedible products such as lubricants, creams, and lotions can also be produced depending on the processing of the oils. Natural oils are composed of long chains of fatty acid esters called triglycerides, as shown in Figure 9.4. The triglyceride chains are often polyunsaturated, the degree of which influences their stability against oxidation in air.

Catalytic hydrogenation of the double bonds improves the stability against air and raises the melting point such that solids can be produced from the extracted liquid oils. Thus, the precursor to chocolate candy, margarine, or a cake mix is liquid oil that upon hydrogenation becomes an edible solid at room temperature. The more the double bonds hydrogenated (the more saturated) higher the melting point, the greater the stability toward air, but for food products saturated fats are injurious to our health by deposition of cholesterol in blood vessels. The goal of a good catalyst and process is to produce a reasonably healthy product with the desired melting point range with sufficient air stability to permit good shelf life.

Oils are classified by the length of the glyceride chain and degree of poly-unsaturation. Typically, nature produces oils with each chain length between 12 and 22 carbons with up to three unsaturated bonds per chain, usually all in the *cis*-form. Triglycerides with 18 carbons per chain length and three double bonds at positions 9, 12, and 15 counting from the first carbon in the ester group are called linolenic and designated C18:3 with a melting point of $-24\,°C$. This structure is shown in Figure 9.6. The iodine value (IV) is the number of grams of I_2 to react with $100\,g$ of oil and roughly is a measure of the number of double bonds in the oil. For linolenic, the IV $= 261$. The outermost double bond is so reactive toward air that oils with three double bonds in the alkyl chain are rare. Therefore, the most prevalent in nature have double bonds at positions 9 and 12 and are referred to as linoleic (C18:2) with a melting point of $-13\,°C$ and IV $= 173$. Its reactivity is about half that of linolenic. The least reactive is oleic (1/20 that of linolenic) with only one double bond per chain

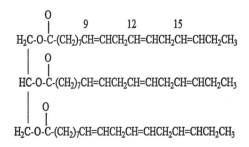

Figure 9.6   Linolenic oil shown as an example of an unsaturated fat molecule.

length at position 9 (C18:1) and a melting point of 5.5 °C and IV = 86. Stearic (IV = 0) is the term used for glycerides with all bonds saturated (C18:0) and melting at 73.1 °C. Not surprisingly, steric has virtually no reactivity toward air and can be thought of as animal fat.

The source of the oils plays a major role in producing a product with the desired melting point, stability, and health consequences. Cotton, sunflower, corn, and soy bean oils are a mixture of the four basic triglycerides with 50–70% C18:2 being the most dominant followed by 20–30% C18:1 with less than 1% C18:3. Less than 10% are other saturated oils such as C16:0. Palm kernel and coconut oils have almost 80% saturated triglycerides (C12:0, C14:0, and C16:0), have high melting points, and are stable against air. They are used for protecting the skin against excessive sun exposure. In contrast, olive oil has up to 80% C18:1 and is therefore relatively healthy.

The most common oil hydrogenation catalysts are 20–25% Ni on Al_2O_3 and SiO_2. Nickel salts are either impregnated or co-gelled with a carrier precursor such as a soluble Al or Si salt. The catalytically active state of Ni is the reduced (metallic) form. The reduction activation step is performed during manufacture at which time the catalyst is coated with a fatty gel to protect it from air oxidation during shipment. This protective layer causes an induction period in the reactor as it slowly dissolves prior to the reaction commencing.

The reaction profile is generally sequential with hydrogenation first occurring on the most active double bonds followed by those less active. Time distribution in Figure 9.5 shows the linoleic (C18:1) form decreasing as the oleic form (C18:1) increases. After extended reaction, time stearic (C18:0) begins to form as the oleic is slowly hydrogenated. As the temperature is raised from 140 to 200 °C, the rate of disappearance of C18:2 increases (lower product distribution) while C18:1 is shown to increase. The amount of C18:0 produced decreases with temperature because C18:1 converts to *trans* rather than the fully saturated C18:0. Thus, one can design the process to control the product distribution in a predictable manner. These sequential reactions are represented by in Figure 9.7.

By knowing the rate/time dependence of the reaction, one can tailor the product to the desired degree of saturation, melting point, and health index.

Less than 1% by weight of catalyst is added to a batch reactor where the fatty protective gel slowly dissolves and the hydrogenation reaction commences. Temperatures of 100 °C and H_2 pressures of 3–5 atm are used to ensure adequate dissolution of the H_2 into the feed stream. Reactions are carried out between 100 and 160 °C.

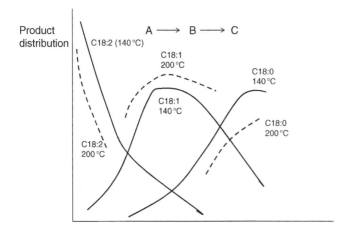

Time

Figure 9.7   Sequential reactions at 140 and 200 °C.

Stirring is vigorous to maximize bulk gas diffusion to the catalyst surface. The catalyst particles are small to minimize pore diffusion resistance and increase liquid–solid mass transfer area. Increasing temperatures favor the hydrogenation of the C18:2 oil producing greater amounts of C18:1 and the *trans*-isomer. Pressure favors more C18:0. Controlling the time on stream, temperature, and pressure is a critical parameter necessary to avoid excessive formation of *trans*-isomer and saturation of all the double bonds. Cu supported on Al_2O_3 is less active than Ni and this property is used to "brush" hydrogenate. Only minimum hydrogenation occurs maintaining the melting point, but is sufficient to improve stability against air. This is sometimes used for producing salad oils.

Hydrogenation of the linoleic form with a melting point of −13 °C will produce the oleic product with a melting point of 5.5 °C very suitable for consumption. During the partial hydrogenation process, it is most desirable to minimize isomerization to the *trans*-isomer (the hydrocarbon groups are *trans* to each other across the remaining unsaturated bonds) because this structure raises the "bad" cholesterol or LDL (low density lipids). A partially hydrogenated *cis*-structure may have a melting point of 6 °C, whereas its *trans*-isomer melts at 40 °C. The *trans*-isomer is more readily formed with acidic catalysts, at high reaction temperatures, at high Ni catalyst loadings, and at low hydrogen pressures (low concentration of H_2 at the catalyst surface). For this reason, SiO_2 is often used as the carrier for Ni since it has lower acidity than Al_2O_3, which favors the more desired *cis*-isomer. Pt-containing oil hydrogenation catalysts produce considerably less *trans*-isomer than Ni; however, the high activity causes excessive hydrogenation of the double bonds. To minimize this effect, NH_3 is intentionally added to the feed or catalyst to poison its activity toward hydrogenation. By so doing, low *trans*-oil is produced without excessive saturation of the double bonds. In 2006, the U.S. Food and Drug Administration required labels that report the amount of *trans*-components present in edible products. In some places in the United States, *trans*-containing food products are heavily regulated.

Catalyst deactivation is mainly caused by mechanical attrition due to the rigorous stirring. In most cases, adsorption guard beds are used upstream to remove most of the impurities such as sulfur and phosphorous often found in the feed. Recognizing that some poisons may break through, the catalyst has an average pore size sufficiently large to admit the triglycerides but smaller than the average size of the organic compounds containing P and S. The spent catalyst is separated from the product by filtration. Given the increasing cost of Ni, it is recovered, refined, and used to make fresh catalyst.

9.3.2 Hydrogenation of Functional Groups

Function groups on organic molecules are often hydrogenated to make specialty chemicals for the pharmaceutical, polymer and petrochemical industries. Some of the more common examples are $-NO_2$, $-HC=O$, $-C=O$, $-COOH$, $-CN$, $C=C$, and other groups. Precious metals are commonly used for specialty chemicals, but reduced base metals such as Ni, CuCr, Co, and Cu are also used.

High surface area γ-Al_2O_3 is a common carrier for precious and base metals. For some specialty hydrogenation reactions, natural and synthetic activated carbons are commonly used with very interesting activity and selectivity results. Carbons are harvested from plants, shells, nuts, wood, peat, coal, and so on and contain various inorganic ash components (SiO_2, Al_2O_3, alkali, etc.) indicative of their origin. They are activated by solvent washing to remove undesirable ash components. They are treated at high temperatures in steam and/or carbon dioxide as well as other proprietary methods. Such activations enhance the internal surface area, but more importantly provide functional groups on the surface that impart special chemistry for adsorption of impurities and the catalytic reactions. These functional groups are dependent on the carbon source and can vary in acidity and basicity and can be hydrophilic or hydrophobic. An important key factor is their insolubility in many solvents and pH conditions in which Al_2O_3 and SiO_2 are not stable. For this reason, carbons are commonly used in air and water filtration and purifications. For many organic catalytic hydrogenation reactions, the carbons disperse the active metals but can also adsorb reactants and thereby create a synergistic effect with the catalytic metals enhancing the rate and selectivity of reactions. In the simplest case, the carbon functional group adsorbs the organic reactant, while the H_2 molecule is dissociated on the catalytic metal producing active H atoms where they "spill over" and complete the reaction. This model is only intended as an example of the dual functionality of carbon-supported reactions and is not a universal mechanism. Broadly speaking, this field is highly empirical and catalyst companies maintain highly secretive formulations and process conditions with metal–carbon combinations for a given reaction. Some general examples will be shown below.

Let us consider the catalytic hydrogenation of nitrobenzene for which a variety of products can be obtained depending on the metal, the carbon, and the process conditions. Selectivity is critical for these reactions. These reactions are not balanced as written, but serve the purpose of demonstrating different selectivity for different metal catalyst, carbons, and process conditions. The catalysts are powdered and the reactions all occur in the slurry phase.

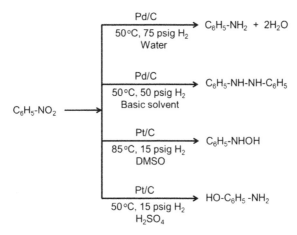

Now consider the hydrogenation of a nitro group on a chlorinated aromatic:

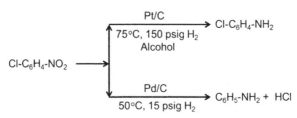

Dechlorination occurs with Pd/C under the conditions shown, while hydrogenation of the nitro group producing aniline occurs with Pt/C.

Saturation of the aromatic ring (aniline) to the cyclohexane occurs with catalytic metals on different carbons and process conditions.

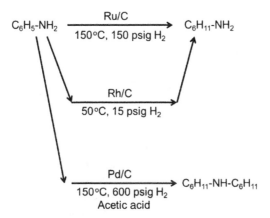

Rh/C requires less severe conditions (lower temperatures and pressures) to produce the cyclohexane-substituted amine than Ru/C. Coupling occurs with Pd/C, but at high pressures.

Reductive alkylation (used in the rubber industry, for example, tires) is the addition of a ketone group to a functionalized aromatic such as nitrobenzene:

$$C_6H_5-NO_2 + R-CO-R' + H_2 \xrightarrow[\substack{50\,°C,\ 250\ psig\ H_2 \\ Alcohol}]{Pt/C} C_6H_5-NH-RCHR'$$

Aromatic aldehydes can be hydrogenated to aromatic alcohols or completely to the corresponding methyl aromatics without saturation the ring:

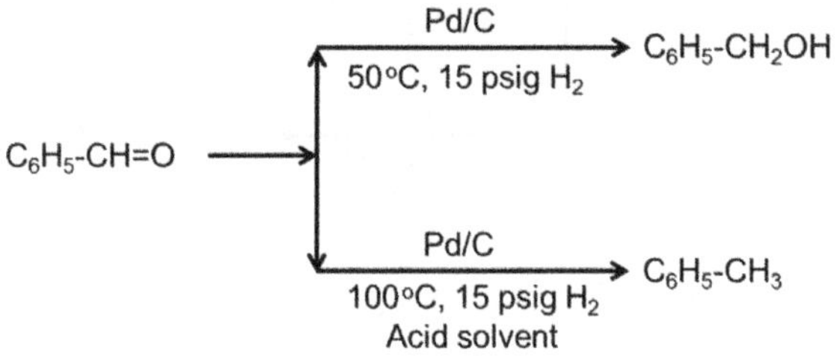

It should be understood that these carbons are all from different sources and activation procedures. It is clear that they must be insoluble in different solvents. One additional advantage is the ease of separation from the metal once the catalyst is deactivated. One simply burns the carbon leaving an ash rich in the catalytic metal.

Slurry-phase production of H_2O_2 is achieved by the hydrogenation of anthroquinone to hydroxyl hydroanthroquinone followed by the noncatalytic oxidation of hydroxyl hydroanthroquinone, producing H_2O_2, and the reactant quinone shown schematically below:

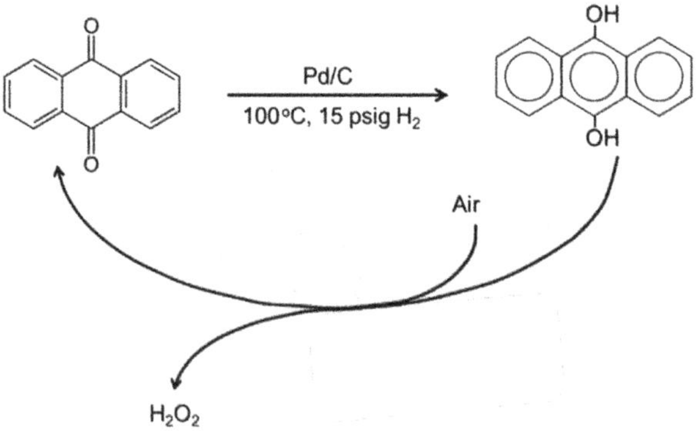

Terephthalic acid ($HOOH-C_6H_4-COOH$) is a monomer for the production of polyesters. It is produced by the air oxidation of dimethyl benzene with traces of aromatic aldehyde impurities (formyl benzoic acid (colored body) imparting a yellow color that must be removed to produce an acceptable white polyester. This is accomplished by a fixed bed hydrogenation of the aldehyde using Pd on a granular carbon to methyl benzoic acid. The latter remains in solution as the terephthalic acid is crystallized and separated. The process is carried out with the liquid feed entering the fixed bed upflow and the recycled H_2 staged into the fixed bed of catalyst. The catalytic reaction is pore diffusion controlled, so the Pd is edge coated on the granular carbon. The carbon imparts very unique performance and its characteristics and source are maintained by proprietary catalyst suppliers.

$$HOOC\!-\!C_6H_4\!-\!CHO \xrightarrow[\text{50 °C, 600 psig } H_2]{\text{Pd/C (granular)}} HOOC\!-\!C_6H_4\!-\!CH_3$$

9.3.3 Biomass (Corn Husks) to a Polymer

Furfural is extracted from cornhusks, bran, wheat, and oats and is a feedstock for furfural alcohol, which is a monomer for production of corrosion-resistant resins and coatings. It is an example of the use of the hemicellulose (cell wall portion) of nonedible biomass to make a useful product. It utilizes a reduced $CuO\!-\!Cr_2O_3$ catalyst powder in a slurry-phase reactor. It operates at very high H_2 pressures at about $200\,°C$.

9.3.4 Comparing Base Metal and Precious Metal Catalysts

Precious metals are rare and expensive, but have remarkable catalytic activity, selectivity, and life for many different types of reactions, many of which are described in this book. Base metals such as Ni, Co, and Cu are also good catalyst materials, but are considerably less expensive and much more abundant than precious metals. Given this, why select an expensive precious metal when a base metal would clearly be more economical? The reason is often precious metals allow the same reaction but at much less severe conditions. It should be understood that precious metals are recycled after use by a variety of separation methods. The simplest of which is combustion of the carbon. For the carriers such as Al_2O_3, leaching with acid or base is used. Also, for special cases, pyrometallurgical methods are used.

An example of a comparison of a base metal and precious metal is presented below for the hydrogenation of acetophenone:

Another successful application of biomass to an edible product is the hydrogenation of corn sugar (glucose) to sorbitol (an artificial sweetener).

This reaction is carried out with a Ni-based catalyst in a fixed bed with Ni/Al_2O_3 or in the slurry phase with an activated unsupported Ni catalyst, referred to historically as Raney nickel, named after the inventor.

Raney nickel is prepared by melting Ni and Al metals, thus producing an alloy that is cooled and then selectively leached in NaOH dissolving the Al and leaving a porous structure of high surface area NiO and in solution $NaAlO_x$. Promoters such as Zn and Co are sometimes added to the melt to enhance performance. This catalyst is very attrition resistant and, therefore, structurally stable for rapidly stirred slurry-phase reactions. Since it does not contain Al_2O_3, it is less acidic than supported catalysts minimizing undesired side reactions such as isomerization.

Glucose   Sorbitol

Raney nickel is also used in fixed bed processes such as hydrogenation of benzene to cyclohexane that is oxidized to adipic acid used in the production of nylon. These reactions are highly exothermic, so the reactor must be cooled externally. Typically, the reactor is surrounded by cooling jackets of a suitable fluid. The diameter of the reactor is small to enhance radial heat transfer.

Benzene   Cyclohexane   Adipic acid

Adipic acid is used as a monomer for polyamides.

There are reactions where reduced base metal oxides are more selective than precious metals. In the example below, Pt hydrogenates only the double bond while the reduced CuO—Cr_2O_3 hydrogenates both functional groups producing an alkyl alcohol (used as plasticizer).

9.4   DEHYDROGENATION

Olefins, such as ethylene and propylene, are in high demand for producing polyethylene and polypropylene. The monomer styrene (for polystyrene) is also in high demand produced through dehydrogenation of the ethyl benzene. Other applications such as alkylation where olefins are combined with other hydrocarbons to produce larger molecules (e.g., detergents and gasoline) are also in high demand. Traditionally, olefins were produced by naphtha cracking (thermal process); however, now in the United States, with the high availability of natural gas and its associated higher hydrocarbon liquids (the fracking process), ethane and propane are inexpensive feedstocks for polymers such as polyethylene and polypropylene. In many cases, the fluidized catalytic cracking process is modified to enhance production of propylene at the expense of gasoline when the demand for the latter is decreased.

Dehydrogenation is a highly endothermic process and uses heterogeneous catalysts to facilitate the reaction. The reactor and process design must be optimized for maximizing heat input and the inevitable formation of coke from olefins. The thermodynamics favor high temperatures, which also catalyze the formation of diolefins (dienes) and ultimately coke. Several commercial process technologies exist for the dehydrogenation of paraffins to olefins, each using different catalysts, reactor designs, and operating conditions.

One commercial example is the CATOFIN technology practiced by ABB Lummus for the conversion of propane to propylene (Figure 9.8). The catalyst for the CATOFIN process is primarily 20% Cr_2O_3 supported on Al_2O_3 with other proprietary promoters. In this process, gas-phase dehydrogenation occurs over a fixed bed

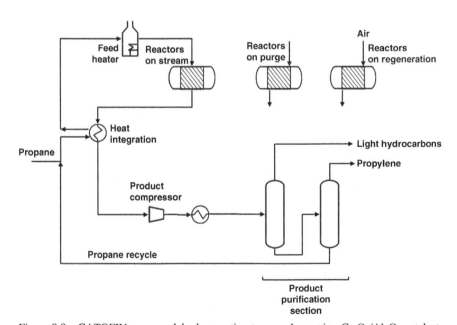

Figure 9.8   CATOFIN propane dehydrogenation to propylene using Cr_2O_3/Al_2O_3 catalyst.

catalyst. Coking is managed by using multiple reactors in parallel. While one set of reactors are "on stream," the others are regenerated by oxidizing the coke. The cycle time between "on stream" and regeneration is on the order of minutes, allowing the use of the enthalpy from coke oxidation to be used to drive the endothermic dehydrogenation reaction.

$$CH_3CH_2CH_3 \longrightarrow CH_2{=}CHCH_3 + H_2$$
$$\longrightarrow CH_2{=}C{=}CH_2$$
$$\longrightarrow Coke$$

The selectivity to mono-olefins is maximized by operating the adiabatic fixed reactor at ~50% conversion per pass with feed recycle and intermittent reheating (Figure 9.6). The feed enters the reactor at about 600 °C (preheated from the product heat exchanger plus additional heat) and rapidly cools to about 500 °C due to the endothermic heat requirement. The feed–product mixture is then reheated to 600 °C for additional conversion. The olefin is continuously condensed (propylene condenses at −42 °C) and separated from the propane and H_2 and recycled along. However, most of the H_2 is exported for other uses. The H_2 reduces coke formation, but shifts the equilibrium opposing dehydrogenation. Due to the excessive coke formation, the process operates with a swing reactor; one reactor is removing coke by regeneration, while the others are in the process stream generating olefins. The catalyst cycle is short requiring regeneration almost every 10 min. The coke is removed by carefully controlling the exotherm by limiting the amount of air entering the reactor to minimize sintering.

Biodegradable detergents are produced by UOP (one example of the Oleflex process) by sulfation of the olefin groups in alkyl benzenes with H_2SO_4. The alkyl benzene is dehydrogenated to the corresponding olefin over a fixed bed at 500 °C of $Pt/\gamma\text{-}Al_2O_3$ + alkaline earth promoters (e.g., CaO) added to neutralize the acid sites on the $\gamma\text{-}Al_2O_3$, which catalyzes di-olefin formation and ultimately coke. Additives such as As, Sn, and Ge are also added to promote the reaction. The process operates similar to that shown in Figure 9.8 with feed–product recycle and preheating after each pass.

The polymer polystyrene is used in cups, insulation, and packaging and is made from its monomer styrene by dehydrogenating ethyl benzene (EB) in a fixed bed reactor. The low surface area catalyst is a 1/8 in. diameter extrudate composed of Fe_2O_3, Cr_2O_3, K_2O, and CaO. The K_2O is added as K_2CO_3, which when decomposed creates some large pores to minimize pore diffusion. It also eliminates acidity minimizing coke formation. Styrene–butadiene polymers are used to make tires.

$$C_6H_5{-}CH_2CH_3 \rightarrow C_6H_5CH_2{=}CH_2 + H_2, \quad \Delta H° = 120\,\text{kJ/mol}$$

In the process (Figure 9.9), superheated steam at 900 °C is added to minimize cracking and coke formation of the EB to bring a final feed temperature of 600 °C where it enters the first adiabatic reactor. The steam also maintains the Fe_2O_3 in the proper oxidation state for catalytic effectiveness. The reactant/product is reheated to 600 °C

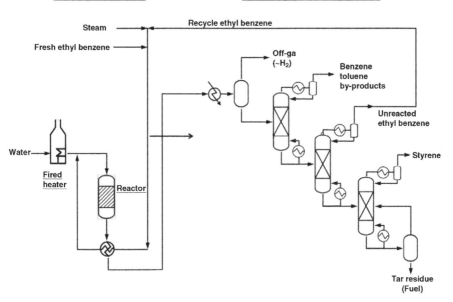

Figure 9.9   Flow diagram for dehydrogenation of ethyl benzene to styrene.

to achieve about 70% conversion (low equilibrium constant $= 0.376$). The styrene product can be separated from the styrene by cooling (styrene condenses at $-30\,°C$, while the EB at $-95\,°C$). The balance of unreacted EB is recycled to the front end of the process.

There is an autothermal reforming process where a small amount of O_2 is added to supply the endothermic heat of reaction by oxidizing some of the H_2 produced. The catalyst is supported Pt/Al_2O_3 pellets. This simplifies the supply of heat, but consumes valuable H_2.

9.5   ALKYLATION

Ethyl benzene is produced by adding ethylene to benzene by what can be considered the reverse of catalytic cracking. The most common catalysts are mineral acids (Brønsted and Lewis) such as HF, H_2SO_4, BF_3, or $AlCl_3$, but solid acids such as Nafion and faujasite Y-zeolite (Lummus—UOP process) are being slowly introduced to minimize the environmental and corrosive hazards of the liquid mineral acids.

$$C_6H_6 + CH_2\!\!=\!\!CH_2 \rightarrow C_6H_5CH_2CH_3 \quad \Delta H° = -100\,kJ/mol$$

Since the reaction is exothermic, the equilibrium constant is favored at low temperatures. For example, the $K_e = 2.4 \times 10^4$ at 225 °C, while at 500 °C the $K_e = {\sim}10$. The liquid-phase process is carried out at pressure in excess of 600 psig H_2 at about 250 °C with addition of ethylene to the liquid benzene at the inlet of both the first and second stages with cooling before the second stage. EB is distilled as the product.

QUESTIONS

1. **a.** What is the reaction and conditions for the purification of terephthalic acid using a Pd/C catalyst?

    b. What undesired by-products would you predict if some Ru were present as an impurity in the catalyst?

2. **a.** In the hydrogenation of $CH_2=CHCHCl_2$, predict the products if you use

    1. Pt/C

    2. Rh/C

    b. In the hydrogenation of an aromatic ketone, predict the products using

    1. Pd/C

    2. Rh/C

3. Explain the important process steps in the ethyl benzene (EB) conversion to styrene? How is coke minimized? What catalyst regeneration methods are used and show the reactions?

4. Can you suggest any alternative processes to produce styrene by dehydrogenation of ethyl benzene that might simplify the existing process and perhaps improve selectivity?

5. Consider the comparison of Ni/Al_2O_3 and Pd/C as powdered catalysts in the slurry-phase hydrogenation of benzylaldehyde to methylbenzene (toluene). The nickel catalyst is less expensive but much less active than Pd.

    a. On the same plot, compare the qualitative rate versus weight for both catalysts.

    b. What physical properties of the catalyst are important for recycle of the catalyst for additional batches?

    c. Show how the entire plot of rate of reaction versus weight of the Pd catalyst will vary with increased agitation.

    d. Explain physically what happens when agitation is increased. Why is bubble size of the H_2 important?

6. What environmental disadvantages are associated with the current catalysts used for alkylation?

7. Describe the process steps utilized in the hydrogenation of vegetable oils to make desired edible products.

8. In the hydrogenation of an aromatic ketone, predict the products using

    1. Pd/C

    2. Rh/C

Aromatic ketone

9. You are trying to hydrogenate butylene to butane. Both Pd- and Ni-based catalysts are well known to catalyze this reaction. In general, what are the advantages and disadvantages associated with precious metal and base metal catalysts that will influence your decision as to which catalyst to use?

BIBLIOGRAPHY

Allen, R. (1982) Hydrogenation, in *Baily's Industrial Oil and Fat Products*, Vol. 2, 4th edn (ed D. Swern), John Wiley & Sons, Inc., New York.

Bartholomew, C. and Farrauto, R. (2006) *Fundamentals of Industrial Catalytic Processes*, 2nd edn, John Wiley & Sons, Inc., Hoboken, NJ, Chapter 7, pp. 488–554.

Bricker, J. (2012) Advanced catalytic dehydrogenation technologies for production of olefins. *Topics in Catalysis* 55 (19–20), 1309–1314.

Gallezot, P. (2003) Hydrogenation: Heterogeneous, in *Encyclopedia of Catalysis*, (ed Horvath,), Vol. 17–55, Wiley-Interscience, Hoboken, NJ, pp. 239–298.

Resasco, D. (2003) Dehydrogenation: Heterogeneous, in *Encyclopedia of Catalysis*, (ed I. Horvath,), Vol. 17–55. John Wiley & Sons, Inc., Hoboken, NJ, pp. 49–79.

Roberts, G. (1976) The influence of mass and heat transfer in the performance of heterogeneous catalysis in gas/liquid/solid systems. *Catalysis for Organic Synthesis*, Academic Press, New York.

Rylander, P. (1983) Catalytic processes in organic conversions, in *Catalysis, Science and Technology* (eds J. Anderson and M. Boudart), Springer, Berlin, Germany.

Rylander, P. (1985) *Hydrogenation Methods*, Academic Press.

CHAPTER 10

PETROLEUM PROCESSING

10.1 CRUDE OIL

In nature, the conversion of biomass to crude oil is a very complex process involving many factors. As a result, the composition of crude oil varies considerably depending on its origin. According to year 2011 market statistics, Russia provides about 13% (~10 million barrels per day or MBBL/day), Saudi Arabia 13% (~10 MBBL/day), and the United States about 10% (~7.5 MBBL/day) with Iran, China, Canada, Iraq, UAE, Venezuela, Mexico, Kuwait, Brazil, Nigeria, and Qatar each contributing smaller amounts to comprise the balance. As an example of the variation in composition by region, Table 10.1 compares oil composition of three extreme sources from around the globe: North Sea, Saudi Arabia, and the Oil Sands of Canada. It should be understood that within each region the compositions can vary considerably.

What is highly desirable is "sweet" crude containing low sulfur (<0.42%) and low content of Ni and V compounds with a high content of the light liquid fraction boiling below 200 °C for gasoline applications. The S, Ni, and V are present as large molecular weight organic compounds and the higher their content the more severe hydrotreating processing conditions are needed for use in transportation or heating applications. Roughly 30% of the oil from Algeria and Libya is a light liquid fraction boiling below 200 °C, which is highly desirable for gasoline blends, while Venezuela oil contains only 5–10% light crude. Most crude contains higher boiling components with widely different sulfur, Ni, and V contents. Heavy gas oil (350–550 °C) from Venezuela contains 1.78% sulfur, while Mexican crude from the Maya region contains almost 5% sulfur. Vacuum residues (boiling point >550 °C) from Kuwait contain 3.66% sulfur, 33 ppm Ni, and 87 ppm V, while the Maya vacuum residues contain almost 6% sulfur, 130 ppm Ni, and almost 700 ppm V. Consequently, processing to remove the sulfur, Ni, and V and to reduce the boiling ranges and molecular weight is an important step in the process.

An even more dramatic difference is demonstrated when comparing oil from the North Sea with Canadian oil (tar) sands as shown in Table 10.1. The former contains only 0.4% sulfur, a combination of Ni and V of 10 ppm, 29% of the oil boiling in the gasoline range, and 33% in the diesel range. Oil sands contain almost 5% sulfur, a combination of Ni and V of 350 ppm, a fraction in the gasoline range, and only 17% in the diesel range. The predominant fraction in oil sands is undesirable 55% vacuum

Introduction to Catalysis and Industrial Catalytic Processes, First Edition. Robert J. Farrauto, Lucas Dorazio, and C.H. Bartholomew.
© 2016 John Wiley & Sons, Inc. Published 2016 by John Wiley & Sons, Inc.

TABLE 10.1   Typical Properties of Crude Oil Sources

Property	North Sea	Saudi Arabia	Oil Sands of Canada
Density (kg/m^3)	834	893	1010
Sulfur (wt%)	0.4	2.8	4.8
V + Ni (ppm)	10	100	350
Gasoline (wt%)	29	17	0
Diesel (wt%)	33	28	17
Vacuum gas oil (wt%)	18	26	28
Vacuum residue (wt%)	18	26	55

residue (BP > 550 °C). These must be processed by cracking or hydrocracking to produce the more desirable C_5–C_{18} fractions.

When sulfur-containing compounds are combusted, highly acidic and corrosive SO_x (SO_2 and SO_3) are formed that will poison the atmosphere. Removing sulfur is an important petroleum processing step, which will be discussed in detail in this chapter. Organic compounds of Ni and V will also lead to abrasion of mechanical equipment in their use and thus require special processing (hydrotreating) for removal, which will also be discussed.

10.2   DISTILLATION

The ultimate goal of the petroleum refinery is to maximize the transformation of crude oil into high-value chemical products, most significantly for transportation, including gasoline, diesel, and aviation fuels. Crude oil is a mixture of a wide variety of hydrocarbons and the required chemical upgrading varies with each class of hydrocarbons. Heavy hydrocarbons require different chemical processing from the lighter ones. As a result, the first step in petroleum processing is separating crude oil into chemically similar streams through a continuous process called distillation. Distillation is a process by which hydrocarbons can be physically separated by selective evaporation and condensation, making use of the fact that the boiling temperature varies between different hydrocarbons. In this process, the crude oil is heated at atmospheric pressure to as high a temperature as possible without thermally decomposing the oil, which is roughly 400–500 °C. At atmospheric pressure, this temperature is not high enough to boil the heavy hydrocarbons (>C_{40}) contained in crude oil. The "bottom" of the atmospheric distillation column is fed to a second distillation column operated under vacuum conditions (0.1 atm absolute pressure). These conditions are sufficient to boil even the heavy wax (<C_{70}) hydrocarbons, which are called vacuum oils. Very large hydrocarbons (>C_{70}) remain as liquid called bitumen or asphalt, which is the "bottom of the barrel" of crude oil. Figure 10.1 illustrates the distillation process, the typical hydrocarbon fractions obtained, and the chemical upgrading performed on each fraction. It is noted that a desalting process, before the crude is distilled, is necessary to dissolve the inorganic constituents present in crude oils.

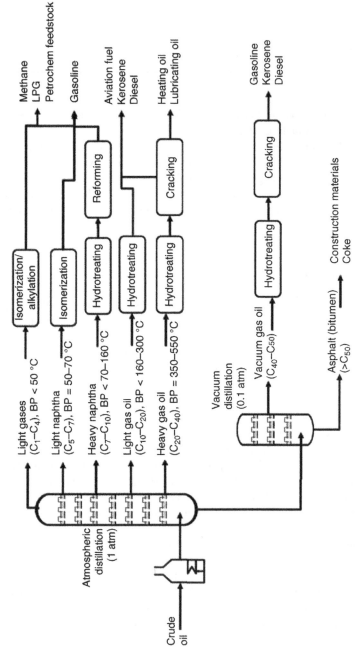

Figure 10.1 Simplified illustration of the crude oil refining process. The desalting process (removal of inorganic components in the crude using a water wash) is not shown.

The lightest hydrocarbons (C_1–C_4, BP < 70 °C) are recovered from the top of the column. Light hydrocarbons (C_5–C_7, BP ~ 70 °C) and heavy hydrocarbons (C_7–C_{10}, BP = 70–160 °C) collectively representing gasoline-range molecules (commonly referred to as naphtha) are next to be recovered. As we continue down the column to heavier higher boiling hydrocarbons, the next fraction typically collected is C_{10}–C_{16}, which is collectively called kerosene used for aviation fuel. The next fraction is light gas oil (C_{14}–C_{20}, BP = 160–300 °C) used for diesel fuel. The heaviest fraction recovered from the atmospheric column is the heavy gas oil (C_{20}–C_{40}, BP = 350–550 °C), which is the hydrocarbon liquid typically used to fuel ships or power stations.

Hydrocarbons containing metals (Ni and V) and sulfur are heavier and become more concentrated in the gas oils. Thus, the heavier hydrocarbons require additional "hydrotreating" processing to demetalize and desulfurize the oil prior to chemical upgrading. The most valuable chemicals are high-octane gasoline and diesel fuel. Thus, downstream chemical upgrading will focus on maximizing these fractions. The lightest hydrocarbons may undergo alkylation or isomerization in order to yield higher octane products that are added to the gasoline pool. Naphtha fractions will undergo reforming reactions to transform low-octane molecules into higher octane molecules. The heavy oils and vacuum gas oils can undergo cracking reactions to lighter hydrocarbons suitable for gasoline or diesel fuels. Once fractionated, a variety of chemical processes are available to transform the various hydrocarbons in order to maximize the yield of the desired chemicals.

10.3 HYDRODEMETALIZATION AND HYDRODESULFURIZATION

The metals and sulfur contained in the heavy fractions (such as molecules illustrated in Figure 10.2) must be removed from the hydrocarbon before further upgrading can be performed. The selective removal of metal or sulfur from the hydrocarbon structure is performed using a process called hydrotreating. The process must be economical, with little unnecessary saturation of hydrocarbons (e.g., aromatics) with minimum consumption of H_2. Hydrotreating is a catalytic process in which the metal is extracted from the hydrocarbon and physically

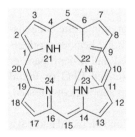

Nickel porphyrin   Thiophene

Figure 10.2   Examples of metal-containing (nickel porphyrin) and sulfur-containing (thiophene) species typically found in crude oil.

incorporated into the catalyst while selectively hydrogenating the hydrocarbon at the site where the metal existed in the structure (Equation 10.1). The catalyst is typically (3% Co + 15% Mo)/Al$_2$O$_3$ with particles a few mm in diameter. Sometimes, Ni, which is more active than Co, is used when additional hydrogenation activity is required. Although sulfur is usually a poison for most catalysts, the hydrodemetalization/hydrodesulfurization (HDM/HDS) catalysts are all presulfided to decrease activity toward excessive consumption of H$_2$ that leads to unwanted saturation of aromatic molecules and excessive light gas formation. Thus, selectivity is enhanced by intentionally poisoning the catalyst with sulfur. The activity is reduced in the sulfided catalyst decreasing the large hydrogen consumption toward undesired products. So activity is sacrificed for improved selectivity.

$$R\text{-}M + \tfrac{1}{2}H_2 \xrightarrow{Co,Mo(S)/Al_2O_3} M/Al_2O_3 + R\text{-}H \qquad (10.1)$$

$$R\text{-}S + \tfrac{3}{2}H_2 \xrightarrow{Co,Mo(S)/Al_2O_3} H_2S + R\text{-}H \qquad (10.2)$$

The hydrotreating process is carried out at 450–500 °C and pressures in excess of 30–100 atm in a fixed bed reactor. The conditions vary with the quality of the feed. Heavy crudes, with high concentrations of metals and sulfur, are processed at pressures approaching 100 atm. A simplified process flow diagram is shown in Figure 10.3. H$_2$ and oil are compressed and preheated (via heat exchange with the HDS reactor) and fed to the first HDM reactor. The first upstream catalyst (S–3% Co, S–15% Mo on Al$_2$O$_3$) is designed to remove the high metal content by using large pore structured carrier such as α-Al$_2$O$_3$ that allows deep penetration of the R-M species (R = general symbol used to designate the organic molecule) into the depths of the pore structure where the metals remain deposited (M/Al$_2$O$_3$) and separated from the R-H species produced. The liquid hourly space velocity (LHSV) is typically 0.5–2 h^{-1}, where the lower value is for high-metal-containing feeds. The second bed will treat a feed with less metal and smaller molecules, so its pore size is that of θ-Al$_2$O$_3$ onto which the sulfided Co/Mo is deposited to allow for conversion of the remaining R-M and some HDS. The sulfided Co/Mo catalyst in the final bed is supported on high surface area γ-Al$_2$O$_3$ and is designed to perform almost complete hydrodesulfurization of the R-S compounds generating H$_2$S gas and R-H. The Ni and V species, deposited in the first two catalyst beds, are often recovered by leaching along with the Co/Mo, which can be recycled.

The products are depressurized and cooled to allow separation of the condensable liquids from the light gases. The liquids are distilled with the heavier products recycled to the feed. The H$_2$S is removed through a ZnO bed and a depropanizer separates the H$_2$ from the light hydrocarbon gases C$_1$–C$_4$ (called LPG or liquid petroleum gas). The remaining H$_2$ is combined with a makeup supply, compressed, and recycled to the feed.

The pore structure of the used HDM catalysts contains V and Ni metal and coke deposits. Coke is a general term for hydrogen-deficient high-boiling hydrocarbons. The continuous deposition of metals and coke causes catalyst deactivation (in 6–12

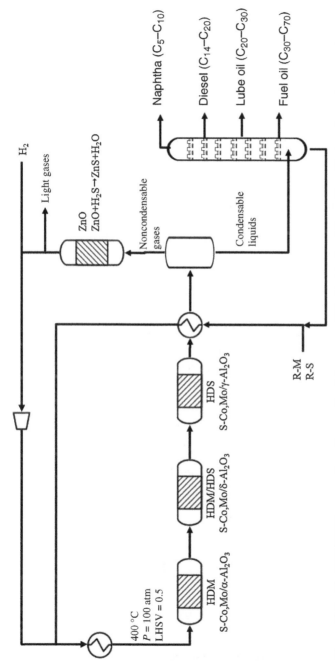

Figure 10.3 The HDM/HDS process flow diagram. Inset shows presulfided catalyst and its positive effect on decreasing excessive gas make and hydrogen consumption.

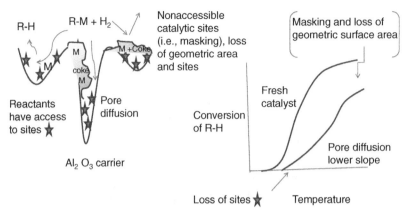

Figure 10.4   Catalyst deactivation by HDM metal deposition (masking) and coking.

months) due to loss of accessible catalytic sites resulting in masking and pore diffusion. These are illustrated in Figure 10.4 (left). Masking decreases the geometric surface area (outer perimeter area of the catalyst particle), consequently reducing the maximum achievable conversion in the mass transfer controlled regime. Pore diffusion is characterized by a decrease in the slope of the conversion–temperature plot due to its reduced activation energy relative to chemically controlled reactions. The inlet temperature to the feed gas must be increased to compensate for the loss in activity. These phenomena are illustrated in the simplified conversion–temperature plot shown in Figure 10.4 (right).

The coked catalyst is regenerated by carefully adding small amounts of diluted O_2 in N_2 while monitoring the bed temperature to avoid high-temperature runaway reactions. This process is continued until the exotherm is very low indicating coke has been removed. A schematic is shown in Figure 10.5.

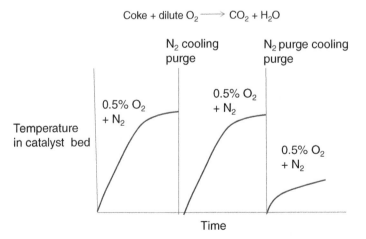

Figure 10.5   Controlled O_2 addition in coked catalyst regeneration.

The deposited V and Ni can be selectively leached leaving the Co/Mo catalyst on Al_2O_3 more or less intact and a candidate for reuse.

10.4   HYDROCARBON CRACKING

The yield of the various transportation fuel hydrocarbon fractions can be significantly increased by breaking the large hydrocarbons into smaller lighter hydrocarbons. This becomes particularly critical as heavier sources of crude oil are used, such as oil sands where more than 80% of the oil is heavier than diesel fuel and gasoline content is effectively zero. The breaking of large hydrocarbons into smaller fragments is accomplished through various chemistries called catalytic cracking. This process is carried out after significant HDM and HDS has occurred.

10.4.1   Fluid Catalytic Cracking

The first cracking catalysts were amorphous SiO_2–Al_2O_3 combinations where the inclusion of Al (+3) in the tetrahedral structure of SiO_2 (Si = +4) generates an imbalance of charge requiring a positive charge to maintain electrical neutrality. This positive charge is a proton that imparts acidic properties to the combination. The more Al (+3) included in the structure, the greater the number of acidic sites present. Both Si and Al are bonded to four O ions as indicated by the vertical lines below.

$$\text{-O-}\overset{|}{\underset{|}{\text{Si}}}\text{-O-}\overset{|}{\underset{|}{\text{Si}}}\text{-O-}\overset{|}{\underset{\overset{|}{H^+}}{\text{Al}}}\text{-O-}\overset{|}{\underset{|}{\text{Si}}}\text{-O-}$$

These catalysts are produced by reacting sodium silicate with a water-soluble salt of Al followed by thermal treatment. The Na^+ is replaced via ion exchange with an ammonium salt, which displaces the Na^+ cation with the ammonium cation, NH_4^+. The exchanged catalyst is then calcined in air where the ammonium cation decomposes liberating ammonia leaving behind the H^+ as the charge balancing cation becoming the catalytically active site for cracking.

The second-generation cracking catalysts were also SiO_2–Al_2O_3 in composition but prepared in a manner that results in important physical and chemical properties, the most unique being their controlled well-defined crystal structures and controlled pore sizes in a range of molecular aperture sizes of about 0.4–1.3 nm (4–13 Å). They were first found in nature but many are now produced synthetically. This unique pore structure is responsible for separating molecules in accordance with their cross-sectional area (kinetic area). A molecule smaller than the aperture can enter the interior but the larger ones are excluded. Hence, the term molecular sieve is often used to describe zeolites. They are usually identified by the Si/Al ratio that determines their structure, acidity, and stability.

Faujasite Y zeolite (designated HY) is a widely used zeolite for cracking catalysts. It belongs to a family of many different zeolite structures that vary in

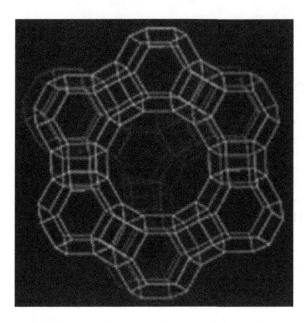

Figure 10.6   Faujasite zeolite.

composition, structure, pore size dimension, and acidity. It has a Si/Al ratio of 1.0–1.5 with a pore or aperture size of 0.74 nm or 7.4 Å composed of 12 oxygen anions as shown in Figure 10.6. The midpoint of each line represents an O^{2-} bonded to Si^{4+} or Al^{3+}. It is the AlO^- site that requires a metal cation for charge balance. For cracking catalysts, these sites are H^+. The higher the Al content (lower Si/Al ratio), the greater the number of acid sites but the lower its hydrothermal stability. This is important for the life of the catalyst since it is continuously regenerated to remove coke at elevated temperature. The zeolite is embedded within an amorphous SiO_2–Al_2O_3 (also called the matrix structure) that not only initiates the cracking of the large molecules but also captures impurities, such as traces of organic metal compounds (R-M, Ni, and V compounds) not completely removed during HDM, that will severely deactivate the zeolite.

The matrix, having its own acid sites, also cracks large molecules into smaller sizes that can enter the zeolite (HZ), which can "polish" them to desired products. Catalyst particle sizes vary between 50 and 100 μm depending on the fluidization dynamics of the process. Modern cracking catalysts are a combination of 75% amorphous SiO_2–Al_2O_3 and 25% faujasite. MgO–SiO_2 is blended into the matrix to react and immobilize the residual vanadium, present in the feed, preventing it from migrating to and destroying the zeolite structure. Some Sb compounds are also added to alloy and deactivate the deposited Ni that would catalytically dehydrogenate the feed hydrocarbons, leading to excessive H_2 generation with subsequent formation of olefins that are coke precursors.

During the fluidized catalytic cracking (FCC) process, a C—C bond is broken and a proton is transferred from the catalyst to the molecule forming a positively

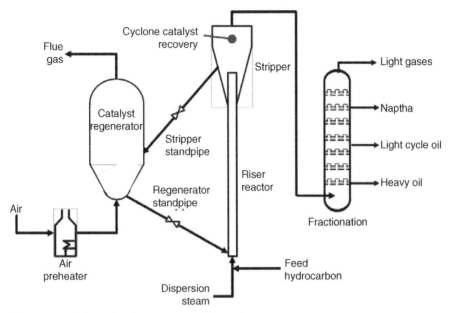

Figure 10.7   Schematic of FCC reactor with catalyst regenerator.

charged carbocation. This ion can react with other hydrocarbons transferring its proton generating new carbocations. Ultimately, the large molecule is cracked to a smaller alkane and alkene (olefin) with the regeneration of the protonated zeolite completing the catalytic cycle.

$$C_{16}H_{34} \xrightarrow{\text{HZ}} C_8H_{18} + C_8H_{16} \tag{10.3}$$

The olefins are highly reactive toward polymerization leading to long chains of high-boiling hydrocarbons (e.g., coke) that block the pores of the catalyst preventing access of the feed molecules leading to a loss in activity.

Cracking is carried out in a fluid bed reactor shown in Figure 10.7. Catalyst particles are mixed with feed and fluidized with steam upflow in a riser reactor where the endothermic cracking reactions occur at around 500 °C. The fluidized process allows for excellent heat transfer. The active life of the catalyst is only on the order of a second because of excessive coke deposition. The deactivated catalyst particles are separated from the product in a cyclone and transported into a separate reactor where they are regenerated with a limited amount of air. The regenerated catalyst is mixed with the incoming feed that is preheated by the heat of combustion of the coke. Coking, by olefin polymerization reactions, has activation energy of 40 kJ/mol, while the desired cracking reaction has activation energy two times higher (80 kJ/ mol). Thus, coking is kinetically favored and must be factored into the overall process. This is accomplished by utilizing the heat of combustion of the coke to preheat the feed. The gaseous products are distilled according to their boiling ranges as shown in Figure 10.7.

10.4.2 Hydrocracking

This is usually used for heavy feeds such as vacuum oils and oil sands that are highly deficient in H_2 that must be reduced in size and increased in hydrogen content. Multiringed polyaromatic hydrocarbons are to be cracked to monoaromatic or substituted cyclohexane molecules with an increase in hydrogen content to be used for transportation or heating. Two different types of catalysts are used: (1) (Co, Mo)/Al_2O_3 and (2) Pt and/or Pd ion exchanged onto zeolites (faujasite, mordenite, or ZSM-5). The presence of high-pressure hydrogen significantly reduces coke formation and thus extends the time between regenerations.

10.5 NAPHTHA REFORMING

In the gasoline-fueled internal combustion engine, maximum power occurs when the fuel–air mixture combusts at maximum compression. This occurs when the piston reaches the top of the cylinder. By design, combustion occurs by action of the spark plug, which ignites the mixture at maximum compression. However, the fuel and air mixture temperature increases as it is adiabatically compressed. Depending on the composition of the gasoline, the fuel/air mixture can autoignite prior to the piston reaching the top of the cylinder, which decreases engine performance. The "octane number" is an empirically derived number that is associated with a molecule's ability to resist combustion during compression. Reference compounds that bound the octane scale are *iso*-octane (octane number $= 100$) and heptane (octane number $= 0$). Aromatics and branched hydrocarbons have high octane numbers and are ideal for gasoline. The goal of the catalytic reforming process is to convert low-octane hydrocarbons into high-value, high-octane fuels.

The most widely used reforming catalyst is 0.3% Pt $+$ 0.3% Re on (1% Cl$^-$) chlorinated Al_2O_3 particles (3–5 mm diameter). It is a dual function catalyst since different components favor different reactions. The Pt is the active component primarily for dehydrogenation and aromatization reactions, while the Cl$^-$ adds to the acidity of the carrier and is the active site for isomerization. The Re is believed to minimize coke formation. Dehydroisomerization requires both metal and acid functions. Some reactions are endothermic (dehydrogenation and dehydroisomerization) and others are exothermic (isomerization and dehydroaromatization).

For example, the dehydrogenation of cyclohexane to benzene and H_2 is catalyzed by the Pt component, and increases the octane number from 75 to 106.

$$C_6H_{12} + \text{heat} \xrightarrow{\text{Pt}} C_6H_6 + 3H_2 \tag{10.4}$$

Acid-catalyzed isomerization of *n*-butane to *iso*-butane increases octane from 94 to 101.

$$CH_3CH_2CH_2CH_3 + \text{heat} \xrightarrow{Al_2O_3(Cl^-)} \overset{\displaystyle CH_3}{\underset{\displaystyle CH_3CH_2CH_3}{|}} + \text{heat} \tag{10.5}$$

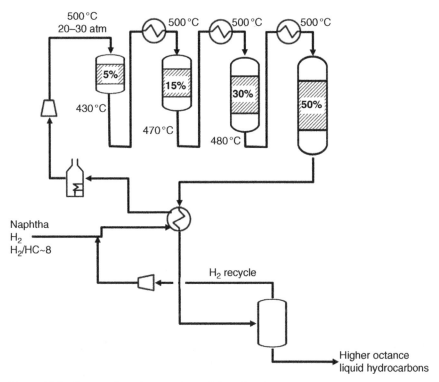

Figure 10.8   Process flow diagram for naphtha reforming.

Heptane has a defined octane number of 0 and when dehydroaromatization occurs toluene is formed with an octane number of 124. This reaction requires a dual function in acid and metal catalytic components.

$$CH_3(CH_2)_5CH_3 + heat \xrightarrow{\text{Pt-Al}_2\text{O}_3(\text{Cl}^-)} C_6H_5CH_3 + 4H_2 \qquad (10.6)$$

The reforming process operates with three or four reactors in series, shown in Figure 10.8. The feed is delivered to the first reactor at 500 °C that is charged with the smallest amount of catalyst (5% of the total and the highest space velocity) to perform the easy but highly endothermic dehydrogenation reactions. To minimize coke formation, a small amount of H_2 is recycled from the product. The products and unreacted feed are then reheated to 500 °C and fed to a second bed containing about 15% of the total catalyst charge where endothermic isomerization reactions occur. After a small decline in bed temperature, the unreacted feed and product are then reheated again to 500 °C where the more difficult dehydroisomerization reactions take place with 30% of the total catalyst charge. The final reactor contains 50% of the total catalyst charge and performs the most difficult dehydrocyclization reactions. Swing reactors are in place to allow the process to continue as each bed is being regenerated by coke burn-off. After coke burn-off, the Pt has sintered and Cl^- is lost but can be

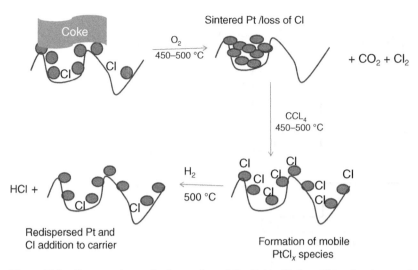

Figure 10.9 Regeneration and rejuvenation of (Pt, Re)/γ-Al$_2$O$_3$ + Cl$^-$ reforming catalyst.

rejuvenated by the addition of chloride (CCl$_4$). The addition of Cl$^-$ leads to the formation of mobile PtCl$_x$ complexes that break apart the sintered Pt agglomerates and upon H$_2$ reduction redispersion of the Pt occurs increasing the catalytic surface area while restoring the lost Cl$^-$ from the carrier. A cartoon showing regeneration and rejuvenation is provided in Figure 10.9.

QUESTIONS

1. What physical and chemical properties make zeolites so important in molecular separations and catalysis?

2. How is coke on the FCC catalyst integrated (utilized) into the process? What technique would you use to establish the optimum conditions for regenerating the catalyst?

3. Describe the role of the SiO$_2$–Al$_2$O$_3$ matrix (non-zeolite), with its additives, in the FCC catalyst.

4. **a.** How are the different structures of Al$_2$O$_3$ used in the HDM/HDS process? What instrumental technique in the laboratory would you use to determine the effectiveness of each of the Al$_2$O$_3$ to maximize capacity of metals?

    b. How is the catalyst modified to minimize unwanted cracking reactions and to maximize selectivity for short times on stream?

5. Refer U.S. Patent 3,415,737 (available through "Google") entitled "Reforming a sulfur-free naphtha with a platinum–rhenium catalyst" by Harris Kluksdahl. In your answers to the claims, use your own words rather than copying what is stated.

    a. How did the inventor determine the chemical relationship between the Pt–Re catalyst? What characterization method was used to determine this conclusion?

b. What catalyst preparation and process conditions are used to improve reforming performance?

c. How did the inventor use a model compound to demonstrate the invention for different amounts of Pt and Re and reduction temperature?

d. Compare Claims 1 and 3.

e. Compare Claims 1 and 6.

f. What is stated in Claim 7?

g. Contrast Claims 1 and 2.

6. In 1978, catalyst engineers and scientists at Mobil (now ExxonMobil) published an important parametric study using a process development unit (PDU) in the laboratory to provide the proper operating conditions for a new plant to make gasoline from methanol. Please secure this paper (Chang et al. (1978) Process studies on the conversion of methanol to gasoline. *Industrial Engineering Chemical Process Development* 17 (3), 255–260) and answer the question asked.

   a. What reasons did the authors give for the use of ZSM-5 (also called pentasil) zeolite to control the product distribution? Describe some of the unique properties stated in our book about ZSM-5.

   b. Explain what is meant by alkylate in Table 1. (*Hint:* Alkylation is the opposite of cracking, see the book.) How is alkylation used to increase the yield of gasoline?

   c. Why is it necessary to control the heat of reaction (both reactions are exothermic)?

   d. Write the two reactions ((1) dehydration of CH_3OH to dimethyl ether (DME) and (2) DME to gasoline using the ZSM-5 zeolite catalyst). What properties of the methanol dehydration catalyst have we studied that are also important for another reaction (name?) we studied?

   e. Explain the operation of the PDU unit. Note they use ^0F. How do they control the production of light gas products? What is the deactivation mode for the DME to gasoline catalyst? Is there anything that can be done to regenerate it?

   f. What was the difference in performance between pure methanol and crude methanol?

7. Why are large-pore catalysts required for hydrodemetalization?

8. Why do we not use a fixed bed reactor for fluid catalytic cracking?

9. A $PtRe/Al_2O_3$ reforming catalyst loses its ability to isomerize straight-chain alkanes to branched alkanes but still maintains its ability to dehydrogenate alkanes to olefins. What can you conclude about the deactivation mode?

BIBLIOGRAPHY

Aitani, A. (1996) Chapter 13: Catalytic reforming processes, in *Catalytic Naphtha Reforming: Science and Technology* (eds G. Antos, A. Aitani, and J. Parera), Marcel Dekker, New York.

Babich, I. and Moulijn, J. (2003) Science and technology of novel processes for deep desulfurization of oil refinery stream: a review. *Fuel* 82 (6), 607–631.

Bartholomew, C. (1994) Catalyst deactivation in hydrotreating of residue: a review, in *Catalytic Hydroprocessing of Petroleum Distillates* (eds M. Oballa and S. Shih), Marcel Dekker, New York. pp. 1–32.

Bartholomew, C. and Farrauto, R.J. (2006) Chapter 9, in *Fundamentals of Industrial Catalytic Processes*, 3rd edn, John Wiley & Sons, Inc., Hoboken, NJ, pp. 635–699.

Dight, L. and Kennedy, J. (1971) *Cracking catalyst*. U.S. Patent 4,171,286.

Fromet, G. and Marin, G. (1987) Hydrocracking: science and technology. *Catalysis Today* 1 (4), 367–473.

Gates, B., Katzer, J., and Schuit, G. (1993) Chapters 3 and 5, in *Chemistry of Catalytic Processes*, McGraw-Hill, New York.

Maxwell, I. (1987) Zeolite catalysis in hydroprocessing technology. *Catalysis Today* 1 (4), 385–414.

Olsen, T. (2014) An oil refinery walk-through. *Chemical Engineering Progress* 110, 34–40.

Rahimi, N. and Karimzadeh, R. (2011) Catalytic cracking of hydrocarbons over modified ZSM-5 zeolites to produce light olefins. *Applied Catalysis A: General* 398, 1–17.

CHAPTER 11

HOMOGENEOUS CATALYSIS AND POLYMERIZATION CATALYSTS

11.1 INTRODUCTION TO HOMOGENEOUS CATALYSIS

Historically, heterogeneous catalysts were the first to be widely used commercially and have dominated industrial practice for many decades. However, enzymes contained in yeast, which is a homogeneous catalyst, have been used in fermentation processes for centuries. Nearly 200 years ago, nitrogen oxides served as a homogeneous catalyst in the production of sulfuric acid in the lead chamber process. However, it was not until the 1960s that homogeneous catalysts began to play a more significant role in large-scale commercial processes, where they were used in the production of specialty chemicals, polymers, food products, and pharmaceuticals. Over the past five decades, their use in chemical processes has increased dramatically. By the beginning of this century, homogeneous catalysts accounted for roughly one-third of the catalyzed industrial chemical processes.

As introduced in Chapter 1, a homogeneous catalytic process is one in which all reactants and catalysts are present in the same phase. Homogeneous catalysts include simple molecules or ions such as HF and H_2SO_4 or complex molecules such as metal–ligand complexes, organometallic complexes, macrocyclic compounds, and enzymes, all of which are soluble in the reacting fluid phase.

They provide several inherent advantages over heterogeneous catalysts, including allowing reaction pathways not possible or very difficult on heterogeneous catalyst surfaces, higher activity or selectivity for particular reactions, as a result of higher activity the ability to conduct reactions at more favorable conditions, and the absence of pore diffusion limitations. However, there are several inherent disadvantages, including their poor thermal stability requiring relatively mild reaction conditions and the added processing required separating the catalyst from the reaction products.

Homogeneous catalysis is a complex topic in catalysis and could easily be the subject of an entire book. This chapter is not intended to provide an exhaustive review of homogeneous catalysis but to give the reader an introduction to homogeneous

Introduction to Catalysis and Industrial Catalytic Processes, First Edition. Robert J. Farrauto, Lucas Dorazio, and C.H. Bartholomew.

catalysis by reviewing some of the key homogeneous catalysis processes currently practiced.

11.2 HYDROFORMYLATION: ALDEHYDES FROM OLEFINS

Olefins are in high demand especially as the market for polyethylene and poly-propylene (plastics) continues to expand at a high rate. In addition to polymers, they are also starting reagents for production of short- and long-chain linear alcohols for use in plasticizers and surfactants, detergents, paints, and even automobile wind-shields. Cobalt carbonyls were discovered in Germany in 1938 and found to be excellent homogeneous catalysts for hydroformylation reactions where H_2 and CO can be added to the olefin increasing the carbon chain of an aldehyde by one (Figure 11.1). The aldehydes are then hydrogenated to the corresponding alcohol. The hydroformylation reaction is carried out at high pressure where the carbonyl catalyst is stable and soluble in liquid propylene. Often a solvent is added, such as methylcyclohexane.

$$CH_3=CH\text{-}CH_2 + H_2 + CO \xrightarrow[\substack{200-300 \text{ atm} \\ 140-180\,^\circ C}]{HCo(CO)_4} CH_3CH_2CH_2CH=O \quad + \quad \substack{CH_3 \\ | \\ HC\text{-}CH=O \\ | \\ CH_3}$$

$$\xrightarrow[\substack{Pd/Al_2O_3 \\ H_2 \\ 20 \text{ atm} \\ 100\,^\circ C}]{} CH_3CH_2CH_2CH_2OH$$

$$(11.1)$$

The linear aldehyde is most desirable and therefore catalyst and process conditions are optimized to maximize its production. The aldehyde is then hydroge-nated to produce a linear alcohol. The cobalt carbonyl catalyst $HCo(CO)_4$ requires 200–300 atm to maintain the CO complexed to the cobalt coordination sphere. The reaction is operated between 140 and 180 °C yielding a 4:1 ratio of linear to branched aldehydes. The hydrogenation reaction occurs at moderate conditions mostly with a heterogeneous catalyst such as Pd on Al_2O_3 at modest conditions.

Advancements have been made (Shell) by adding phosphine ligands to the cobalt coordination sphere (cobalt triorganophosphine) with sufficient hydrogenation activity to produce the alcohol during hydroformylation. This allows a one-step procedure to generate the alcohol directly from the olefin.

The discovery of rhodium triphenylphosphine homogeneous catalysts has dramatically reduced the process conditions with greater selectivity to the linear aldehydes. The homogeneous catalyst (often referred to as the Wilkinson catalyst) written in its fresh state as $RhCl(PPh_3)_3$, where P is phosphorous and Ph_3 represents three phenyl groups coordinated to the Rh coordination sphere, produces a 20:1 ratio

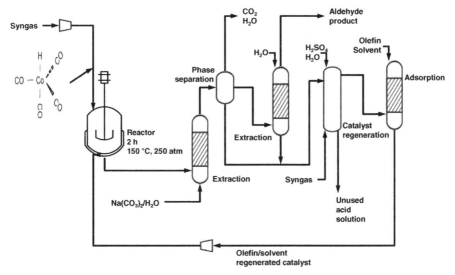

Figure 11.1   Hydroformylation process using a cobalt homogeneous catalyst.

of linear to branched aldehydes at less than 25 atm and 125 °C. The process described in Figure 11.2 is carried out by Dow Chemical (formerly Union Carbide) in concert with Davy McKee.

Detergent alcohols (C_{12}–C_{18}) are also made via hydroformylation. For these applications, the most active olefins are those with a terminal double bond. The process operates at 180 °C and as high a pressure as 250 atm. Hydrogen and carbon monoxide (1:1 H_2/CO) are fed to the reactor, which adsorb and react with the liquid-phase olefin. The fresh catalyst is $HCo(CO)_4$. The average residence time in the continuous stirred tank reactor (CSTR) is about 2 h. The products are treated in a

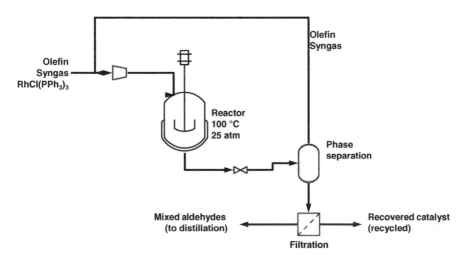

Figure 11.2   Dow (Davy McKee) LP Oxo Selector process using the Rh catalyst.

second reactor with Na_2CO_3/H_2O, which extracts the catalyst as $NaCo(CO)_4$ separating it from the aldehyde products. The extracted catalyst is treated with H_2SO_4 forming the original catalyst that is dissolved in the olefin feed and the process continued.

The catalyst is present in an ampule when mixed with a 1:1 mixture of H_2, CO, and olefin. The combination is compressed to 25 atm where the ampule is broken exposing the catalyst to the reaction mix at about 100 °C. The process is conducted in a CSTR. The pressure of the gas/liquid product is reduced at the exit separating the unreacted H_2, CO, and olefin from the aldehyde and catalyst (solid). The light gases are recycled to the compressor and fresh feed. The catalyst is filtered from the product and regenerated. The aldehydes are purified (linear from the branched species) by distillation.

Dow (and Davy McKee) claims that the costs of the Rh catalyst and phosphine ligands are not key issues given the less severe operating conditions relative to cobalt catalyst processes. However, proprietary methods have been developed allowing for catalyst recovery. The general mechanism is believed to be the insertion of the olefin into the coordination sphere of the homogeneous catalyst with addition of CO and H_2 from the catalyst to the olefin producing the aldehyde. The CO and H_2 replace the H and CO from the catalyst.

11.3   CARBOXYLATION: ACETIC ACID PRODUCTION

The first production of acetic acid using a homogeneous catalyst was pioneered by BASF in 1913 using methanol-soluble CoI in a carbonylation reaction in which CO is added to methanol.

$$CH_3OH + CO \xrightarrow[\substack{25\,atm \\ 175\,°C}]{Rh(CO)_2I_2} CH_3\overset{\overset{\displaystyle O}{\|}}{C}\text{-OH} \tag{11.2}$$

In the 1970s, Monsanto developed a new RhI precursor catalyst that in the presence of HI in methanol decreased the process severity (25 atm CO and 175 °C). The actual catalyst was formed in the reaction mix and believed to be $Rh(CO)_2I_2$. The catalyst and process was 99% selective to acetic acid.

CO, CH_3OH, H_2O, and the water-soluble catalyst feed is compressed to 25 atm. The reaction is carried out below 200 °C in a CSTR (Figure 11.3). The pressure at the exit is reduced allowing the unreacted CO and CH_3OH (BP = 65 °C) to be separated from the higher boiling acetic acid (118 °C) and recycled to the inlet of the reactor. The acetic acid was dried and separated from the undesired by-product (propionic acid, BP = 141 °C) and catalyst by distillation. Careful control of the acid is necessary to avoid destabilizing the Rh catalyst and to minimize reactor corrosion.

The Cativa™ process by BP utilizes an iridium complex in combination with a Ru carbonyl complex and other promoters (CH_3I) and is claimed to be more effective than Rh catalysts for acetic acid production. BP claims that the Ir complex is more stable than Rh over a wider range of CO pressures. The process operates at

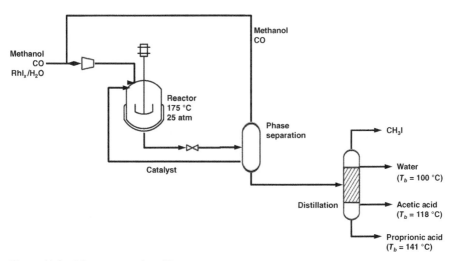

Figure 11.3   Monsanto acetic acid process.

less than 200 °C and a pressure of 28 atm. The catalyst is more soluble and thus less water is needed and most importantly is much more selective with less undesired acetaldehyde and propionic acid. The process design is similar to that used with the Rh catalyst.

11.4 ENZYMATIC CATALYSIS

The most well known use of enzymes (nature's catalysts) is the fermentation of starch and sugars (using yeast) for the production of alcohols. Let us also remember how our bodies utilize hundreds and thousands of enzymes that assist in performing metabolic functions. Fermentation and production of ethanol will be briefly discussed in the alternative energy chapter; however, enzymes are also used in other industrial applications not as well known. What makes enzymes so unique is a great mystery of nature, but biochemists have successfully studied and applied them to a few important reactions. They are proteins with high biological molecular recognition forming selective bonds with reactant sites. They are active at temperatures <50 °C (unlike most inorganic catalysts that require much higher temperatures to initiate reactions). They are present in bacteria cells but can permeate through the cell walls. The pharmaceutical companies isolate the enzymes using filtration or a GC column. Sometimes, the entire cell is added to the reaction mixture allowing the enzymes to permeate through the cell walls.

Enzymes catalyze reactions lowering activation energies. For example, the hydrolysis of urea, for the release of NH_3 in the application of fertilizers, occurs with activation energy of <125 kJ/mol. This allows substantial rates consistent with farming. The decomposition of H_2O_2 using the enzyme catalase has an activity 107 times that of the same reaction catalyzed by Pt.

One important medical application is the production of a drug effective in the treatment of leukemia. L-Aspartic acid is produced by the catalytic reaction of fumaric acid by the enzyme L-aspartase.

$$HOOC-CH=CHCOOH \longrightarrow HOOC-CH_2\overset{\overset{\displaystyle H}{|}}{\underset{\underset{\displaystyle NH_3}{|}}{C}}{}^*-COOH \qquad (11.3)$$

L-Aspartic acid is known as a chiral compound since the C^* is bonded to four different groups. This produces a compound that is nonsuperimposable upon its mirror image creating molecular specificity by bonding to specific sites in the DNA interrupting the growth of cancer cells.

An interesting commercial application was introduced in the 1970s by Corning Corporation demonstrating the feasibility of bonding enzymes to porous glass and avoiding the difficulty of separation of a homogeneous process. They were able to convert corn starch to glucose using an enzyme (glycoamylase) bonded to porous glass. The glucose (low sweetness) is then isomerized to fructose (high sweetness) using glycose isomerase bonded also to porous glass.

11.5 POLYOLEFINS

Specially prepared plastics are rapidly replacing traditional metal components because of their strength, transparency, resilience, lighter weight, and greater corrosion resistance. The largest volume products are polyethylene and polypropylene. Each has its own contributions to the marketplace where the former is primarily used for low-strength applications such as milk and food containers. Polypropylene is used when enhanced strength, higher melting temperatures, and greater resistance to chemicals such as chemical holding tanks and automobile bumpers are required.

11.5.1 Polyethylene

Polyethylene ($-CH_2-CH_2-)_n$ has a molecular weight of about 5×10^5 with a melting point of 135 °C. It is used extensively for many household products, including food bags and milk containers.

A slurry-phase process (Phillips) for its production utilizes chromium oxide deposited on SiO_2 dispersed in a solvent such as cyclohexane at 100 °C and an ethylene pressure between 30 and 40 psig. The active site is believed to be Cr^{2+} produced by the reduction of Cr^{6+} by ethylene. The catalyst is contained in an ampule to avoid contact and oxidation with air. The ethylene feed is dissolved in cyclohexane and with the catalyst is fed to a loop reactor at a flow rate of about 5 m/s with a residence time of up to 2 h. The loop reactor (Figure 11.4) provides sufficient turbulence and heat removal without buildup of polymer on the reactor walls. The polymer in the solvent is typically about 25% by weight. The solvent in the product is flashed and recycled leaving the polymer. The catalyst is retained in the polymer since its concentration is extremely low. The operating conditions are adjusted to produce both high- and low-density polyethylene. The reaction mechanism proposed is that

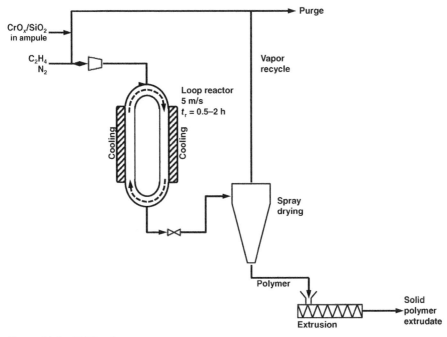

Figure 11.4   Phillips loop reactor.

the polymer coordinates with one of the Cr^{2+} sites and the incoming ethylene coordinates with another site. Insertion of the ethylene into the double bond of polymer propagates its growth.

A second method of production (Figure 11.5) utilizes the Ziegler–Natta $TiCl_4$ deposited on $MgCl_2$ as the catalyst. The $TiCl_4$ is reduced to the $TiCl_3$ in an activation step using ethyl benzoate. The $MgCl_2$ helps to fracture the catalyst that is retained in the final product. A liquid cocatalyst such as an alkyl aluminum halide is added along with other olefins such as butane to modify the density and flexibility of the final product. This is a reactive catalyst that must be prepared at the exclusion of air and water, so after its preparation and activation it is protected in an ampule. The alkyl group of the cocatalyst coordinates with the Ti^{3+} sites. The polymer grows by insertion of the alkyl group into the double bond of the adsorbed polymer creating another empty coordination site on the Ti^{3+}, which provides a new site for the next ethylene molecule.

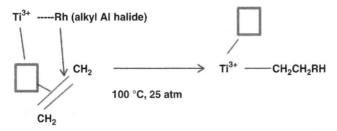

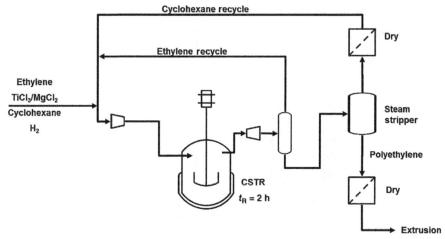

Figure 11.5 $TiCl_3/MgCl_2$ process for polyethylene.

The process feeds gaseous ethylene (and often with butane as a comonomer) and liquid cyclohexane with the catalyst (ampule) and compressed to 25 atm and heated to 100 °C. H_2 is also added to control the chain length of the polymer. After about 2 h of residence time in a batch reactor, the gas/liquid product goes to a separator where the used ethylene is recycled to the feed and the polymer, cyclohexane, and catalyst go to a steam stripper where the cyclohexane is separated, dried, and recycled to the front end of the process. The polymer is further dried and pelletized.

11.5.2 Polypropylene

$$n[CH_2=CHCH_3] \longrightarrow [\text{-}CH\text{-}CH_2\text{-}CH\text{-}CH_2\text{-}]_n \atop \quad\quad CH_3 \quad\quad CH_3$$ (11.4)

Polypropylene can have an isotactic structure in which all —CH_3 groups are on the same side of the polymer chain. This is the most important of all the isomers with a molecular weight of 500,000, a density of 0.9, and a melting point of 170 °C. It is transparent and rigid and hence it is extensively used for pipes, tubes tanks, and so on.

The syndiotactic isomer has the —CH_3 groups positioned on alternate sides of the polymer chain with a molecular weight of 300,000 and a melting point of 135 °C. It is used in soft packaging such as snack bags. The least important is the atatic isomer with random orientation of the —CH_3 groups.

Modern production for polypropylene uses the Zeigler–Natta catalyst, which is also used for polyethylene production. The catalyst is $TiCl_4$ supported on $MgCl_2$ (in an ampule) along with the aluminum alkyl halide cocatalyst such as diethyl aluminum fluoride, $(CH_3CH_2)_2AlF$. The $MgCl_2$ is milled to a very disordered but active structure and the $TiCl_4$ is deposited on it. Production of the polypropylene is carried out at 100 psig of H_2 and polypropylene at 65 °C in a slurry-phase reactor similar to that used for polyethylene. The polyethylene, H_2, and isobutene are separated from the polymer and solvent via a gas–liquid separator (pressure reduction), dried by passing through a zeolite, and recycled to the front end of the process. The desired isotactic polymer (containing traces of catalyst) is insoluble in xylene and can be separated from the other soluble isomers. Antioxidants and UV stabilizers are added to the polymer before it is extruded into pellets.

A fluidized gas-phase reactor is also used. The process is carried out at 50 and 100 °C (avoiding melting of the polymer during reaction) and 100–600 psig. The polypropylene product is mixed in a separate reactor with ethylene to make a block polymer with enhanced mechanical properties.

Water, CO, and O_2 are the most significant poisons and are carefully removed upstream of the process.

The catalyst preparation and the process are far more complicated than presented here, so the reader is encouraged to refer to more detailed references.

QUESTIONS

1. Polypropylene:

   a. How is the length of the polymer chain (its molecular weight) controlled during the slurry process?

   b. What is the most important commercial isomer of polypropylene and how does its structure differ from less the important isomer?

   c. How is the catalyst separated from the final polymer

2. What are the advantages/disadvantages of homogeneous catalysts compared to heterogeneous catalyst?

3. What would be a worthy goal of research in homogeneous catalytic reactions?

4. What the advantages/disadvantages of using base metal compared to precious metal homogenous catalysts?

5. What are the limitations of using enzymes as catalysts for large scale industrial chemical production?

6. What processes are typically used for homogenous reactions?

7. What is the most common use of enzymes and why are they important?

BIBLIOGRAPHY

Enzymes

Achle, W. (ed.) (2004) *Enzymes in Industry*, Wiley-VCH Verlag GmbH, Weinheim, Germany.

Chaplin, M. and Bucke, C. (1990) *Enzyme Technology*, Cambridge University Press, New York.

Chen, W., Brulmann, F., Lee, K., and Deshusses, M. (2003) Whole cell catalysis, in *Encyclopedia of Catalysis*, Vol. 6 (ed I. Horvath), John Wiley & Sons, Inc., Hoboken, NJ, pp. 658–680.

Dordick, J. (1991) Chapter 1: An introduction in industrial biocatalysis, in *Biocatalysts for Industry* (ed J. Dordick), Plenum Press, New York.

Monter, S. (2006) *Alcohol Fuels*, Taylor & Francis, Boca Raton, FL.

Homogeneous Catalysis

Anonymous (2013) Hydroformylation's Diamond Jubilee. *Chemical and Engineering News* 22, 38–40

Crabtree, R. (2003) Homogeneous catalysts and catalysis, in *Encyclopedia of Catalysis*, Vol. 3 (ed I. Horvath), John Wiley & Sons, Inc., Hoboken, NJ, pp. 479–492.

Evans, D., Olson, J., and Wilkinson, J. (1968) Hydroformylation of alkenes by use of rhodium complex catalysts. *Journal of the Chemical Society A* 12, 3133–3136.

Jane, J. (2000) The Cativa™ process for the manufacture of acetic acid. *Platinum Metals Review* 44 (3), 94–105.

Kaminsky, W. (2000) Polymerization catalysis. *Catalysis Today* 62, 23–34.

Keep, A. (2003) Catalyst preparation—homogenous, in *Encyclopedia of Catalysis*, Vol. 2 (ed. I. Horvath,), John Wiley & Sons, Inc., Hoboken, NJ, pp. 379–387.

Polymerization

Henrici, G. and Olive, S. (1981) Mechanism of Ziegler–Natta catalysis. *ChemTech* 11, 746–752.

Levine, I. and Karol, F. (1977) *Preparation of low and medium density polyethylene in fluid bed reactor.* U.S. Patent 4,011,382.

McDaniel, M. (1988) Controlling polymer properties with the Phillips chromium catalyst. *Industrial & Engineering Chemical Research* 27, 1559–1564.

Stevens, M. (1999) *Polymer Chemistry: An Introduction*, Oxford University Press, Oxford, NY.

Xie, T., McAuley, K., Hsu, J., and Bacon, D.W. (1994) Gas phase ethylene polymerization: production processes, polymer properties and reactor modelling. *Industrial & Engineering Chemical Research* 33, 449–479.

CHAPTER 12

CATALYTIC TREATMENT FROM STATIONARY SOURCES: HC, CO, NO$_x$, AND O$_3$

12.1 INTRODUCTION

One of the key elements in any new manufactured product is the life cycle analysis and a key element of the analysis is the effect on the environment. This includes the sources of contaminants to the water system as well as the air we breathe. Of course, it is difficult to manufacture products without some waste streams. Wastewater treatment, a prime consideration in the 1960s and 1970s in the United States, has met with success in cleaning up many rivers and lakes. It is continuously being addressed in the United States and certainly is receiving worldwide attention as it is becoming a serious issue on a global scale. Probably more common, but many times less noticeable, are the emissions of organic compounds from manufacturing of the multitude of consumer products used every day. In most manufacturing processes, for the raw materials, intermediates, or the finished product, organic materials are present as chemicals, solvents, release agents, coatings, decomposition products, pigments, and so on that eventually must be disposed. In such manufacturing, there is usually a gaseous effluent that contains low concentrations of organics vented to the atmosphere. Commonly associated with volatile organic compounds (VOCs) are emissions of CO and oxides of nitrogen (NO$_x$ = NO, NO$_2$, and N$_2$O). Examples of commercial processes having VOC and carbon monoxide emissions are chemical plants, petroleum refineries, pharmaceutical plants, automobile manufacturers, wire coating operations, painting facilities, and so on. Power plants are a large contributor to NO$_x$ as well as hydrocarbons, depending on the fuel source and CO.

The structure and composition of the VOCs determine the degree to which they undergo photochemical reactions with oxides of nitrogen (NO$_x$) forming hazardous ozone, also referred to as smog, in a series of complicated atmospheric reactions. Consequently, they are regulated by local and federal laws. They are also referred to as reactive organic gases (ROGs). Their approximate reactivity as a function of molecular type is indicated below.

oxygenates > aromatics > olefins > saturated molecules (large > small)

Introduction to Catalysis and Industrial Catalytic Processes, First Edition. Robert J. Farrauto, Lucas Dorazio, and C.H. Bartholomew.
© 2016 John Wiley & Sons, Inc. Published 2016 by John Wiley & Sons, Inc.

Methane is excluded from current U.S. regulations due to its low photochemical reactivity, which means it does not participate in ozone-forming reactions. However, this must be reconsidered for future regulations given its contribution to greenhouse effects.

The organics present depend on the process and can include ketones, aldehydes, aromatics, paraffins, olefins, acids, chlorinated hydrocarbons, fluorinated hydrocarbons, and higher molecular weight organics, which are often present as aerosols. In coal-fired power plants, inorganic particulates (ash such as SiO_2, Al_2O_3, Hg, and As) are also emitted together with the other pollutants.

A number of methods are available for abating these emissions, but catalytic abatement is commonly used especially when the pollutants are in low concentrations. For concentrations above about 1% liquid or solid absorbents, scrubbing, condensation, membranes, thermal flares, and electrostatic precipitators (for removal of ash) may be used. This chapter will focus on catalytic abatement.

12.2 CATALYTIC INCINERATION OF HYDROCARBONS AND CARBON MONOXIDE

The catalytic incineration method has become most popular because, in many cases, it is more versatile and economical for the low concentrations of organic emissions (i.e., <5000 vppm).

The basic catalytic oxidation reaction of an organic molecule is shown below.

$$C_xH_y + \left(x + \frac{y}{4}\right)O_2 \rightarrow xCO_2 + \frac{y}{2}H_2O \tag{12.1}$$

CO, although not a VOC, often accompanies the VOCs and thus it too is oxidized.

$$CO + \tfrac{1}{2}O_2 \rightarrow CO_2 \tag{12.2}$$

For heteroatom VOCs such as halide (X) compounds,

$$3CH_3X + 5O_2 \rightarrow 3CO_2 + 4H_2O + HX + X_2 \tag{12.3}$$

For perfluoro compounds,

$$C_2F_6 + 3H_2O \rightarrow CO + CO_2 + 6HF \tag{12.4}$$

For oxygenates,

$$C_xH_yO + \left(x + \frac{1}{4}y - \frac{1}{2}\right)O_2 \rightarrow xCO_2 + \frac{y}{2}H_2O \tag{12.5}$$

Carbon monoxide is relatively easy to catalytically oxidize, so the conditions for complete abatement are determined by the organic constituent. The actual operating

TABLE 12.1   Catalytic Oxidation Using Pt on a Monolith (M) Versus Thermal Combustion Relative to Flammability Ranges

1% Feed concentration	T_{in} (°C), $Pt/Al_2O_3//M$	Flammability range (wt%)	Thermal combustion (°C)
CO	100	12.5–74	700
Benzene	210	1.4–7.1	500
Formaldehyde	150	7–7.3	430
Propane	400	2.1–9.5	500
Ethanol	300	4.3–19	500

temperature and amount of preheat needed depend on the organic molecule, space velocity, composition of feed (i.e., contaminants, water vapor, and so forth), and organic concentration. One way of comparing thermal versus catalytic abatement is to look at the energy required (air preheat temperature) to obtain quantitative removal of a given hydrocarbon species. The operating temperatures shown in Table 12.1 are well below the corresponding temperatures necessary to initiate thermal (noncatalytic) oxidation. The catalyst initiates reaction at lower temperatures by lowering the activation energy. This demonstrates the major advantage of catalyzed processes, that is, they proceed faster than noncatalytic reactions, allowing lower temperatures for the same amount of conversion. This translates directly to improved economics for fuel use and less expensive reactor construction materials, since corrosion is greatly reduced. Catalyst can also operate outside of the flammability range and thus does not require additional fuel to elevate the temperatures as is usually the case with flaring.

Selection of the catalytic material for various organic pollutants has been the subject of many studies. Both base metal oxides and precious metals, and their combinations are used for both hydrocarbons and chlorinated hydrocarbons. As a rule, precious metals (especially platinum and/or palladium dispersed on carriers (γ-Al_2O_3) deposited on monolithic (honeycomb) structures) are preferred because of their activity, resistance to deactivation, and their ability to be regenerated. The monolith offers low pressure drop substrate for the catalyst. When the catalyst is deposited onto the carrier and deposited onto the monolith walls, it is referred to as a washcoat. Platinum appears to be the preferred catalyst for saturated hydrocarbons and higher molecular weight species. Palladium is preferred for methane and low molecular weight olefins. Metal oxides are less frequently used except for those conditions where the feed gas is relatively free of contaminants such as sulfur. An exception is the use of a Cr_2O_3 catalyst in a fluidized bed process in which abrasion renews the catalytic surface and minimizes the retention of poisons.

It is worth noting in Table 12.1 that a catalyst can completely oxidize materials outside of the flammability range at relatively low temperature while thermal combustion requires additional fuel to raise the temperature.

Table 12.2 shows some relative light-off temperatures for various organic families of molecules using a monolithic Pt catalyst.

TABLE 12.2 Light Temperatures for Various Families of Molecules Using a Pt Monolith Catalyst

Hydrocarbons: relative light-off temperatures (°C)

Pt/Al$_2$O$_3$//monolith

C$_{4+}$ (150) < C$_{3=}$ (200) < C$_{2=}$ (250) < C$_3$ (300) < CH$_4$ (500)

Oxygenates: relative light-off temperatures (°C)

Pt/Al$_2$O$_3$//monolith

Alcohols (100) < ketones (220) < acetates (230)

12.2.1 Monolith (Honeycomb) Reactors

The preferred support for abatement catalysts is the monolith or honeycomb structure either ceramic (cordierite) or metallic (aluminum or stainless steel). The monolith structure offers the major advantage of low pressure drop due to its high open frontal area (OFA) (i.e., 70–90%). This was briefly discussed in Chapter 2.

For VOC catalysts, the best operating region is where the reaction is controlled by bulk phase mass transfer. When bulk mass transfer (BMT) limits the reaction rate, the abatement process is effectively passive to temperature fluctuations. Here the most important design parameter is the degree of turbulence (i.e., Reynolds number), the geometric surface area (GSA) of the support, and the reactor cross-sectional area. All these must be selected at the lowest possible pressure drop. The monolith support satisfies these criteria. In some limited cases, other catalyst structures such as screens and beads are used.

When a reaction is in the bulk mass transfer control regime, the rate or conversion has a low sensitivity to temperature (i.e., low activation energy). For the condition of bulk mass transfer control, the intrinsic activity of the catalyst must be greater than the rate of mass transfer of the pollutant to the catalyst surface (washcoat). Furthermore, pore diffusion must be fast relative to bulk mass transfer. When the rate of a reaction is controlled bulk mass transfer, it generates a flat portion of the conversion versus temperature plot described in Chapter 1. The rate of reaction is proportional to the mass transfer coefficient (k_{MT}) (which increases with turbulence or Reynolds number), the geometric surface area (a_s) of the catalyst, and the pollutant concentration.

$$r = k_{MT}a_s C_{A,bulk} \tag{4.22}$$

For a monolith support, the geometric surface area is defined as the total wall surface area (TSA) of the monolith. The mass transfer coefficient increases with an increase in linear velocity (LV = flow rate (STP)/frontal area of monolith) and with a decrease in the channel size. To increase bulk mass transfer, one can also take advantage of the entry region where the boundary layer has not yet fully developed. At the leading edge of the monolith, the boundary layer is thin and gradually thickens until it becomes fully developed a short distance from the leading edge. There are cases where the reactor design is composed of thin slices of catalyst monolith with small gaps in-between. By using thin slices, one can prevent the flow from ever becoming fully developed and reduce the average thickness of the boundary layer in the overall reactor. In reality, the length of the slice is greater than entry region due to strength requirements for the monolith slice. However, the average boundary layer thickness is still lower than the equivalent length of continuous monolith. This approach provides a very inexpensive simple means to induce turbulence in the reactor and increase the rate of bulk mass transfer. This design is used in ozone abatement catalysts described later in this chapter.

The disadvantage of increasing the bulk mass transfer rate is the negative effect on pressure drop. Increasing turbulence and increasing gas linear velocity both result in higher pressure drop. Thus, an optimal balance must be struck between pressure drop and reaction (bulk mass transfer) rate.

12.2.2 Catalyzed Monolith (Honeycomb) Structures

Ceramic monoliths are typically made of cordierite ($2MgO-2Al_2O_3-5SiO_2$) specifically designed with low thermal expansion to resist cracking during rapid temperature swings often referred to as thermal shocks. They are extruded and calcined to temperatures in excess of $1000\,^{\circ}C$ and can have varying channel diameters and wall thicknesses to meet GSA requirements and as low a pressure drop as possible even after the catalyzed washcoat is applied. Specific catalyst compositions are kept proprietary; however, most commonly, 0.1–0.5% Pt on high surface area γ-Al_2O_3 is used, mostly as a washcoat on a monolith. In limited cases, small amounts of Pd or Rh are added to promote a particular reaction where the Pt may be deficient. The washcoat is composed of the high surface area carrier impregnated with catalytic precursor salts in aqueous solution. The carrier may first be ball milled to $<10\,\mu m$ in slightly acidified water producing slurry with 30–40 wt% solids content. The monolith is then dipped into the carrier slurry. Excess slurry is blown out of the channels with air (air knife), dried, and calcined to produce an adherent bond to the monolith walls. It is then dipped into a solution containing the component salts, dried at about $120\,^{\circ}C$, and calcined at $500\,^{\circ}C$ in air to decompose the salts generating catalytic metals or metal oxides. Alternatively, the carrier may first be impregnated with precursor salts of the catalytic components. Precipitating agents are then added to fix the metals or metal oxides to the carrier. The combination is then ball milled to produce the slurry into which the monolith is dipped, channels blown free of excess slurry, dried, and calcined to remove the combustible salts. This final step produces a washcoat with an adherent bond with the monolith.

The use of metal monoliths can offer many advantages such as larger open frontal area (~90% thinner walls) for a given GSA, improved thermal characteristics, and the potential for complex flow patterns within the monolith designed to increase the bulk mass transfer rate. Adhesion of the catalyst washcoat to the surface of the metal surface is critical and many metals do not provide a suitable surface. One well-known metal that does provide a suitable surface is a Fe–Cr–Al high-temperature alloy commercially named Fecralloy. Fecralloy composition is roughly 70% Fe, 20% Cr, 5–10% Al, and trace metals including Y and Zr. Fecralloy thermal resistance comes from the aluminum migrating to and oxidizing on the surface, forming a protective layer to prevent the oxidation of the Fe. For catalyst washcoating, one can take advantage of the very uniform roughened Al$_2$O$_3$ layer as a good surface for catalyst adhesion. For catalyst washcoating, the Fecralloy surface is given an 800 °C air pretreatment to generate a roughened Al$_2$O$_3$-rich surface to which the washcoat will adhere. Care must be taken to ensure an adherent bond since the metal expands to a greater extent than the washcoat. If care is not taken, adhesion is lost and the washcoat may attrite from the surface especially during large temperature changes in use. Monolith geometries vary depending on the nature of the feed and pressure drop restraints. Table 12.3 lists a number of different ceramic monoliths (honeycombs) with varying cell densities (i.e., 400 cpsi), wall thicknesses (i.e., 6.5 mils = 0.0065 in.), and geometric surface areas. In general, the higher the cpsi, the greater the pressure drop.

Feed gases containing dust or ash (especially from coal-fired power plants) usually require larger diameter honeycomb channels (i.e., 100 cpsi). Larger channels decrease pressure drop but also have lower GSA per given volume, which decreases bulk mass transfer conversion. The larger the open frontal area, the lower the pressure drop. Therefore, some compensation in size and properties of the total monolith is required to give a specific conversion with an acceptable pressure drop. The volume of catalyst used also depends on the degree of conversion desired. For example, 30,000 l/h space velocity will typically give 99% conversion for a 200 cpsi honeycomb, while 60,000 l/h gives 90% conversion. Trade-offs of this type require that the plant engineers work closely with the catalyst manufacturer.

12.2.3 Reactor Sizing

In many environmental applications, the reactant exists in ppm level concentrations and the oxygen is in great excess. As described in Section 4.4.4, this special condition results in the reaction effectively being first order with respect to the pollutant, and zero order with respect to oxygen. However, in most cases, these reactions are bulk mass transfer controlled, where the rate of reaction is described by Equation 4.22. It is also reasonable to assume that the reactor is isothermal since the reaction enthalpy is very small (due to pollutant concentration being very small) compared with the thermal mass of the gas mixture. Given these assumptions, we can integrate the expression given for the rate of bulk mass transfer (Equation 4.22) to yield Equation 12.6 that allows laboratory tests for use in designing the environmental abatement reactor.

$$r = k_{MT}a_s C_{A,bulk} \tag{4.22}$$

$$\ln\left(\frac{C_b^i}{C_b^o}\right) = \frac{k_{MT}a_s}{SV} \tag{12.6}$$

The ratio of inlet and outlet concentrations (degree of conversion) depends on the environmental regulations.

The mass transfer coefficient can be roughly estimated in a small-scale laboratory reactor from a conversion (of the pollutant to be abated) versus temperature plot using a catalyzed monolith (with a given cpsi based on pressure drop) to be used in the final design. In the laboratory test, the monolith will have a much smaller diameter than that in the plant, so the laboratory inlet flow must be adjusted to give a

TABLE 12.3 Nominal Properties of Standard and Thin-Wall Cordierite Substrates

Ceramic cell density (cell/in.^{2})/wall thickness (6.5 = 0.0065 in.)	400/6.5	470/5	600/3.5	600/4	600/4.3	900/2.5	1200/2.5
Substrate diameter (mm)	105.7	105.7	105.7	105.7	105.7	105.7	105.7
Substrate length (mm)	98	88	76	76	76	76	35
Substrate volume (l)	0.86	0.77	0.67	0.67	0.67	0.67	0.31
Material porosity (%)	35	24	35	35	35	35	35
OFA	0.757	0.795	0836	0.814	0.800	0.856	0.834
GSA (m^2/l)	2.74	3.04	3.53	3.48	3.45	4.37	4.98
TSA (m^2)	2.35	2.35	2.35	2.32	2.30	2.91	1.53
Hydraulic diameter (mm)	1.10	1.04	0.95	0.94	0.93	0.78	0.67
Flow resistance (l/cm^2)	3074 (105%)	3274 (100%)	3780 (100%)	3990	4122	5412	7589
Bulk density (g/l)	395	390	267	303	324	235	269
Heat capacity @ 200 °C (J/(K1))	352	348	238	270	289	209	240
Heat capacity @ 200 °C (J/K)	302	269	159	180	193	140	74
Substrate mass (g)	339	301	178	202	216	156	83

Reproduced from Chapter 7 of Heck, R.M., Farrauto, R.J., and Gulati, S.T. (2009) *Catalytic Air Pollution Control: Commercial Technology*, 3rd edn, John Wiley & Sons, Inc., New York.

linear velocity approaching that expected in the plant by giving a similar degree of turbulence (Reynolds number). The diameter of the monolith in the final design is dictated by the vent into which it will be located. By applying Equation 12.6 and the estimated mass transfer coefficient, one can then determine the approximate length (i.e., SV) of the monolith to be used from the known flow rate.

This method should be applied using a catalyst that has experienced the real feed gas since some catalyst deactivation due to temperature excursions and/or poisoning by feed contaminants may occur. To address this reality, it is preferred that a slipstream test be conducted where a small portion of the real exhaust is passed over a small-sized sample of the catalyst for 3–6 months. The conversion versus temperature profile before and after the test provides critical insight into how much additional catalyst may be needed to meet the emission standards. Likely a lower space velocity or high inlet temperature will be needed to meet the environmental requirements. The advantage of slipstream testing is presented in Section 12.2.4.

For feeds that are known to contain large amounts of sulfur (i.e., greater than 50 ppm), less reactive carriers such as TiO$_2$ or α-Al$_2$O$_3$ are used for the Pt. Those carriers are relatively inert to the formation of sulfates compared with high surface area γ-Al$_2$O$_3$. For feeds without sulfur, nonprecious metal catalysts such as CuO or Co$_2$O$_3$ are also used on the γ-Al$_2$O$_3$ carrier. Of course, it is also possible to operate the catalyst at high enough temperatures so that the sulfur does not adsorb on the catalytic sites, thus preventing deactivation by poisoning. However, this requires additional energy and a fuel penalty. Furthermore, base metal oxides such as CuO and Co$_2$O$_3$ undergo sintering, oxidation state change, and reaction with the carrier at elevated temperatures.

The abatement system is usually designed for maximum heat recovery, as shown in Figure 12.1a and b. The igniter is needed only to preheat the pollutant-laden inlet gas to initiate the oxidation reaction. If the inlet gas temperature becomes sufficiently higher than that necessary for "light-off," no external heat source is needed to sustain catalytic oxidation. The catalyst shown is a precious metal deposited on a high surface area carrier such as γ-Al$_2$O$_3$ washcoated onto the walls of the monolith structure. The inlet air containing pollutants is preheated through the exit heat exchanger. If the exotherm for the combustion reaction is sufficiently large, a second heat exchanger, in series, may be used to export process heat.

12.2.4 Catalyst Deactivation

Sintering is not usually a problem for VOC applications, since modern catalyst materials have been developed from mobile source applications (i.e., automotive catalysts) to resist high-temperature degradation. Deactivation usually occurs by *fouling* or *masking* in which a residue from the process stream deposits on the catalyst surface. This residue may be dust from the manufacturing process or an organic char. Another example of this mechanism is via aerosols that contain a metalorganic compound that, when decomposed, leaves a residue of inorganic material or ash. A common source is lubricating oil, which decomposes leaving Zn, P, and Ca compounds on the catalyst surface. When these materials deposit, they can block access to the pores and/or react with the catalytic sites, rendering them less active. A physical representation of these phenomena is shown in Figures 5.4 and 5.5 (see Chapter 5).

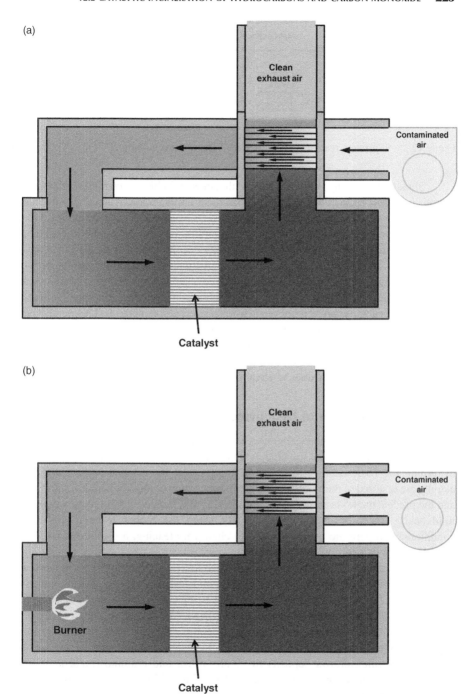

Figure 12.1   (a) VOC abatement process with heat integration. (b) VOC abatement with supplemental heating.

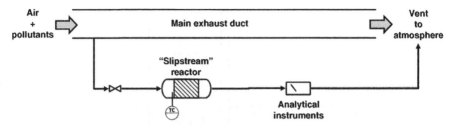

Figure 12.2   Slipstream reactor concept used for VOC abatement design.

Slipstream testing (Figure 12.2) is the most reliable procedure to size and establish process control for stationary abatement applications. A small percentage (1–2%) of the actual effluent stream is passed over a washcoated monolith catalyst for some prolonged length of time such as ~3 months. By comparing the conversion versus temperature plot generated in the laboratory of the fresh and aged (slipstream) catalysts, one can determine the extent of deactivation expected in the final application. This allows catalyst sizing, inlet temperatures, and space velocity to be adjusted to meet the required conversion. From a process point of view, the contaminants present on the catalyst can be determined by characterization methods (Chapter 3) that will provide insight into minimizing their effect. For example, it is common for process ash (oil constituents, contaminants in the air, and so on) to accumulate on a catalyst. The extent of contamination can be addressed by either upstream filtration or a regular maintenance schedule to clean or regenerate the catalyst.

12.2.5   Regeneration of Deactivated Catalysts

Various catalyst regeneration techniques can be utilized. However, the most practiced technique involves chemical washing, which selectively dissolves impurities without disturbing the basic catalyst material. In many cases, the catalyst surface and the activity are restored to their initial state. The composition of the regeneration solutions and the precise methods employed are proprietary but include careful treatment with mild acids, bases, or chelating agents. By using the proper reactor, operating conditions, and periodic maintenance with chemical cleaning, a useful catalyst life of 5–10 years is common.

The usual procedure in developing regeneration technology is to remove a catalyst sample from the commercial reactor and test for activity in a laboratory unit. Using a model compound similar to that being converted in the actual plant, various chemical washing procedures can be tried. Depending on the severity of deposits, using the correct proprietary aqueous solutions various extents of regeneration can be obtained. Analysis of the catalyst using optical emission spectroscopy often shows deposits of alkali, silica, and alumina (present in air intake) causing reductions in surface area from 15 to 3.55 m^2/g. The deposits were masking the pores of the catalytic surface, and the alkaline wash substantially removed them and restored the surface area to 9.60 m^2/g or 65% of the original value. Using such regeneration procedures has resulted in substantially longer life for organic abatement catalysts.

12.3   FOOD PROCESSING

Restaurant cooking, especially those that fry food, emits significant amounts of visible fine particulate matter (i.e., smoke) and VOCs into the air, many of which impart a distinct odor. On cold days, the VOCs, especially those derived from cooking oils, can condense on the cold walls of the chimney and ultimately lead to chimney fires.

Restaurant equipment that contributes to these emissions includes charbroilers, griddles, and deep fat fryers. According to a recent study conducted by the University of California, Riverside, College of Engineering, Council for Environmental Research and Technology, charbroiling is the major contributor to emissions from restaurants and contributes more than 80% of all the cooking emissions (i.e., VOC and particulate matter). At present, Rule 1138 only requires the addition of an emission control device for the chain-driven charbroiler. An extensive study was conducted by the Southern California Air Resources Board to evaluate the performance of various available and emerging control techniques for chain-driven charbroilers. Among all the assessed technologies that are capable of reducing both particulates and VOCs (i.e., catalytic oxidizer, fiber bed filter, and incineration), the catalytic oxidizer is the only device that is cost effective.

Unlike the traditional catalytic VOC abatement technology, the effect of both catalyst and converter design on control of cooking emissions is not well studied. The catalytic converter containing catalyst is mounted onto a ventilation shroud above the cooking surface as shown in Figure 12.3. The hot cooking emissions flow through

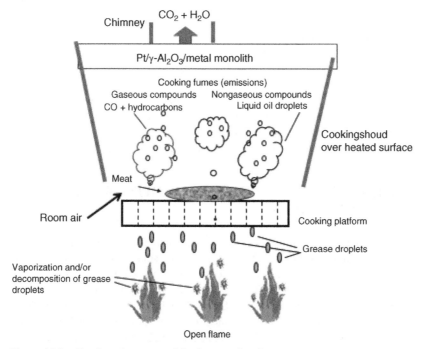

Figure 12.3   Catalyst abatement of food processing fumes.

the ventilation shroud by natural convection upward to the converter. A thin-wall metal monolith, designed for low pressure drop and light weight, washcoated with a Pt/γ-Al$_2$O$_3$-containing catalyst destroys the odor-bearing compounds and the potential for chimney fires. Thus, the unnecessary maintenance of the chimney due to organic deposits provides a cost savings to the restaurant.

The catalytic converter unit operates at a nominal temperature range of 530–650 °C while cooking food, and 370–430 °C during idle or warm-up mode. The treated cooking exhaust is emitted from the converter, mixes with room air, and is then vented outside the room through the ventilation hood installed above the broiler.

The system design is a trade-off between mass transfer and pressure drop. This is the classic catalytic reactor design problem and the optimum converter design for a charbroiler requires high VOC and particulate removal efficiency with a minimum pressure drop loss. A metal monolith with nonparallel channels gives better mass transfer limited conversion rate with lower pressure drop loss. The skewed channels of the monolith cause a disruption of the boundary layer and therefore higher mass transfer performance creates a higher but acceptable pressure drop.

12.3.1 Catalyst Deactivation

Like many other catalytic pollution abatement technologies, the activity of food service catalyst can be affected by fouling. The most common deposits found on food service catalysts are unburned grease particles and/or partially burned organic char. The grease particles are normally found on areas where the catalyst temperature is not high enough to initiate oxidation. The organic char is the result of the dehydrogenation of the deposited grease particles. As the grease builds up on the catalyst surface (masking), it hinders the diffusion of oxygen to the catalyst surface. This favors "coking" of the grease particles rather than catalytic oxidation since the oxygen is blocked from reaching the catalytic surface. The deposits from a chain-driven charbroiler are typically analyzed by XPS and several inorganic compounds such as P, K, Na, and Si were found. These compounds are believed to be from various sources, such as the ingredients used for food preparation, chemical agents used to clean the cooking appliances, and decomposition of the phospholipids contained in the meat. An in-house washing method was developed using mild alkaline solution to dissolve the impurities.

12.4 NITROGEN OXIDE (NO$_X$) REDUCTION FROM STATIONARY SOURCES

NO$_x$ is a pollutant that is generated in any process that uses air to combust a fossil fuel. Nearly half of urban environment's NO$_x$ emissions are sourced from the transportation sector and how to treat this was discussed in the mobile section. The other half is sourced from stationary sources. NO$_x$ is formed by the high-temperature thermal reaction of N$_2$ and O$_2$ used in combustion. Some of the big stationary emitters include coal-, fuel-, or natural gas-fired power plants, commercial or industrial boilers, or waste incinerators. The treatment of NO$_x$ can be handled by modifying a combustion process to prevent the formation of thermal or flame generated species or it can be converted back to N_2 with a separate reducing chemical. By modifying the combustion process, the thermodynamics and kinetics of NO_x formation are being shifted to less favorable conditions.

Lower flame temperatures can be achieved by decreasing the temperature of the incoming combustion air or by introducing steam into the flame zone. Steam injection will also dampen the amount of radicals created in the flame zone that leads to NO_x formation. Furthermore, it is useful in reducing unburned hydrocarbons leading to particulate species. Another means to lower flame temperature is to lower the oxygen (i.e., air/fuel ratio). As will be shown in gasoline (Chapter 13) and diesel engines (Chapter 14), the hottest flame is achieved when the combustion process operates at stoichiometric conditions, the point at which the exact amount of oxygen is present to completely react with all of the fuel. Moving away from the stoichiometric point will not only lower the flame temperature, but also release more partially combusted products (i.e., CO or VOCs). A way to lower oxygen concentration and simultaneously drop the flame temperature is by using a process called flue gas recycle. The flue gas of the combusted products is depleted of oxygen and enriched in nitrogen, steam, and carbon dioxide. Recycling a portion of the exhaust not only lowers the oxygen concentration but the excess gases act as a heat sink to decrease the flame temperature. Techniques used to lower flame temperatures or excess oxygen can result in 50–80% lower NO_x emissions depending on the specifics of the design.

12.4.1 SCR Technology

The stationary source catalyst technology for controlling NO_x at the outlet of power plants is called selective catalytic reduction (SCR). The first SCR catalysts were Pt-based catalysts; however, the temperature window for operation is too narrow and requires careful process control. Since about 1980, V_2O_5 has been mostly used. It relies on the selective reduction of NO_x by ammonia over the catalyst.

The major desired reactions shown below are all catalyzed by V_2O_5 supported on TiO_2 on a ceramic or metal monolith structure. Gaseous ammonia is injected ($\sim 1.1\, NH_3/NO$) into the grid and acts as the reducing agent forming N_2 over the catalyst. The first three reactions are desired and dominate below about 450 °C. Above this temperature, the undesired decomposition of NH_3 to N_2 and/or NH_3 oxidation to NO_x occur explaining the maximum followed by the decline in NO_x conversion. Metal-exchanged zeolites (Fe/Beta) can also be used, especially at temperatures in excess of 450 °C where they are stable and selective toward N_2; however, most power plant exhausts do not reach these temperatures. Comparative performance is shown in Figure 12.4. Furthermore, the added expense is not justified when compared with the V_2O_5 catalysts and the convenience of controlling the temperature suitable for the V_2O_5 catalyst.

$$4NH_3 + 4NO + O_2 \rightarrow 4N_2 + 6H_2O \tag{12.7}$$

$$4NH_3 + 2NO_2 + O_2 \rightarrow 3N_2 + 6H_2O \tag{12.8}$$

$$2NH_3 + NO + NO_2 \rightarrow 2N_2 + 3H_2O \tag{12.9}$$

$$2NH_3 + \tfrac{3}{2}O_2 \rightarrow N_2 + 3H_2O \tag{12.10}$$

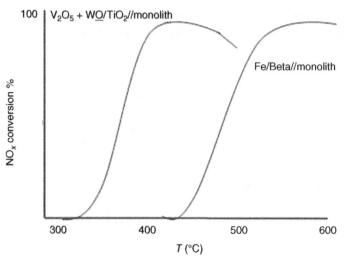

Figure 12.4   SCR with V_2O_5 and a metal-exchanged zeolite: 1.1 NH_3/NO and zeolite.

The catalyst temperature is controlled between 300 and 450 °C. If the NH_3 to NO ratio exceeds about 1.1, NH_3 slip occurs at the exit and therefore its injection must be carefully controlled and matched to the NO$_x$ content in the exhaust.

Another important concern with SCR systems is their reactivity with sulfur. Since some sulfur is present in power plants especially those fired by coal, the carrier for the V_2O_5 used is high surface area TiO_2 (anatase structure) that is much less reactive to SO$_x$ than Al_2O_3. If sulfur is present in the exhaust stream, often as SO_2, the SCR catalyst can oxidize this to SO_3. SO_3 will go on to react with water to make sulfuric acid, a major contributor to acid rain, or react with ammonia to form ammonium sulfates that can create fouling on downstream pipes or release as particulate aerosols.

$$2SO_2 + O_2 \rightarrow 2SO_3 \tag{12.11}$$

$$SO_3 + H_2O \rightarrow H_2SO_4 \tag{12.12}$$

$$NH_3 + SO_3 + H_2O \rightarrow NH_4HSO_4 \tag{12.13}$$

A schematic of an SCR is provided in Figure 12.5. The basic components include the ammonia injection manifold and the catalyst bed. Good mixing of the NH_3 and exhaust gas upstream of the catalyst is important in order to avoid the above-mentioned by-products.

Associated with coal-fired power plants is ash that can block channels of the monolith and/or mask the surface of the catalyst. It is therefore common to use monoliths with large-diameter channels (~100 cpsi) for those fuels where ash is expected. For coal-fired power plants, it is also normal to place an electrostatic precipitator (ESP) upstream from the catalyst to remove the ash before it can cause problems on the catalyst. Since small ash particles can pass uncaptured through the ESP, periodically the catalyst can be purged with air to remove the surface ash as a method of regeneration.

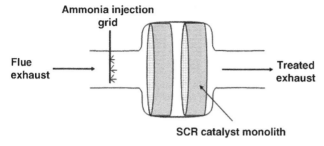

Figure 12.5   SCR reactor schematic. It would be worth mentioning that the widening, that is, lower velocity, increases contact time.

12.4.2   Ozone Abatement in Aircraft Cabin Air

Wide-body aircraft fly over polar routes en route to Europe or Asia from the United States at altitudes above about 30,000 ft in order to limit fuel consumption. At these altitudes, they penetrate the ozone layer (1–4 ppm) depending on the flight pattern. Makeup air brought into the cabin contains a few ppm of O$_3$ (lung irritant) that adds to the discomfort of passengers and crew. In 1980, the FAA established regulations for reduction of ozone entering the airplane. A number of solutions were considered including adsorption on carbon filters; however, due to high weight, pressure drop, and poor capacity, they were disqualified. The solution came from the use of ozone abatement catalysts on low pressure drop monolith structures housed in the heating and air conditioning ducts. The ozone-containing makeup air was bled from one compression stage of the jet engine at a temperature between 120 and 200 °C (the compressor discharge temperature). Therefore, the catalyst had to be active and stable within this temperature range at a space velocity of about 500,000 h^{-1}. After considerable laboratory and flight testing, a Pd/Al$_2$O$_3$//monolith (ceramic) in a unique reactor design was found to satisfy all requirements.

$$2O_3 \xrightarrow{\text{Pd/Al}_2\text{O}_3\text{//monolith}} \rightarrow 3O_2 \qquad (12.14)$$

Later metal monoliths with lower weight, thinner walls, and thus lower pressure drop than the ceramic were used. Since the O$_3$ levels were only a few ppm, bulk mass transfer to the catalyst surface became rate limiting. To enhance BMT, the catalyst was segmented into six pieces each about 1 in. thick (see Figure 12.6). This design provided sufficient turbulence (entrance and exit mixing effects disrupted the boundary layer at the catalyst surface) enhancing O$_3$ contact with the catalyst. This design allowed reduction of the O$_3$ to less than 0.1 ppm.

12.4.3   Deactivation

Deactivation came mostly from land operations especially as the plane waited to takeoff. The air entering the engine and the catalyst from the planes ahead waiting for takeoff contained sulfur from the jet fuel and lubricating oil components (S, P, Zn, and particulates) deposited on the catalyst masking the surface leading to deactivation.

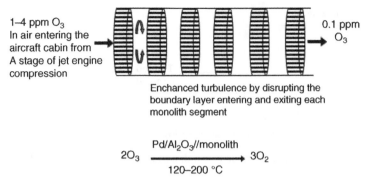

1–4 ppm O$_3$
In air entering the
aircraft cabin from
A stage of jet engine
compression

0.1 ppm
O$_3$

Enchanced turbulence by disrupting the
boundary layer entering and exiting each
monolith segment

$$2O_3 \xrightarrow[120\text{–}200\ °C]{Pd/Al_2O_3//monolith} 3O_2$$

Figure 12.6   Ozone abatement reactor design.

Catalyst companies developed solvents to wash the catalyst free of contaminants during periodic maintenance of the airplane about every 6–12 months.

12.5   CO$_2$ REDUCTION

Preventing the generation of CO$_2$ is perhaps the most straightforward means to address the greenhouse gas problem. To do this, high carbon content fossil fuel combustion needs to be substituted with renewable and alternative forms of energy, especially in the generation of power. This includes the use of hydro, wind, solar, and geothermal forms of energy. Furthermore, it must be decided as a society that nuclear power is an acceptable form of carbon-free energy. These forms of energy can be converted into electricity or used for heating, such as the case with solar thermal technologies. Furthermore, renewable sources could be used to electrolyze water to form hydrogen and carbon monoxide used for the synthesis of higher hydrocarbons in a Fischer–Tropsch process (Chapter 7). The CO can also be processed via water gas shift and PSA to high-quality H$_2$, which is fed to a highly efficient fuel cell for the direct production of electricity.

For mobile applications that rely on diesel or gasoline combustion engines, alternative engines can be used. Hybrid gasoline/electric engines have been on the market for nearly a decade. All-battery electric vehicles are now available commercially; however, recharging generates CO$_2$ at the power plant. The electric energy used to drive the vehicles is in part lower due to the greater efficiency achieved with electric motors. Furthermore, when the electricity that is used to charge the on-board batteries is sourced from renewable technologies, then the total carbon footprint can be reduced. Fuel cell vehicles are now available in selected locations and offer significant reductions in CO$_2$; however, the hydrogen infrastructure needs to be built to provide the H$_2$ fuel to all locations.

Renewable fuel derived from biological sources, such as ethanol or biodiesel, is being used in gasoline and diesel engines as a carbon neutral technology. The concept here is that the CO$_2$ that is generated when a bioproduct is combusted is balanced by the CO$_2$ that is consumed during the photosynthetic growth of the crop. It should be noted that validity of carbon neutrality of the production of biofuels is currently under debate and the demand for biofuels directly competes with society's demand for food products.

Finally, the obvious way in which CO_2 emission is being handled is by becoming more energy efficient. For mobile applications, this means for every pound of carbon burned, more of that energy be directed into powering the vehicle. The ultimate efficiency is set by the thermodynamic cycles, but getting closer to this limit through engine and design improvements will result in improved fuel economy. Moving from heat engines to fuel cells is another approach that is being developed for improving energy efficiency. Heating demands are decreasing as commercial and residential buildings are making use of more advanced insulation and building materials. Devices running on electrical power are being made more efficiently, such as energy star appliances or low energy consuming fluorescent light bulbs. Finally, the easiest way in which carbon emissions can be lowered is through continued conservation efforts supported by communities and governments alike.

QUESTIONS

1. Secure the following paper and answer the related questions: Zhang, X., et al. (2005) Preparation and characterization of CuO/CeO_2 catalysts and their applications in low-temperature CO oxidation. *Applied Catalysis A: General* 295, 142–149.

   a. What did the HRTEM results (Figure 1) tell the readers about CeO_2-D compared with CeO_2-A, -B, and -C?

   b. What did XRD analysis of CeO_2-D relative to CeO_2-A indicate about their crystallite sizes? How would this result be expected to influence the catalytic activity of the CuO/CeO_2 catalysts? Write the equation used to generate the crystallite sizes and give the definition of the terms in the equation.

   c. What did XRD show about the CuO/CeO_2 catalysts calcined at different temperatures? What does noncrystalline mean and what is its significance in catalysis?

   d. Was there an optimum condition for preparing the most active catalyst? What is the likely rate-limiting step for the catalyst calcined at 800 °C? What did XPS (Figure 9) tell the authors about the oxidation states of the Ce and Cu?

   e. Temperature-programmed reduction (TPR) is used to understand the interaction between the catalytic component and the carrier. (Refer to Figure 7 in the paper.) Explain TPR and how it is performed. What do the peaks at low temperatures (134–140 °C) relative to those at higher temperature (219 °C) mean? Describe how this is important for catalytic activity.

   f. Have the authors performed enough experiments to permit CuO/CeO_2 catalysts to replace Pt or Pd in an automotive application of oxidation of pollutants? What experiments would you suggest to accomplish their goal of a new automobile oxidation catalyst application?

2. Obtain the following paper and answer the related questions: Duran, F.G., et al. (2009) Manganese and iron oxides as combustion catalysts of volatile organic compounds. *Applied Catalysis B: Environmental* 92, 194–201.

   a. Why did the authors combine iron oxides with manganese oxides and what did the authors suggest happened? What characterization technique was used to differentiate the different compounds formed?

b. MnO$_x$ can have many different oxidation states all of which have different structures and catalytic activity. What characterization technique would you use to understand the transformation temperatures and associated energy with conversion from MnO$_2$ to Mn$_2$O$_3$?

c. In Table 1, which catalyst and preparation method gave the best results? (T_{50} corresponds to the temperature required to obtain 50% conversion of the ethanol and ethyl acetate.) What might influence the data reported for $T = 80\%$?

d. What can you conclude from reviewing Tables 1 and 2, where one catalyst material has a higher activity even though its surface area is lower than the other?

e. Examine Figure 1. Why are the peaks broader for FeMn 3:1 (500) compared with FeMn 3:1 (650)?

f. What are your thoughts on comparing normalized activity (or activity per surface area) for catalysts as was reported in Figure 8? (Activity per catalytic site is referred to as a turnover number that reports the activity of the individual site.)

g. What is TPR and how is it measured? What does the presence of Fe do to the TPR of Mn$_2$O$_3$?

3. If you were the highly paid engineer responsible for selecting a catalyst for abatement of undesired ethanol emissions from your plant, what additional data would you want to see from the catalyst company that may not be reported in this paper? What information would you provide to the catalyst company?

4. **a.** You are employed as an environmental engineer and have been given a task to design a stationary oxidation catalytic abatement system for a chemical plant that emits unacceptable amounts of hexane (solvent) and carbon monoxide from an electrical wire coating operation. The coating is designed to provide insulation for the metal wires. What information must you consider before you start the design process? Expect that your colleagues in the plant will be cooperative to your questions and will be needed to participate in the final plan. Consider the most important monolith designs to meet the requirements, especially pressure drop and BMT conversion.

b. What tests would you have performed in the laboratory to assist in first design? It is only necessary to provide a qualitative plan outline?

5. Secure U.S. Patent 5,580,535: Hoke, J.B., Larkin, M.P., Farrauto, R.J., Voss, K.E., Whiteley, R.E., and Quick, L.M. (1996) System and method for abatement of food cooking fumes.

a. How was TGA/DTA used to evaluate candidate catalysts?

b. What is the significance of Table II?

c. Would you suggest a ceramic or metal monolith for the abatement catalyst? Why?

d. Contrast Claims 1–7. How are they different from each other?

e. Contrast Claims 1, 8, and 9.

6. Questions (a), (b), and (c) apply to wide-body aircraft such as Boeing and Airbus jumbo jets. The ozone abatement catalyst (Pd/γ-Al$_2$O$_3$//metal monolith) is installed in the heat and air conditioning air intake to improve the quality of air in the cabin and cockpit. Questions (d) and (e) apply to ozone abatement catalyst deposited on the radiator of ground level vehicles.

a. Why is the catalyst segmented into five or six separate monoliths?

b. Describe the typical duty cycle (duty cycle = its modes of operation during its life) for the airplane ozone abatement catalyst and how it affects its life? What is the best way to determine the duty cycle? What characterization tools would be helpful in determining deactivation mechanisms? Explain why you have selected these methods.

c. Why is a catalyzed metal monolith preferred over a ceramic monolith? Why is a small amount of Pt added to the Pd/Al$_2$O$_3$//monolith catalyst?

d. What is the nature of the flow in car radiators and how does this impact the rate-limiting step for decomposition of ozone? What is the catalyst?

e. What is the main deactivation mechanism for the O$_3$ catalyst on the car and how did catalyst companies design the catalyst to minimize this mode of deactivation? What characterization tools would be helpful in determining all of the deactivation mechanisms? Explain the reasons for your selections.

7. What are the key advantages of catalytic abatement compared with thermal combustion? What are its disadvantages?

8. What is a VOC? What is the most significant issue caused by CO and VOCs entering the atmosphere?

9. A straight channel metal monolith is used in a CO/HC oxidation catalyst in a power plant. The straight channel design is replaced with a herringbone channel design (herring bone = zig zag channels, parallel to flow). The effective channel diameter is the same, as well as the GSA and Residence time. However, the conversion is observed to increase with the new design.

   a. What process is rate controlling?

   b. Why does the conversion increase? Be specific.

10. If one were to explain a plot of NO$_x$ formation versus air/fuel ratio during combustion, you would see NO$_x$ formation to reach a maximum near stoichiometric A/F condition, and NO$_x$ formation to be lower at lean (high A/F) and rich (low A/F) conditions. Why does NO$_x$ increase, reach a maximum, and then decrease?

11. Consider the formation of NO$_x$ in a gas turbine. Why does NO$_x$ formation increase with residence time? Residence time is the amount of time combustion gases spend in the combustion chamber of the turbine system.

BIBLIOGRAPHY

Environmental Oxidation Catalysis

Bonacci, J., Farrauto, R., and Heck, R. (1989) Catalytic incineration of hazardous wastes, in *Encyclopedia of Environmental Control Technology*, Vol. 1 (ed. P. Cherminisoff), Gulf Publishing, pp. 130–178.

Duprez, D. and Cavani, F. (2014) *Advanced Progress in Oxidation Catalysis*, World Scientific Publishers, London, UK.

Farrauto, R., Hobson, M., Kennelly, T., and Waterman, E. (1992) Catalytic chemistry of supported palladium for combustion of methane. *Applied Catalysis A: General* 81, 227–234.

Farrauto, R., Lampert, J., Hobson, M., and Waterman, E. (1995) Thermal decomposition and reformation of PdO catalysts: support effects. *Applied Catalysis B: Environmental* 6, 263.

Forzatti, P. (2000) Environmental catalysis for stationary applications. *Catalysis Today* 62, 51.

Heck, R.M., Farrauto, R.J., and Gulati, S. (2009) Chapters 11, 14, and 15, in *Catalytic Air Pollution Control: Commercial Technology*, 3rd edn, John Wiley & Sons, Inc., Hoboken, NJ.

McCarty, J. (1995) Kinetics of PdO combustion catalysts. *Catalysis Today* 26, 283.

Spivey, J. (1987) Complete catalytic oxidation of volatile organics. *Industrial and Engineering Research* 26, 2165–2180.

Spivey, J. and Butt, J. (1992) Literature review: deactivation of catalysts in the oxidation of volatile organic compounds. *Catalysis Today* 11, 465–500.

Selective Catalytic Reduction of NO$_x$

Bosch, H. and Janssen, F. (1988) Catalytic reduction of nitric oxides: a review of the fundamentals and technology. *Catalysis Today* 2, 369–521.

Cohn, G., Steele, D., and Andersen, H. (1961) *Nitric acid tail gas abatement*. U.S. Patent 2,975,025.

Heck, R.M., Farrauto, R.J., and Gulati, S. (2009) Chapter 12, in *Catalytic Air Pollution Control: Commercial Technology*, 3rd edn, John Wiley & Sons, Inc., Hoboken, NJ.

Ozone Abatement

Carr, W. and Chen, J. (1982) *Ozone abatement catalysts having improved durability and low temperature performance*. U.S. Patent 4,343,776.

Heck, R.M., Farrauto, R.J., and Gulati, S. (2009) Chapters 10 and 16, in *Catalytic Air Pollution Control: Commercial Technology*, 3rd edn, John Wiley & Sons, Inc., Hoboken, NJ.

Heck, R.M., Farrauto, R.J., and Lee, H. (1992) Commercial development and experiences with catalytic ozone abatement in jet aircraft. *Catalysis Today* 13, 43–85.

Food Processing Catalysts

Hoke, J., Larkin, M., Farrauto, R, Voss, K., Whitely, R., and Quick, M. (1998) *System and method for abating food cooking fumes*. U.S. Patent 5,756,053.

CHAPTER 13

CATALYTIC ABATEMENT OF GASOLINE ENGINE EMISSIONS

13.1 EMISSIONS AND REGULATIONS

13.1.1 Origins of Emissions

The development of the four-cycle spark-ignited combustion engine permitted the controlled combustion of gasoline that provides the power to operate the automobile. Gasoline, which contains a mixture of paraffins and aromatic hydrocarbons, is combusted with controlled amounts of air producing complete combustion products of CO_2 and H_2O. Equation 13.1 uses isooctane as an ideal and simple model for the hydrocarbons combusted in the gasoline IC engine during the power stroke:

$$C_8H_{18} + 25/2O_2 \rightarrow 8CO_2 + 9H_2O + \text{Heat} \qquad (13.1)$$

No combustion is 100% complete due to nonhomogeneous mixing and, therefore, there are undesired combustion products of CO and unburned hydrocarbons (UHC). The CO levels range from 1 to 2 vol%, while the unburned hydrocarbons are from 500 to 1000 vppm. During the combustion process, very high temperatures are reached due to diffusion burning of the gasoline droplets, resulting in thermal fixation of the nitrogen in the air to form NO_x, which is a combination of NO, NO_2, and N_2O. Levels of NO_x are in the 100–3000 vppm range. The quantity of pollutants varies with many of the operating conditions of the engine, but is influenced predominantly by the air/fuel ratio in the combustion cylinder. Figure 13.1 shows the engine emissions from a spark-ignited gasoline engine as a function of the air/fuel ratio.

The term used to describe the air/fuel (A/F) weight ratio is λ, defined as the actual air/fuel ratio divided by the air/fuel ratio at the stoichiometric point or $\lambda = (A/F)_{\text{actual}}/(A/F)_{\text{stoichiometric}}$. The stoichiometric point $\lambda = 1$ is the precise amount of air (O_2) required to oxidize all of the fuel, which for gasoline is approximately 14.6 (wt/wt). For an A/F that is rich (insufficient air), $\lambda < 1$ and for an A/F lean mixture (excess air), $\lambda > 1$. This will become a critical control value for three-way catalysts (TWCs) to be discussed later in the chapter.

When the engine is operated rich of stoichiometric, the CO and HC emissions are highest while the NO_x emissions are depressed. This is because complete burning of the

Introduction to Catalysis and Industrial Catalytic Processes, First Edition. Robert J. Farrauto, Lucas Dorazio, and C.H. Bartholomew.
© 2016 John Wiley & Sons, Inc. Published 2016 by John Wiley & Sons, Inc.

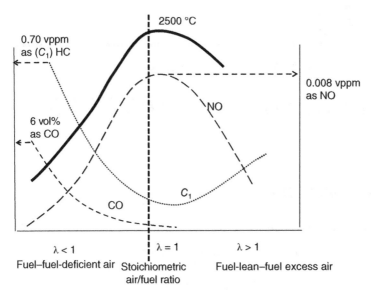

Figure 13.1   Gasoline-relative engine emissions and temperature as a function of air/fuel ratio.

gasoline is prevented by the deficiency of O_2. The level of NO_x is reduced because the adiabatic flame temperature is reduced and the available unreacted CO and HC can undergo reactions generating H_2 allowing the reduction of NO to N_2. On the lean side of stoichiometric, the CO and HC are reduced since nearly complete combustion dominates. Again, the NO_x is reduced since the operating temperature is decreased. Just lean of stoichiometric operation, the NO_x is maximum, since the adiabatic flame temperature is the highest. At stoichiometric, the adiabatic flame temperature is lowered slightly because of the heat of vaporization of the liquid fuel gasoline. The actual operating region of combustion for the spark-ignited engine is defined by the lean and rich flame stability, beyond which the combustion is too unstable.

Within the region of operation of the spark-ignited engine, a significant amount of CO, HC, and NO_x is emitted to the atmosphere. The consequences of these emissions have been well documented, but, briefly, CO is a direct poison to humans (reacts with hemoglobin in blood), while HC and NO_x undergo photochemical reactions in the sunlight leading to the generation of smog and ozone. NO_2 (the thermodynamically stable form of NO_x at temperatures below about 400 °C) is also a component of acid rain (HNO_3) poisoning rivers and stream.

13.1.2   Regulations in the United States

The necessity to control automobile emissions in the United States came in 1970 when the U.S. Congress passed the Clean Air Act. The requirements under the Clean Air Act were changing as the technology was being evaluated. As a point of reference, the 1975–1976 Federal (49 states) requirements were 1.5 g/mile HC, 15.0 g/mile CO, and 3.1 g/mile NO_x. The Environmental Protection Agency (EPA) established a Federal

Test Procedure (FTP) simulating the average driving conditions in the United States in which CO, HC, and NO_x would be measured. The FTP cycle was conducted on a vehicle dynamometer and included measurements from the automobile during three conditions: (i) cold start, after the engine was idle for 8 h, (ii) hot start, and (iii) a combination of urban and highway driving conditions. Separate bags would collect the emissions from all three modes, and a weighing factor applied for calculating the total emissions. Complete details on the FTP are available. Typical precontrolled vehicle emissions in the total FTP cycle were 83–90 g/mile of CO, 13–16 g/mile of HC, and 3.5–7.0 g/mile of NO_x. A number of changes in engine design and control technology were implemented to lower the engine out emissions; however, the catalyst was still required to obtain greater than 90% conversion of CO and HC by 1976 and to maintain performance for 50,000 miles.

Amendments in the early 1990s to the Clean Air Act have set up more stringent requirements for automotive emissions. It should be mentioned that methane was excluded from hydrocarbon regulations (non-methane hydrocarbons (NMHCs)) since it does not participate in photochemical smog-generating reactions. This will change in the future as its greenhouse gas impact will be regulated. Today, the catalyst is warranted for 150,000 miles in the United States. This figure demonstrates how the technology has improved since the first converters were introduced.

As the automobile engine became more sophisticated, the control devices and combustion modifications have proven to be very compatible with catalyst technology, to the point where today engineering design incorporates the emission control unit and strategy for each vehicle.

In 2010, 800 million passenger cars in use worldwide are equipped with catalytic converters. Annual worldwide production of new cars is exceeding 100 million. In addition, there are about 40% more passenger vehicles represented by trucks. The majority of these vehicles (automobiles and trucks) use a spark-ignited gasoline engine to provide power and this has become the most frequent form of transportation. Gasoline blends still remains a mixture of paraffins and aromatic hydrocarbons that combust in air at a very high efficiency. But other changes to fuel quality have contributed to improved catalyst durability as discussed below.

Oxygenates, mostly alcohols derived from crops such as corn or sugarcane, are now however added to the fuel to decrease fossil fuel usage. Ethanol has a high octane rating and thus replaces some aromatics. Octane is a measure of the ability of the fuel to resist precombustion during the compression stroke in the spark-ignited engine. Various blends of ethanol (10–20%) are currently in use in the United States and mixtures of 85% ethanol and 15% gasoline (E85) are now available to further decrease the use of fossil fuel-derived gasoline. Furthermore, the decrease in sulfur content in the gasoline has significantly improved catalyst durability. The removal of traces of lead compounds (originally added to gasoline to increase octane) has also lengthened catalyst life. Emissions from alternative fuel vehicles and the effectiveness of catalysts are currently being investigated.

Furthermore, advancements in engine strategies will also impact the emissions and put every increasing demand on the catalysts and control strategies. It is clear that automotive emission control is a dynamic technology to be researched well in the future as all electric and fuel cell vehicles very slowly replace the internal combustion engine.

13.1.3 The Federal Test Procedure for the United States

The standard performance tests for three-way catalysts are based on the 1975 Federal Test Procedure. This procedure was also used for the first oxidation catalysts.

The 1975 FTP uses a vehicle on a chassis dynamometer. The test procedure measures emission from a driving cycle through Los Angeles, California and has the following characteristics:

• Cycle length:	11.115 miles
• Cycle duration:	1877 s plus 600 s pause
Bag 1	0–505 s
Bag 2	505–1370 s
Hot soak	600 s
Bag 3	0–505 s
• Average speed:	34.1 km/h
• Maximum speed:	91.2 km/h
• Number of hills	23
• Number of modes	112

Emissions are measured using a constant volume sampling system. The test begins with a cold start phase 1 or bag 1 (at 20–30 °C) after a minimum 12 h soak at constant ambient temperature. After 505 s, the vehicle is driven at the speeds indicated in the phase 2 or bag 2 hot stabilized portion. The vehicle is then run idle for 600 s, after which the phase 3 cycle or bag 3 (hot start) is implemented. This phase is identical to the speeds and accelerations indicated in phase 1. Additions to the test protocol have been further updated to include high-speed and aggressive driving.

13.2 CATALYTIC REACTIONS OCCURRING DURING CATALYTIC ABATEMENT

The basic operation of the catalyst is to perform the following reactions in the exhaust of the automobile:

Oxidation of CO and C_yH_n (hydrocarbon) to CO_2 and H_2O:

$$C_yH_n + (y + n/4)O_2 \rightarrow yCO_2 + n/2H_2O \tag{13.2}$$

$$CO + 1/2O_2 \rightarrow CO_2 \tag{13.3}$$

$$CO + H_2O \rightarrow CO_2 + H_2 \tag{13.4}$$

Equation 13.4 occurs mainly when the engine is operated rich $\lambda < 1$.

Reduction of NO/NO_2 to N_2 under fuel-rich operating conditions:

$$NO(\text{or } NO_2) + CO \rightarrow 1/2N_2 + CO_2 \tag{13.5}$$

$$NO(\text{or } NO_2) + H_2 \rightarrow 1/2N_2 + H_2O \tag{13.6}$$

$$C_yH_n + nNO \rightarrow (n/2)N_2 + yCO_2 + n/2H_2O \tag{13.7}$$

Hydrogen may also be present in the exhaust and this reacts very rapidly essentially at room temperature and is not usually depicted in the exhaust emissions.

13.3 FIRST-GENERATION CONVERTERS: OXIDATION CATALYST

During the early implementation of the Clean Air Act, the catalyst was required to abate only CO and HC. The NO_x standard was relaxed, so engine manufacturers used exhaust gas recycle (EGR) to meet the NO_x standards. With EGR, a small portion of the exhaust, rich in high heat capacity containing H_2O, CO_2, and N_2, is recycled into the combustion chamber, thereby lowering the combustion flame temperature, which results in less thermal NO_x formation. The engine was operated just rich of stoichiometric to further reduce the formation of NO_x and secondary air was pumped into the exhaust gas to provide sufficient O_2 for the catalytic oxidation of CO and HC on the catalyst.

During this period, many catalytic materials were studied and the area of high-temperature stabilization of alumina was explored. It was known that the precious metals, Pt and Pd, were excellent oxidation catalysts; however, the cost and supply of these materials were bothersome. Therefore, many base metal candidates were investigated, such as Cu, Cr, Ni, and Mn. They were less active than the precious metals, but substantially cheaper and more readily available. However, studies in the earlier to mid-1970s showed that the base metal oxides were very susceptible to sulfur poisoning with thermal stability and were, therefore, eliminated from consideration. Precious metals were therefore the catalysts of choice, which is still the case and to date no base metal oxides are used as primary catalytic components for gasoline vehicle emission control. Some historical data are shown in Equation 13.4. This situation is somewhat different for modern diesel fuels since they now contain low sulfur and the exhaust experiences much lower temperatures, as will be discussed in Chapter 14.

Therefore, the first-generation oxidation catalysts were a combination of Pt and Pd and operated in the temperature range of 250–600 °C, with space velocities varying during vehicle operation from 10,000 to 100,000 1/h, depending on the engine size and mode of the driving cycle (i.e., idle, cruise, or acceleration). Typical oxidation catalyst compositions were Pt and Pd in a 2.5:1 or 5:1 ratio ranging from 0.05 to 0.1 troy oz/car (a troy oz is about 31 g). The catalytic converter was positioned in the exhaust piping physically mounted in a metal canister and engineered for the proper

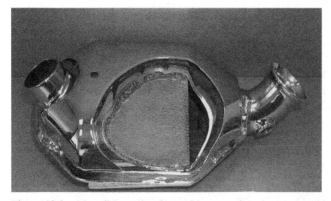

Figure 13.2   Monolith catalyst housed in a metal canister secured in the exhaust. (Reprinted with permission from BASF.)

flow dynamics. The catalyst was secured to avoid mechanical failure and insulated to retain the heat. Figure 13.2 shows a typical monolith catalyst in the canister that is welded into the exhaust.

13.4 THE FAILURE OF NONPRECIOUS METALS: A SUMMARY OF CATALYST HISTORY

Even today there is a strong imperative to utilize transition metals (often referred to as base metals) and metal oxides as the catalysts for gasoline emission control given their availability and low price relative to the precious metals. In particular, the oxides of Cu and Cr were good candidates for catalyzing the oxidation of both CO and HC. In laboratory experiments using propane and propylene as reasonable model compounds for exhaust HCs, a mixture of CuO and $CuCr_2O_4$ had very good initial activity, but once exposed to the anticipated maximum temperature expected in the exhaust (i.e., 800 °C), the structure converted to $Cu_2Cr_2O_4$ with lower activity (as measured by an increase in T_{50}, the temperature necessary for 50% conversion). The decrease in activity was accompanied by a change in physical characteristics such as a reduction in BET surface area and growth in XRD crystal size due to sintering.

$$\underset{\text{(active)}}{CuO} + CuCr_2O_4 \xrightarrow{800\ °C/\text{air}} \underset{\text{(inactive)}}{Cu_2Cr_2O_4} + 1/2O_2 \qquad (13.8)$$

In order to disperse and enhance the number of catalytic sites available to the reactants and to stabilize the catalytic components against sintering, a high surface area $\gamma\text{-}Al_2O_3$ material was introduced as a carrier for the Cu and Cr oxides. $\gamma\text{-}Al_2O_3$ is composed of pores in the 2–50 nm range, which are large enough for the gaseous reactants to penetrate and undergo surface reaction with the catalytic components dispersed on its internal surface. Although initial activity of the supported Cu/Cr catalyst was good, thermal treatment resulted in loss of activity due to the formation of a $CuAl_2O_4$, which had considerably less activity than the fresh catalyst.

$$\underset{\text{high activity}}{CuO/\gamma\text{-}Al_2O_3} \xrightarrow{800\ °C/\text{air}} \underset{\text{low activity}}{CuAl_2O_4} \qquad (13.9)$$

In addition to thermal treatments, catalysts were also exposed to oxides of sulfur (e.g., SO_2) since gasoline contained up to 0.1% (1000 ppm) of sulfur in the mid-1970s. Typically, catalyst performance declined after exposure to sulfur oxides due to formation of inactive sulfates. Equation 13.10 shows the formation of catalytically inactive $CuSO_4$ when CuO is exposed to SO_2/SO_3. Cobalt-containing catalysts also suffered from the same type of chemical reaction (Equation 13.8).

$$\underset{\text{active}}{CuO} + SO_2/SO_3 \xrightarrow{500\ °C/\text{air}} \underset{\text{inactive}}{CuSO_4} \qquad (13.10)$$

$$\underset{\text{active}}{Co_2O_3} + SO_2/SO_3 \xrightarrow{500\ °C/\text{air}} \underset{\text{inactive}}{Co_2(SO_4)_3} \qquad (13.11)$$

In contrast, $CuCr_2O_4$ had been shown to adsorb SO_2/SO_3 forming a chemisorbed layer rather than a compound of sulfate. Although this reduces the catalytic activity, it allows regeneration by a simple water wash, which is not practical for a commercial vehicle.

$$CuCr_2O_4 + SO_2/SO_3 \xrightarrow{\text{500 °C/air}} CuCr_2O_4 \cdots SO_2/SO_3 \qquad (13.12)$$
$$\underset{\text{high activity}}{} \qquad \underset{\text{poor activity}}{}$$

It was clear in 1973 that the lack of thermal stability and the vulnerability to oxides of sulfur eliminated base metals from further consideration. In contrast, precious metals deposited on carriers were found to function after such treatments and were the only viable catalyst materials that could be used. Studies attempting to replace precious metals continue, but none have been commercially successful.

13.4.1 Deactivation and Stabilization of Precious Metal Oxidation Catalysts

The precious metal oxidation catalyst was negatively affected by the exhaust impurities of sulfur oxides and tetraethyl lead from the octane booster, both present in the gasoline, and phosphorus and zinc from engine lubricating oil. As the research was ongoing for improved catalyst compositions, Pb present as an octane booster continued to deactivate most severely all the catalytic materials. Poisoning of Pt and Pd by the traces of Pb (about 3–4 mg/gallon of Pb were in "unleaded" gasoline) was caused by formation of a low-activity alloy:

$$\text{Pt or Pd} + \text{Pb} \xrightarrow{\text{air,900 °C}} \text{PtPb or PdPb} \qquad (13.13)$$

The impact of lead poisoning on all catalysts (and humans) contributed to the decision that led the Federal Government to mandate its removal from gasoline. This proved a benefit for autoemission control. The use of catalysts was now more feasible in meeting the 50,000-mile performance requirements in 1975. It is interesting to note that removal of lead from the gasoline was still the first step in implementing automobile emission control technologies in the emerging countries such as India and China.

However, the operating environment of the catalyst was still hostile in that phosphorus (P) and zinc (Zn) derived from lubricating oil and sulfur (S) from the fuel itself were present, as well as the severe temperature transients and possible temperature exposure of 800–1000 °C maximum. Desorption of sulfur compounds, adsorbed on the catalytic sites, occurs at >700 °C, but any residual sulfur decreases catalyst performance. The P and Zn in lubricating oil form a polymeric film on the Al_2O_3 masking the sites, but it too is partially reversed at higher temperatures where the polymer decomposes.

The combinations of Pt and Pd dispersed onto high surface area γ-Al_2O_3 particles were found to have reasonably good fresh activity. After high-temperature aging (900 °C in air/steam to simulate engine exhaust conditions), the catalyst usually lost some of its activity, as evidenced by increased temperatures for 50% conversion of both CO and HC. Characterization of the partially deactivated catalyst by BET surface area measurements and X-ray diffraction patterns showed that the γ-Al_2O_3 had undergone severe sintering to a lower surface area, more crystalline phase such as α-Al_2O_3 (see Chapter 5). The high-area pore structure of the γ-Al_2O_3 effectively collapses and occludes the active catalytic species, making them inaccessible to the reactants (see Chapter 5). Naturally, this results in a loss of catalytic performance. Since no other carrier materials had all the desirable properties of γ-Al_2O_3, research was directed toward understanding and minimizing the sintering mechanisms of γ-Al_2O_3 under autoexhaust conditions. It was known that certain contaminants such as Na and K acted as fluxes, accelerating the sintering process of γ-Al_2O_3. Thus, preparations had to exclude these elements. In contrast, small amount (1–3%) of different La_2O_3, SiO_2 was conducted at temperatures as high as 1200 °C with and without stabilizers. A typical fresh surface area of γ-A_2O_3 is around 125–150 m^2/g.

Although the precise mechanism for their stabilizing effect is not known, high-resolution surface studies have indicated that these oxides enter into the surface structure of the γ-Al_2O_3 and greatly diminish the rate of the chemical and physical changes occurring normally during sintering. Further studies have shown that the elements primarily act on free (grain) surfaces, rather than in the bulk volume obtained using first-principles atomistic calculations.

The development of thermally stable high surface area γ-Al_2O_3 by the incorporation of oxides was a breakthrough in materials technology, and its use uncovered another problem: agglomeration or sintering of the Pt and Pd during high-temperature exposure in the automobile exhaust. Hydrogen chemisorption and XRD studies revealed that the Pt and Pd, initially well dispersed on stabilized γ-Al_2O_3, had undergone significant crystallization after high-temperature treatment. It was clear that Pd was more stable than the Pt in the high-temperature environment of the first oxidation catalysts.

The first-generation oxidation catalysts were comprised of 2.5:1 weight ratio of Pt to Pd at about 0.05% total precious metal for beads and about 0.12% for honeycombs. Stabilizers such as CeO_2 and La_2O_3 were also included in the formulation to minimize sintering of the γ-Al_2O_3 carrier.

13.5 SUPPORTING THE CATALYST IN THE EXHAUST

To support the catalyst converter in the exhaust, there were two approaches considered: the monolith structure that proved to be the most successful and the particulate, or bead bed, that failed after several years of poor mechanical durability on the road.

13.5.1 Ceramic Monoliths

Monoliths themselves have very little surface area, but have low-porosity walls upon which a catalyst could be deposited. It is a unitary structure composed of inorganic oxides or metals in the structure of a honeycomb with equally sized and parallel channels that may be square, sinusoidal, triangular, hexagonal, round, and so on.

Monoliths are available as ceramic and metal with different channel dimensions and shapes.

Commercial ceramic monoliths (made by extrusion from synthetic cordierite, $2MgO \cdot 2Al_2O_3 \cdot 5SiO_2$) have large pores and low surface areas (i.e., $0.3 \, m^2/g$), so it is necessary to deposit a high surface area carrier containing the catalytic components onto the channel walls. The catalyzed coating is composed of a high surface carrier such as Al_2O_3 impregnated with catalytic components. This is referred to as the *catalyzed washcoat*. Figure 13.3 shows various magnifications of a 400 cells per square inch (cpsi) ceramic monolith with a double washcoat commonly used in automotive applications.

The washcoat can be seen deposited over the entire wall, but it is concentrated at the corners of the square-channel ceramic monolith. The thickness of the "fillet" depends primarily on the geometry of the channel and the coating method. The pollutant-containing gases enter the channels uniformly and diffuse to and through the washcoat pore structure to the catalytic sites where they are converted catalytically. The amount of geometric surface area (GSA), upon which the washcoat is deposited, is determined by the number and diameter of the channels. There is a limit as to how much washcoat can be deposited, since extra deposition results in a decrease of the effective channel diameter, thereby increasing the pressure drop to an unacceptable level.

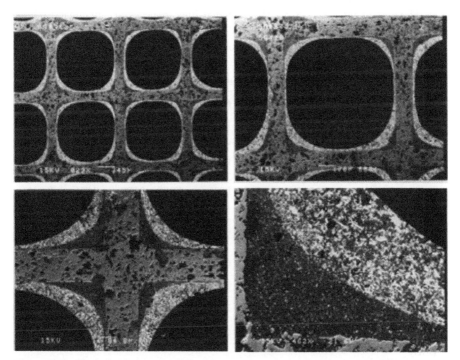

Figure 13.3   Optical micrographs of double-layered washcoated ceramic monoliths. (Reproduced from Chapter 2 of Heck, R.M., Farrauto, R.J., and Gulati, S.T. (2009) *Catalytic Air Pollution Control: Commercial Technology*, 3rd edn, John Wiley & Sons, Inc., New York.)

Monoliths offer a number of engineering design advantages that have led to their widespread use in environmental applications. However, one of the most important is low-pressure drop associated with high flow rates. The monolith has a large open frontal area and with straight parallel channels offers less resistance to flow than a pellet-type catalyst. Low-pressure drop translates to lower compressor costs for stationary applications and greater power saving for mobile sources. Other advantages are excellent attrition resistance, good mechanical and thermal shock properties, ability to make compact reactors, freedom in reactor orientation, and so on. The cordierite material and its engineered shape and structure are designed for low expansion (coefficient $10 \times 10^{-7}/°C$) and resistance to thermal shock leading to cracking. This is critical for automobiles since temperatures in the exhaust can vary significantly as the vehicle accelerates or decelerates during normal driving.

Synthetic cordierite, $2MgO \cdot 2Al_2O_3 \cdot 5SiO_2$, is by far the most commonly used ceramic for monolithic catalyst support applications. The raw materials such as kaolin, talc, alumina, aluminum hydroxide, and silica are blended into a paste and extruded and calcined. Sizes up to about 11 in. (27.94 cm.) in diameter and 7 in. (17.78 cm.) long, with cell densities from about 9 to 1200 cpsi can be made. The required conversion, the physical space available for the reactor, and engineering constraints, such as pressure drop, are considered when designing the monolith size. Some physical properties of selective ceramic monolith structures are given in Table 13.1. Cells per square inch is the first number (i.e., 400/6.5) and 6.5 is the wall thickness in mils (0.0065 in.). High cell densities are necessary for enhanced bulk mass transfer since the GSA is increased favoring this mode of rate control. The higher open frontal area (OFA) is critical for lower pressure drop. Typically, up to 600

TABLE 13.1 Nominal Properties of Standard and Thin-Wall Cordierite Substrates

Ceramic cell density (cell/in.^{2})	400/6.5	470/5	600/3.5	600/4	600/4.3	900/2.5	1200/2.5
Substrate diameter (mm)	105.7	105.7	105.7	105.7	105.7	105.7	105.7
Substrate length (mm)	98	88	76	76	76	76	35
Substrate volume (l)	0.86	0.77	0.67	0.67	0.67	0.67	0.31
Material porosity (%)	35	24	35	35	35	35	35
OFA (open frontal area %)	75.7	0.795	0836	0.814	0.800	0.856	0.834
GSA (geometric surface area) (m^2/l)	2.74	3.04	3.53	3.48	3.45	4.37	4.98
TSA (total wall surface area) (m^2)	2.35	2.35	2.35	2.32	2.30	2.91	1.53
Hydraulic diameter (mm)	1.10	1.04	0.95	0.94	0.93	0.78	0.67
Flow resistance (1/cm^2)	3074	3274	3780	3990	4122	5412	7589
Bulk density (g/l)	395	390	267	303	324	235	269
Heat capacity @200°C (J/(K l))	352	348	238	270	289	209	240
Heat capacity @200°C (J/K)	302	269	159	180	193	140	74
Substrate mass (g)	339	301	178	202	216	156	83

Reproduced from Chapter 7 of Heck, R.M., Farrauto, R.J., and Gulati, S.T. (2009) *Catalytic Air Pollution Control: Commercial Technology*, 3rd edn, John Wiley & Sons, Inc., New York.

cpsi with wall thicknesses of 4 mil are used in gasoline vehicles providing sufficiently low-pressure drop with sufficient bulk mass transfer conversions.

An increase in cell density from 100 to 300 cpsi significantly increases the geometric area from 398 to 660 ft^2/ft^3 (157–260 cm^2/cm^3), but decreases the channel diameter from 0.083 to 0.046 in. (0.21–0.12 cm.). The wall of the ceramic drops in thickness from 0.017 to 0.012 in. (0.04–0.03 cm.). The increase in cell density does cause an increase in pressure drop at a given flow rate. For example, a flow rate of 300 standard cubic feet per minute (SCFM) (8.4 × 10$_6$ cm^3/min) through a monolith of 1 ft^2 (929 cm^2) by 1 in. thick (2.54 cm), the pressure drop for a 100 cpsi is about 0.1 in. (0.254 cm) of water compared to about 0.3 in. for a 300 cpsi monolith. It should be understood that the pressure drop values are for the monolith without the catalyzed washcoat. Applying a washcoat will increase the pressure drop as a function of its thickness and surface roughness.

13.5.2 Metal Monoliths

Monoliths made of high temperature resistant aluminum-containing steels are becoming increasingly popular as catalyst supports, mainly because they can be prepared with thinner walls than a ceramic (Table 13.2). This offers the potential for higher cell densities with lower pressure drop. The wall thickness of a 400 cpsi metal substrate used for automotive applications is about 25% lower than its ceramic counterpart, that is, 0.0015–0.002 in. (0.004–0.005 cm) compared to 0.006–0.008 in. (0.015–0.02 cm), respectively. The open frontal area of the metal is typically about 90 versus 70% for the ceramic with the same cell density. Its thermal conductivity is also considerably higher (about 15–20 times) than the ceramic, resulting in faster heat up. This property is particularly important for oxidizing hydrocarbons and carbon monoxide emissions when a vehicle is cold. Metals substrates also offer some advantages for installation of the converter in that they can be directly welded into the exhaust system.

TABLE 13.2 Nominal Properties of Standard and Thin-Wall Metallic Substrates

Metal cell density (cell/in.^{2})	400/2	500/1.5	500/2	600/1.5	600/2
Substrate diameter (mm)	105.7	105.7	105.7	105.7	105.7
Substrate length (mm)	68	114	114	114	114
Substrate volume (l)	0.60	1.00	1.00	1.00	1.00
OFA (open frontal area) (%)	0.089	0.900	0.880	0.890	0.870
GSA (geometric surface area) (m^2/l)	3.65	4.05	4.00	4.20	4.15
TSA (total wall surface area) (m^2)	2.18	4.05	4.00	4.20	4.15
Hydraulic diameter (mm)	0.98	0.89	0.88	0.85	0.84
Flow resistance (1/cm^2)	2646	3150	3287	3503	3660
Bulk density (g/l)	792	720	864	792	936
Heat capacity @200 °C (J/(K l))	408	371	445	408	482
Heat capacity @200 °C (J/K)	243	371	445	408	482
Substrate mass (g)	473	720	864	792	936

Reproduced from Chapter 7 of Heck, R.M., Farrauto, R.J., and Gulati, S.T. (2009) *Catalytic Air Pollution Control: Commercial Technology*, 3rd edn, John Wiley & Sons, Inc., New York.

A common design is that of corrugated sheets of metal welded or wrapped together into a monolithic structure. In most cases, the washcoat is deposited onto a roughened metallic surface of the already fabricated monolith by a dipping or controlled vacuum deposition. There are some new designs in which the washcoat is first deposited onto a flat, roughened surface prior to final wrapping into the monolith shape.

Adhesion of the oxide-based washcoat to the metallic surface and corrosion of the steel in high-temperature steam environments were early problems that prevented their widespread use in all but some specialized automotive applications. Surface pretreatment (air oxidation at 800 °C converts the Al in the alloy to an Al_2O_3 surface) of the metal has improved the adherence problems, and new corrosion-resistant steels are allowing metals (Fe, Cr, Al alloys referred to as FECRALLOY) to penetrate the automotive markets. They are also used in the close-coupled position for the cold start portion of the FTP in the TWC catalyst exhaust system (to be discussed later) due to their high heat conductivity to ensure rapid heat up. They are currently used extensively for low- temperature applications such as NO_x in power plants, O_3 abatement in airplanes, CO and VOC abatement, and abatement of oil-based emissions from restaurants. Being electrically conductive, they have found use in electrically heated catalytic converters for rapid conversion of emissions during start-up. They are finding greater use in high-performance vehicles where response time is critical during acceleration. Here, the low-pressure drop of the metal monolith is its most desirable property. They are usually more expensive than their ceramic counterpart.

13.6 PREPARING THE MONOLITH CATALYST

The catalyzed carrier (Al_2O_3) is made into an acidified aqueous slurry with a solid content of 30–50%. The mixture is ball milled for at least 2 h to reduce the particle size (typically 5–20 μm) and generate the proper rheology for the subsequent monolith dipping operation.

The preparation of the finished catalyst (mainly for small-scale production) is made by dipping the monolith into the slurry. The monolith generally has some wall porosity or surface roughness to ensure adhesion of the catalyzed washcoat. The excess slurry is air blown to clear the channels and dried at about 110 °C. The final step is calcination, which bonds the catalyzed washcoat securely to the monolith walls and decomposes and volatilizes the excess preparation components. Calcinations are performed in air at temperatures between 300 and 500 °C. Great care must be taken to avoid rapid heat up since H_2O trapped in the micropores can build up sufficient pressure to crack the monolith. Furthermore, exothermic reactions due to decomposing salts can cause localized high temperatures within the catalyst material that can accelerate sintering.

An alternative approach is to first coat the monolithic honeycomb with the uncatalyzed carrier, followed by drying and calcining. It is then dipped into a solution containing the catalytic salts. This method relies on the electrostatic adsorption of the salts on the carrier surface. The supported catalyst is then dried and calcined to its final state.

One manufacturing method uses a vacuum on one end of the monolith to suck the slurry up to a given axial length. It is then turned 180 °C where the uncoated portion of the monolith is similarly coated. This technology allows zone coating where the front of the bed may contain a different formulation than the back.

Some manufacturers that use metal substrates precoat them with the washcoat prior to wrapping or forming the metal into the monolithic structure. It is common to pretreat (air oxidize) the surface of the metal to generate roughness as discussed above. This ensures good bonding to the washcoat. The metal is then calcined to produce a stable bond between the surface and the washcoat. The major advantage is coating uniformity, with no corners containing high localized amounts of washcoat. Automotive ceramic monoliths have well-designed pore structures (of about 3–4 µm) that allow good chemical and mechanical bonding to the washcoat. The chemical components in the ceramic are immobilized, so little migration from the monolith into the catalyzed washcoat occurs.

13.7 RATE CONTROL REGIMES IN AUTOMOTIVE CATALYSTS

At the entrance of the monolith reactor, during a cold start, the reaction is controlled by chemical kinetics rather than by diffusion to or within the catalyst pores structure. For example, in the automobile catalytic converter, the incoming gases are relatively cold when the vehicle has been dormant for an extended period of time, so the reaction rate is governed by the chemical kinetics. In addition to the sensible heat from the gas stream, the catalyst surface heats up due to the heat generated by the oxidation reactions. When sufficiently hot, the rate will be determined by pore diffusion followed by bulk mass transfer. Knowledge of all the rate-controlling steps throughout the entire catalyst duty cycle is essential in designing the catalyst and the reactor to meet stringent environmental regulations.

One important application of the Langmuir–Hinshelwood kinetic reaction model discussed in Chapter 1 can be illustrated in understanding the role of O_2 and CO in the exhaust. The general model predicts kinetic inhibition by CO and O_2 as shown in Equation 13.14:

$$[Rate]_{CO} = kK_{CO}P_{CO} K_{O_2}^{1/2}P_{O_2}^{1/2}/(1 + K_{CO}P_{CO} + K_{O_2}^{1/2} P_{O_2}^{1/2})^2 \qquad (13.14)$$

One approach that was considered in the early days to manage NO_x was to operate the engine rich to decrease its formation. A NO_x reduction catalyst would further reduce it via reactions with high levels of CO. It would then be necessary to add air post NO_x catalyst to oxidize the high levels of CO. In this case, the general model for CO oxidation can be reduced to Equation 13.15 where CO inhibits the oxidation reaction. Thus, by adding air CO is diluted, while increasing the O_2 concentration drives the reaction rate forward:

$$[Rate]_{CO} = kK_{O_2}^{1/2}P_{O_2}^{1/2}/(P_{CO}K_{CO}) \qquad (13.15)$$

13.8 CATALYZED MONOLITH NOMENCLATURE

The common nomenclature is to state the washcoat loading in grams per cubic inches (g/in^3) and in grams of catalytic component (especially for precious metals) per cubic foot (g/ft^3) of monolith. The monolith volume is calculated based on its cross-sectional area and length. It does not take into account the number of channels per square inch. Consequently, the cell density is always stated when describing the finished catalyst dimensions. This is also very important when a space velocity is stated. This definition lacks kinetic rigor; however, it is the nomenclature used since monolith volume reflects available space in the exhaust, which is a premium. True kinetics would have to be calculated based on the total weight of washcoat contained on the walls and/or the amount of precious metal deposited in the monolith.

13.9 PRECIOUS METAL RECOVERY FROM CATALYTIC CONVERTERS

As already stated, automobile catalysts contain expensive precious metals that can be recovered and reused after the catalyst has lost its effectiveness. The hydro-metallurgical procedure simply involves crushing the spent catalyst and treating it with acid to dissolve only the ceramic (Al_2O_3 and monolith) components. This leaves an insoluble precious metal-rich residue that is then further purified by chemical procedures. For catalyzed metal monoliths, the entire structure is dissolved and the precious metals precipitated from the solution.

The pyrometallurgical method utilizes smelting in which the melted oxides of the ceramic and Al_2O_3 float to the top as slag. The highly dense precious metals alloy with an added metal (i.e., Cu or Fe) at the bottom of the smelter. Here they are chemically removed and further purified for reuse.

13.10 MONITORING CATALYTIC ACTIVITY IN A MONOLITH

It is necessary to know the composition of reactants entering the reactor and the products being produced at its exit. What happens inside the monolith is also critical because catalysts, like many products, are designed to expect some loss of performance with use. So even if no apparent loss in performance is observed at the outlet, it is likely that some deactivation occurs along the length and performance is compensated by catalyst downstream from the major reaction front. This has been best illustrated in Figure 13.4, showing the product generated at front moving down the bed as deactivation proceeds from the inlet to the outlet.

The generic illustration is for any catalytic reaction where time is in arbitrary unit with no real significance other than to demonstrate the progress of deactivation. At Time = 0, maximum product formation occurs near the inlet of the axial length.

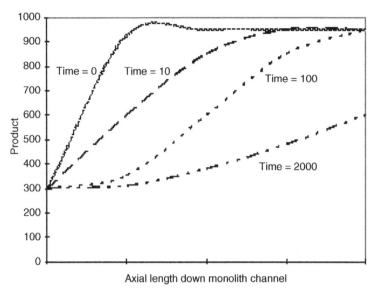

Figure 13.4   Conversion proceeding axially down the channel of a monolith with poisoning. Units of time are arbitrary units. (Reproduced from Chapter 5 of Heck, R.M., Farrauto, R.J., and Gulati, S.T. (2009) Catalytic *Air Pollution Control: Commercial Technology*, 3rd edn, John Wiley & Sons, Inc., New York.)

If the catalyst in the front of the bed is subjected to selective poisoning, the reaction front will begin to move downstream. The rate at which the profile moves downstream increases as the concentration of poisons increases. As poisoning continues, the reaction profile is shifted downstream to a new location at Time = 100 and Time = 1000. Still the outlet concentration of products will not change given one a false sense of catalyst stability. Finally, when the bed becomes sufficiently poisoned, breakthrough occurs at Time = 2000 indicating that the remaining catalyst cannot sustain the degree of desired conversion. This phenomenon is monitored by inserting analysis probes at various axial locations within the monolith channels so that changes could be observed and, if possible, corrective action be taken before the system undergoes complete failure.

For an exothermic reaction along a catalyzed monolith, typical of environmental pollution abatement applications, the change in temperature with length down the channel ($\Delta T/\Delta L$) is initially very high at the inlet and quickly decreases as the pollutant concentration is reduced to close to zero. This is shown in Figure 13.5 for Time = 0. The high peak temperature causes catalyst sintering resulting in a loss in activity and the maximum moves down the channel (Time = 100). Effectively, the exotherm is distributed more uniformly reducing the peak temperature. With time, the peak temperature is further reduced and deactivation slows (Time = 5000) to essentially zero provided no other deactivating mechanisms are operative.

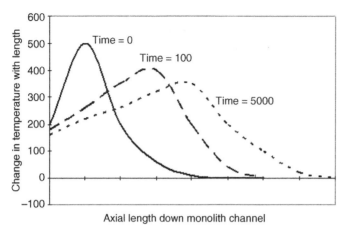

Figure 13.5   Temperature profiles ($\Delta T/\Delta L$) for an exothermic reaction down the axial length of a catalyzed monolith channel caused by sintering. (Reproduced from Chapter 5 of Heck, R.M., Farrauto, R.J., and Gulati, S.T. (2009) *Catalytic Air Pollution Control: Commercial Technology*, 3rd edn, John Wiley & Sons, Inc., New York.)

In any high-temperature process, sintering of the catalytic components and the carrier occurs in the early stages, but then equilibrate to a given value. Any temperature upsets can cause additional sintering and loss of activity.

13.11 THE FAILURE OF THE TRADITIONAL BEADED (PARTICULATE) CATALYSTS FOR AUTOMOTIVE APPLICATIONS

An important question engine manufacturers had to address was how to house the catalyst in the exhaust. The most traditional way was to use spherical particulate γ-Al_2O_3 particles, anywhere from 1/8 to 1/4 in. in diameter, into which the stabilizers and active catalytic components (i.e., precious metals) would be incorporated. These "beads" would be mounted in a spring-loaded reactor bed downstream, just before the muffler. Since the engine exhaust gas was deficient in oxygen, air was added into the exhaust using an air pump. The rationale was simple: Catalysts had been made on these types of carriers for many years in the petroleum, petrochemical, and chemical industries and manufacturing facilities to mass produce them were already in place. There were known reactor designs and flow models that would make scale-up easy and reliable. One major concern was the attrition resistance of the γ-Al_2O_3 particles, since they would experience many mechanical stresses during the lifetime of the converter. The bead bed reactor design for the early oxidation catalysts was of a radial flow design with a large open frontal area to decrease linear velocity and pressure drop.

The beads were manufactured with the stabilizers incorporated into the structure to minimize carrier sintering. The precious metal salts were impregnated into the bead and, using proprietary methods, fixed in particular locations to ensure adequate performance and durability. They were then dried at typically 120 °C and calcined to about 500 °C to their finished state. The finished catalyst usually had about 0.05 wt% precious metals with a Pt/Pd weight ratio of 2.5:1. After 1979, the need for NO$_x$ reduction required the introduction of small amounts of Rh into the second-generation catalysts. The bead catalyst worked initially and gave excellent efficiency regarding removal of the pollutants of CO, HC, and NO$_x$. The problem was durability of the beads themselves as the vibrations of the vehicle eventually ground the beads into a smaller sizes causing settling of the reactor catalyst bed, increased pressure drop, and bypassing of the reactants resulting in poor performance. They were essentially eliminated for use after about 1980 and replaced exclusively with monoliths.

13.12 NO$_X$, CO AND HC REDUCTION: THE THREE-WAY CATALYST

Following the successful implementation of catalysts for controlling CO and HC, the reduction of NO$_x$ emissions in the automobile exhaust to less than 1.0 g/mile had to be addressed. NO$_x$ reduction is most effective in the absence of O$_2$, while the abatement of CO and HC requires O$_2$. The exhaust emanating from the engine can be made sufficiently reducing (rich) so that a catalyst to reduce NO$_x$ could be positioned upstream of the air injection system in the exhaust and oxidizing catalyst. With this arrangement, the H$_2$, produced by reactions of CO and steam (water gas shift) and HC with steam (steam reforming), as later in this chapter, could reduce the NO$_x$, with the assistance from a NO$_x$ reduction catalyst. The remaining CO and HC will be oxidized in a second oxidation catalyst bed.

   A primary catalyst for the reduction reaction was Ru; however, on an occasion when the engine exhaust might be oxidizing and the temperature exceeded about 700 °C, it was found to volatilize by forming a volatile oxide of ruthenium. This approach was dropped from further consideration. If Pt was used instead of Ru, the NO$_x$ was reduced to appreciable amounts of NH$_3$ and not N$_2$. The NH$_3$ would then enter the oxidation catalyst and be reconverted to NO$_x$. In reducing atmospheres, Rh was known to be an excellent NO$_x$ reduction catalyst with virtually no NH$_3$ formation and was thus incorporated into the catalyst.

   If the engine exhaust could be operated close to the stoichiometric, air/fuel ratio $\lambda = 1$, all three pollutants with the right catalyst could be simultaneously converted, and the need for a two-stage reactor with air injection could be eliminated. Figure 13.6 shows such a curve in which the NO$_x$ reduction via reaction with HC and CO occurs readily when the exhaust is rich ($\lambda < 1$), while the CO and HC oxidation reactions are prevented by insufficient O$_2$. Due to insufficient O$_2$, hydrocarbon steam reforming and CO water gas shift reactions generate H$_2$, which is then available for catalytic NO$_x$ reduction over Rh. As the air/fuel ratio approaches the stoichiometric point, there is a narrow window where simultaneous catalytic conversion of all three occurs. On the lean side ($\lambda > 1$), the CO and HC conversions are high, but at the sacrifice of the NO$_x$ conversion since all the CO and HC would be oxidized and little H$_2$ would form from

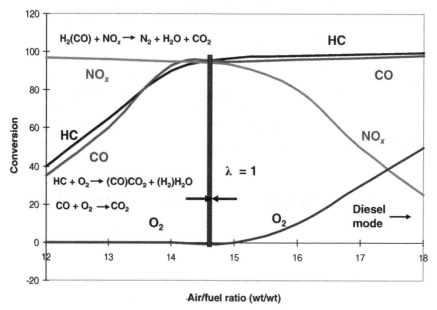

Figure 13.6   Simultaneous conversion of HC, CO, and NO$_x$ for TWC as a function of air/fuel ratio.

the WGS and SR reactions. The optimum catalyst was a combination of 5Pt and Rh dispersed on a γ-Al$_2$O$_3$ washcoated on a monolith structure. The Pt was active for the oxidation of CO and HC, while the Rh was an excellent catalyst for NO$_x$ reduction, all at the stoichiometric ($\lambda = 1$) air/fuel ratio.

The key to advancing this technology was to control the air/fuel ratio of the automobile engine within this narrow air to fuel window at all times. This was made possible by the development of the O$_2$ sensor, positioned immediately before the catalyst in the exhaust manifold. The exhaust gas oxygen (EGO) or lambda sensor was composed of an anionic conductive solid electrolyte of yttrium-stabilized zirconia with electrodes of high surface area Pt. Very few solids are like stabilized ZrO$_2$, which is an oxygen ion conductor. One electrode was located directly in the exhaust stream and sensed the O$_2$ content, while the second was a reference positioned outside of the exhaust in the natural air. The electrode is a catalyst in that it converted the HC and CO at its surface, provided sufficient O$_2$ was present. For correct air/fuel control, the sensor must at all times equilibrate the exhaust gas. If the exhaust was rich, then the O$_2$ content at the electrode surface was quickly depleted. For the condition of a lean exhaust, some O$_2$ remained unreacted, and the electrode sensed its relative high concentration. The voltage (E) generated across the sensor was strongly dependent on the O$_2$ content and is represented by the Nernst equation. The sensor is an oxygen concentration cell.

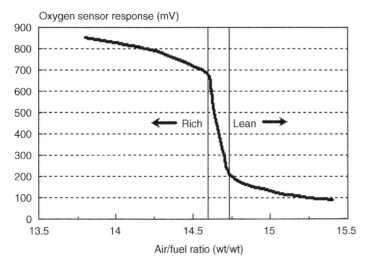

Figure 13.7   Oxygen sensor response output as a function of air/fuel ratio. (Reproduced from Chapter 6 of Heck, R.M., Farrauto, R.J., and Gulati, S.T. (2009) *Catalytic Air Pollution Control: Commercial Technology*, 3rd edn, John Wiley & Sons, Inc., New York.)

$$E = E_o + \mathrm{RT}/nF \ln (P_{O_2})_{\text{reference}}/(P_{O_2})_{\text{exhaust}} \tag{13.16}$$

The voltage signal generated is fed back to the fuel injection control device (i.e., throttle body injector or multipoint injectors), which adjusts the air/fuel ratio. Figure 13.7 shows the response profile for the O$_2$ sensor. Note that it functions similar to a potentiometric titration curve used in aqueous analytical chemistry.

The total device is a very sophisticated electronic control system to maintain the air/fuel ratio within the narrow window, which allows the simultaneous conversion of all three pollutants. This technology, referred to as *three-way-catalysis*, was first installed in large quantities on vehicles in 1979. Even today, the oxygen sensor is the state of the art in air/fuel ratio control in the gasoline internal combustion engine.

Modern sensors have been modified to be more poison tolerant to P and Si found in the engine exhaust. Also, to improve the operating range of the O$_2$ sensor during driving—particularly in cold start—the heated O$_2$ sensor was developed. This is referred to as the heated exhaust gas oxygen (HEGO) type sensor.

Because the control system utilizes "feedback," there is a time lag associated with adjusting the air/fuel ratio. This results in a perturbation around the control set point. This perturbation is characterized by the amplitude of the *A/F* ratio and the response frequency (Hz). Thus, when operating rich, there was a need to provide a small amount of O$_2$ to consume the unreacted CO and HC. Conversely, when the exhaust goes slightly oxidizing, the excess O$_2$ needs to be consumed to return the exhaust to $\lambda = 1$. This was accomplished by the development of the O$_2$ storage component, which liberates or adsorbs O$_2$ during the air to fuel perturbations. Ceria (CeO$_x$), with its defect structure, was found to have the proper redox response and is the most commonly used O$_2$ storage component (OSC) in modern three-way catalytic converters. The defect structure indicates that there is a varying amount of O in the structure generating redox lability due to a mixture of Ce^{4+} and Ce^{3+} necessary for charge neutrality. This imparts the ability of the material to gain and lose electrons by the O_2 content in the gas stream. In the fuel-rich condition, CO is plentiful while O_2 is in excess in the fuel lean condition. The OSC makes its contributions by adjusting the O_2 partial pressure in the exhaust from both chemisorbed O_2 and ceria lattice oxygen.

The reactions are indicated below:

$$\text{Rich condition :} \quad CeO_x + CO \rightarrow Ce_2O_{x-1} + CO_2 \qquad (13.17)$$

$$\text{Lean condition :} \quad Ce_2O_{x-1} + 1/2O_2 \rightarrow CeO_x \qquad (13.18)$$

Another benefit of ceria is that it is a good water gas shift (13.19) and hydrocarbon steam-reforming (13.20) catalyst and thus they catalyze the reactions of CO and HC with H_2O in the rich mode. The H_2 formed then reduces a portion of the NO_x to N_2 over the Rh component in the catalyst.

$$CO + H_2O \rightarrow H_2 + CO_2 \qquad (13.19)$$

$$C_xH_y + 2H_2O \rightarrow (2 + y/2)H_2 + xCO_2 \qquad (13.20)$$

The hydrogen thus formed reduces NO_x via the following:

$$NO_x + H_2 \rightarrow 1/2N_2 + xH_2O \qquad (13.21)$$

Other oxide materials with similar oxygen lability, such as NiO/Ni and Fe_2O_3/ FeO, have also been used as storage components; however, modern OSC materials are CeO_x-ZrO_2 solid solutions that are more thermally stable and have faster kinetics for the redox reactions. Additional promoters may also be added to the OSC to enhance performance. Figures 13.6 and 13.7 show the response of TWCs fully formulated with Ce. If Ce were not present, the CO and HC conversion on the rich side of stoichiometric would decline dramatically since there would be no O_2 available for the oxidation reactions. This is why a stable Ce-containing compound is required to maintain the activity of the oxygen storage for rich-side oxidation reactions.

The three-way catalysts of the late 1980s were primarily composed of about 0.1–0.15% precious metals at a Pt/Rh ratio of 5:1, high concentrations of bulk high surface area CeO_2–ZrO_2 (10–20%), and the remainder being the γ-Al_2O_3 washcoat. The γ-Al_2O_3 is stabilized with 1–2% of La_2O_3 and/or BaO. This composite washcoat is then deposited on a honeycomb with 400 cells/in.[2]. Typically, the washcoat loading is about 1.5–2.0 g/in.[3] or about 15% of the weight of the finished honeycomb catalyst. The size and shape of the final catalyst configuration vary with each automobile company but, typically, they are about 5–6 in. in diameter and 3–6 in. long.

By 2010, the Pt was replaced with Pd mainly due to price considerations. The modern TWC is about a 25:1 ratio of Pd to Rh. Since Pd is not as active for the same oxidation reactions as Pt, its loading was higher. There were additional positive consequence of the use of Pd in place of Pt: (i) It was most active as PdO and was more thermally stable (the active species is PdO) than Pt. (ii) It had some NO_x reduction activity to N_2. (iii) Because of the PdO/Pd, redox couple contributed to OSC allowing reductions in the OSC. Pd is known to be more sensitive to poisons such as sulfur; however, its fuel levels have been significantly reduced (from 100 to <20 ppm of S) allowing it to be used with little negative effect on performance.

This demonstrates how the catalyst and the fuel quality must be optimized together.

13.13   SIMULATED AGING METHODS

The late 1980s and the early 1990s required improvement in technology because of the automobile's changing operational strategies: fuel economy was important, yet operating speeds were higher. This situation resulted in higher exposure temperatures to the TWC catalyst. Higher fuel economy was met by introducing a driving strategy whereby fuel is shut off during deceleration. The catalyst, therefore, is exposed to a highly oxidizing atmosphere that results in deactivation of the Rh function by reaction with the γ-alumina, forming an inactive rhodium–aluminate species.

To simulate these modes of deactivation in the laboratory, engine dynamometer aging cycles were set up to simulate 50,000 miles performance. These aging cycles consisted of repetitive steps of changing the engine speed/load, air/fuel ratio, and exhaust temperature. In some cases, air was injected downstream into the exhaust, while in others the engine was connected to a flywheel and actual fuel cut was simulated. In addition, some cycles were isothermal, while others were exothermal, generating a large temperature rise on the catalyst surface. The aging cycles generating an exotherm on the catalyst surface usually have an inlet temperature of 650–850 °C, with catalyst bed temperatures ranging from 850 to 1100 °C, depending on the concentration of the CO in the engine exhaust. Still other cycles looked at accelerated poisoning of the catalyst by doping the fuel with lubricating oil with high concentrations of phosphorous. Electron microprobe profiles after accelerated tests are shown in Figure 13.8 with locations of S and P.

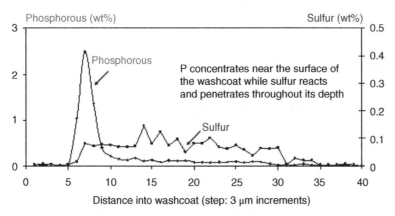

Figure 13.8   Electron microprobe scan of an automotive catalyst contaminated with P and S from lubricating oil.

The concentrations of S and P are much greater in the inlet than the outlet section, indicating the former serves as a filter. The sulfur is uniformly present throughout the washcoat, suggesting an interaction between it and Al_2O_3. The P is concentrated near the outer periphery of the washcoat, but only in the inlet section. The drop in poison concentrations at about 20 μm is at the washcoat–monolith interface.

Several studies have been conducted on the effect these various aging cycles have on the TWC performance. One clear observation is the strong effect of aging temperature and exhaust gas oxygen concentration on performance. Al_2O_3 and PM sintering, undesired interactions of the catalytic components with the Al_2O_3, pore blockage due to poisons from P and Zn present in engine oil, and deactivation by adsorption of fuel sulfur onto the precious metals and formation of $Al_2(SO_4)_3$ all must be factored into the final catalyst design. Reductions in sulfur content in fuels from 1000 ppm (1975) to 20–30 ppm in 2010 have significantly increased the life of the catalysts. A modern TWC will meet the U.S. emission standards after 150,000 miles of driving. Simulating this many miles in the engine laboratory requires a detailed empirical understanding of why gasoline autocatalysts deactivate. Accelerated aging methods utilized in engine laboratories are primarily the results of extensive fleet testing and characterization after thousands of miles of typical driving.

A possible source of deactivation is the loss of the catalytic washcoat due to attrition. This mechanism would be a problem since the gases are flowing at high linear velocities (high turbulence) and exhaust temperatures occur rapidly due to changing driving conditions. The thermal expansion differences between the washcoat and the monolith, especially metal substrates, lead to a loss of both bonding and washcoat. Special surface roughening methods are being used for metal monoliths to provide a more receptive host for generating adherent washcoats. Occasionally proprietary binders containing SiO_2 and/or Al_2O_3 are added to the washcoat formulation to improve the chemical bond between the washcoat and the substrate to ensure an adherent washcoat.

Washcoat loss is observed by preparing a cross section of the honeycomb catalyst and scanning the wall of the channel with either an optical or a scanning electron microscope. The loss of catalytic washcoat material is irreversible and results in a shift in the conversion versus temperature curve to higher temperatures. It also decreases the bulk mass transfer conversion achievable, which is directly proportional to the geometric surface area of washcoat.

13.14 CLOSE-COUPLED CATALYST

During the cold start operation when a vehicle has been dormant, for example, overnight, materials including the engine, the exhaust system, and the catalysts are cold. Thus, when the engine is started before the catalyst gets sufficiently hot to begin catalytic action (usually about 2–3 min for the under-floor position), emissions of CO and HC would be very high resulting in failure of the FTP. This suggested that a close-coupled catalyst (a smaller catalyst position about 1–2 ft from the engine manifold) would get hot sufficiently fast to begin the conversion process. The concept of using a catalyst near the engine manifold or in the vicinity of the vehicle firewall to reduce the heat up time had been published. The practice of "overfueling" or "acceleration enrichment" results in high HC and CO emissions and had to be reduced. Manifold discharge temperatures approached 1000 °C, so this would put a great stress on the catalyst, especially thermally induced sintering. After many other approaches were considered (electrically heated metal monolith-supported catalysts), the close-coupled approach was the least expensive solution; however, the technology had to be developed.

Catalyst manufacturers were once again exploring new carriers capable of retaining high surface areas and metal combinations that resist deactivation due to sintering after high-temperature exposure. A shift in the technology for close-coupled catalyst occurred when a close-coupled catalyst capable of sustained performance after 1050 °C aging was developed and shown to give satisfactory performance in combination with an under-floor catalyst. The close-coupled catalyst was designed mainly for HC removal, while the under-floor catalyst removed the remaining CO and NO_x. The concept inherent in this technology was to have lower CO oxidation activity for the close-coupled catalyst, thus eliminating severe over temperatures when high CO concentrations occur in the rich transient driving cycle. A new design with improved thermally stable carriers and more stable catalytic metals was developed.

Catalyst manufacturers worked on technologies capable of sustained operation and good HC light-off after exposure to 1050 °C. One characteristic of these close-coupled technologies is that Ce is removed. Ce is an excellent CO oxidation catalyst and also stores oxygen, which can then react with CO during the rich transient driving excursions. If the oxygen is stored on the catalyst during severe rich excursions (e.g., fuel enrichment or heavy accelerations), the CO can react at the catalyst surface causing a localized exotherm, resulting in very high catalyst surface temperatures. As a rule, every percent of CO oxidized gives 90 °C rise in temperature. A second factor was to design the catalyst to accept some deactivation. This meant more precious metal and design for loss of activity due to sintering of both the carrier and catalytic components. Thus, some sintering and loss of activity could be tolerated. Predicting the extent of sintering was based on empirical accelerated engine tests. At such high temperatures, poisoning by sulfur, phosphorus, and zinc generally do not occur to any appreciable extent.

One study looked at the geometric effect of the close-coupled catalyst on both performance and light-off. The cross-sectional area and volume of the close-coupled catalyst in the so-called cascade design were studied. It was found that the smaller cross section of close-coupled catalyst with less volume significantly improved the light-off characteristics by reaching light-off temperature faster and this could be combined with a larger cross section under-floor catalyst with more volume to achieve the required emissions. This requirement was the incentive to introduce metal monolithic structures to support the washcoated catalyst. Metal substrates simply get hot faster than ceramics.

The early light-off of the close-coupled catalyst can be accomplished by a number of methods related to the engine control technology during cold start. One of the initial methods was to control the ignition spark retard that would allow

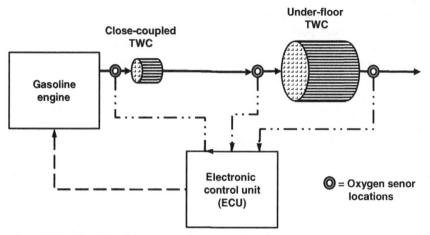

Figure 13.9 Close-coupled TWC catalyst, under-floor TWC, and oxygen sensors connected to electronic feedback to control air/fuel ratio close to stoichiometric ($\lambda = 1$).

unburned gases to escape the engine combustion chamber and continue to burn in the exhaust manifold, thus providing heat to the catalytic converter. These engine control methods have become more sophisticated as illustrated by the studies on actually controlling the degree of exothermic reactions in the exhaust manifold. This is another example of how multidiscipline technologies have contributed to meeting emission standards.

Figure 13.9 shows the catalytic operations along with oxygen sensing and feedback control present in modern TWC exhaust systems.

13.15 FINAL COMMENTS

We have seen steady improvements and dramatic new technologies since 1975 when the catalytic converters were first introduced in the United States. We have seen oxidation catalysts composed of Pt and Pd as washcoats on both ceramic monoliths and impregnated Al_2O_3 beads, removal of Pb compounds in the fuel, new reactor designs for different catalyst structures, introduction of the TWC with Pt and Rh with an electronic feedback control system with the oxygen sensor to control the air/fuel ratio, and the oxygen storage component that is now an intimate part of the TWC. The close-coupled catalyst on metal monoliths was introduced to manage the cold start issue of CO and HC emissions. We have seen high cell density ceramic and metal monoliths with decreasing pressure drop because of thinner walls being used to increase bulk mass transfer conversion with low pressure drop. We have seen that the fuel sulfur levels decreased from almost 1000 ppm (1975) to <20 ppm in 2010. The close-coupled catalyst has been improved to perform TWC in concert with the large-sized under-floor catalyst. Now the TWC catalyst no longer contains Pt, but is a combination of Pd and Rh. We can expect more subtle changes with time especially for the abatement of greenhouse gas emissions, for example, methane and dinitrogen oxide emissions may soon have to be addressed. Clearly, the automotive catalytic converter has been a glowing success for improved air quality. It is a good example of how multidisciplinary teams will be required to meet other challenges in the future.

The question must be asked as to what are the fuels of the future? Abundant natural gas, containing mostly methane, is now being utilized in some vehicles, but currently methane is excluded from hydrocarbon standards. Given its large greenhouse effect, new catalyst and engine technologies will be needed to abate methane emissions. Fuels derived from biomass containing larger amounts of oxygen, such as lingocellulose, relative to the current ethanol additives to gasoline, may have an impact of aldehyde emissions that will have to be evaluated. The slow emergence of H_2 fuel cells could revolutionize the automobile and fossil fuel industry. There are many technology challenges in front of us as we strive for cleaner air with environmental sound sources of fuel. There is no doubt that catalysts will play a major role in the abatement of emissions or better yet in the generation of the renewable fuels for the future.

QUESTIONS

1. What is meant by three-way catalysis or TWC?

2. Sketch the conversion versus λ curve for the TWC catalyst from very fuel rich to very fuel lean.

3. Sketch the region (with and without the OSC present) from $\lambda = 0.99$ to $\lambda = 1.01$ and indicate the reactions occurring.

4. What are the advantages of using Pd + Rh in the TWC instead of Pt + Rh?

5. Describe what happens in the piston when gasoline is used with a lower octane number that is specified by the automobile manufacturer. Indicate what is physically happening during the reactions.

6. Internal combustion engines used for lawn mowers (grass cutters) are a source of many health issues. They were tuned to operate rich of the stoichiometric point. What are the positive and negative consequences of operating the engine rich?

7. Why is the development of the oxygen sensor so important for implementation of TWC technology?

8. Why is Ce important in the utilization of the oxygen sensor HEGO and implementation of TWC technology?

9. What technologies were needed to allow the use of Pd as a NO_x reduction catalyst versus Rh?

10. What operating elements are important in the use of a close-coupled catalyst for cold start performance?

11. Why cannot TWC technologies be used for lean burn engines such as a diesel?

12. What are the major mechanisms for deactivation of TWC technology? Describe how it affects the TWC performance.

13. Design the complete washcoated monolith for low-temperature light-off during the cold start process, for hot start and cruise condition of the FTP. Feel free to explore approaches that may be different from what you have learned in the class.

14. a. Calculate the stoichiometric point ($\lambda = 1$) on a weight–weight basis, for E85 (15% gasoline and 85% ethanol). Assume the vol% of O_2 in air is 21%.

   Assume

$$\text{Gasoline:} \quad C_8H_{18} + 25/2O_2 \rightarrow 8CO_2 + 9H_2O$$

$$\text{Ethanol } C_2H_5OH + 3O_2 \rightarrow 2CO_2 + 3H_2O$$

   b. What are the consequences for fuel economy for using pure gasoline versus E85?

15. Explain the importance of the oxygen sensor. Why does the signal oscillate in amplitude?

16. What are the catalytic reactions for $\lambda = 1$, $\lambda < 1$, and $\lambda > 1$?

17. What alternative materials might be considered as OSC?

18. What laboratory experiments would you perform to determine if these alternative materials were acceptable replacements for CeO_2? When preparing the outline of your plan, ensure to take into account about how you would characterize these materials before and after duty cycle testing.

19. Design a single TWC monolith catalyst for a stoichiometric engine that will successfully pass the cold start, hot start, and steady-state modes of the U.S. FTP.

20. Refer to the TWC U.S. Patent 4157,316 (1979).

   a. Summarize the most important qualitative issues in this patent, and why the choice of catalyst metals (not more than five sentences).

   b. Al_2O_3 is stable as the preferred support.

      (i) What other oxide(s) can be added and why?

      (ii) Which other oxides are preferred additions to Al_2O_3?

   c. What is meant by "fixing" the components in the catalyst?

   d. How is fixing performed?

   e. What is the role of the metal oxides that have multiple oxidation states?

   f. Which metal oxide is preferred?

   g. What alternative mode of conversion of all three components can be used in the exhaust instead of the control of stoichiometric air/fuel ratio?

   h. Why would one want to do that?

   i. What are the problems?

   j. Refer to claims of patent.

      (i) Compare 1b with 2.

      (ii) Compare 1c with 4.

      (iii) Compare 1d with 3.

      (iv) Compare 5 with 6.

      (v) Compare 9e with 15.

      (vi) Compare 1 with 9.

21. You are working for an environmental engineering company. They receive a request from a mower manufacturer to assist in solving the problems associated with the negative pollution aspects of their equipment. Assume you will work with a catalyst company who will also provide information to them. Recommend how all three parties will work together to assist in solving the problem.

BIBLIOGRAPHY

Heck, R.M., Farrauto, R.J., and Gulati, S.T. (2009) Chapter 6, in *Catalytic Air Pollution Control: Commercial Technology*, 3rd edn, John Wiley & Sons, Inc., Hoboken, NJ.

Hochmuth, J., Wassermann, K., and Farrauto, R. (2013) Car cleaning catalyst, in *Comprehensive Inorganic Chemistry II*, Vol. 7 (eds R. Schlogl and H. Niemantsverdriet), Elsevier, The Netherlands.

Thompson, C., Mooney, J., Keith, C., and Mannion, W. Polyfunctional catalysts. U.S. Patent 4,157,316, June 5, 1979.

Oxygen Storage Materials

Cuif, J-P, Deutsch, S., Touret, O., Marczi, Jen, H-W, Graham, G., Chun, W., and McCabe, R. (1998) *High temperature stability of ceria–zirconia supported Pd model catalysts*, SAE 980668.

Poisoning: Aging

Culley, S., McDonnell, T., Ball, D., Kirby, C., and Hawes, S. (1996) *The impact of passenger car motor oil phosphorus levels on automotive emissions control systems*, SAE 961898.

Farrauto, R. and Wedding, B. (1974) Poisoning by sulfur oxides of some base metal oxide auto exhaust catalysts. *Journal of Catalysis* 33 (2), 249–255.

Rokosz, M., Chenb, A., Lowe-Mac, C., Kucherov, A., Benson, D., Peck, M., and McCabe, R. (2001) Characterization of phosphorus-poisoned automotive exhaust catalysts. *Applied Catalysis B: Environmental* 33, 205–215.

Skowron, L., Williamson, W., and Summers, L. (1989) *Effect of aging and evaluation on the three way catalyst performance*, SAE 892093.

Williamson, W., Perry, J., Gandhi, H., and Bomback, J. (1985) Effects of oil phosphorous on deactivation of monolithic three-way catalysts. *Applied Catalysis* 15, 277–292.

Wong, C. and McCabe, R. (1989) Effects of high-temperature oxidation and reduction on the structure activity of Rh/Al_2O_3 and Rh/SiO_2 catalysts. *Journal of Catalysis* 119, 47–64.

Xua, L., Guoa, G., Uyb, D., O'Neill, A., Weber, W., Rokosz, M., and McCabe, R. (2004) Cerium phosphate in automotive exhaust catalyst poisoning. *Applied Catalysis B: Environmental* 50, 113–125.

Zheng, Q., Farrauto, R., Deeba, M. and Valsamakis, I. (2015) A Comparative Thermal Aging Study on the Regenerability of Rh/Al_2O_3 and $Rh/Ce_xO_y-ZrO_2$ as Model Catalysts for Automotive Three Way Catalysts *Catalysts* 5, 1770–1796.

Oxidation Catalysts

Yao, Y.-F. (1975) The oxidation of CO and C_2H_4 over metal oxides. *Journal of Catalysis* 39, 104–115.

CHAPTER 14

DIESEL ENGINE EMISSION ABATEMENT

14.1 INTRODUCTION

14.1.1 Emissions from Diesel Engines

The popularity of diesel engines is derived primarily from their fuel efficiency and long life relative to the gasoline spark-ignited engine. They have high compression ratios and often turbocharged giving rise to their enhanced efficiency relative to gasoline engines. They operate very lean of stoichiometric, ($\lambda \gg 1$) air/fuel ratios greater than about 22 compared to <10 for gasoline engines. They ignite upon compression unlike the spark-ignited gasoline engine. Since the TWC operates at $\lambda \sim 1$, it will not effectively catalyze NO_x reduction in a diesel exhaust. The boiling point (200–340 °C) of diesel (C_{14}–C_{20} mostly paraffinic hydrocarbons) gives rise to particulates (soot) since the fuel is injected as a liquid droplet where combustion occurs from liquid–gas interface inward (core–shell burning). This "diffusion combustion" phenomenon results in the production of significant amounts of NO_x at the fuel liquid–gas interface where the local temperature and oxygen concentrations are high. The inner core is starved of O_2 and when heated pyrolysis occurs giving rise to particulates or soot. Its lean nature results in a cooler combustion with less gaseous NO_x, CO, and HC emissions than its gasoline counterpart. The design of the combustion process, however, results in high particulate emission levels. This is to be contrasted with premixed gasoline–air vapor mixtures (BP = 25–200 °C) that produce virtually no soot. Therefore, new abatement technology for diesels was needed.

Diesels have fuel economies 20–25% greater than gasoline, and therefore produce less CO_2 per mile driven. It is not uncommon for a diesel engine to have a life of 1 million miles, or about 5–10 times that of the gasoline engine. Diesel-fueled vehicles have always been popular in Europe where 50% of new passenger cars sold are diesel powered. Commercial trucks are more or less universally operated with diesel fuel worldwide. Since about 2006 diesel passenger cars started showing increased sales due to improved drivability, cleaner operation, and improved fuel economy. The term cetane number (typically 50 or 60) is used to describe the ability of the diesel fuel to combust when injected into the hot compressed air. This is a

Introduction to Catalysis and Industrial Catalytic Processes, First Edition. Robert J. Farrauto, Lucas Dorazio, and C.H. Bartholomew.

function of the chemistry and structure of the fuel. Aromatics have low cetane numbers, while paraffins are more suitable with higher cetane numbers.

The relationship between particulates (PM) and NO_x is summarized in Figure 14.1 in what is called the NO_x–particulate trade-off. When NO_x emissions are high (e.g., at high combustion temperatures), particulate emissions are low. Likewise, when NO_x emissions are low (e.g., at lower combustion temperatures), the particulate emissions are high.

Also shown on this profile are the emission standards to be met for different countries and years. Regulations require both particulate and NO_x emissions to be close to zero (small box 2010 in bottom left of the graph) in the United States with other countries following closely. Due to the adverse health effects associated with diesel soot and the ozone forming potential of NO_x, both are the major focus for emission regulations. Although CO and HC are also regulated, the amounts produced by diesel engines are generally low enough not to be a major obstacle for meeting emission regulations. This is particularly true for heavy-duty diesel applications.

The solids emitted from diesel engines are essentially dry soot (carbon-rich particles). Liquids composed of unburned diesel fuel and lubricating oils (commonly referred to as soluble organic fraction (SOF)) and gases are also emitted. Some sulfates originating from the combustion of the sulfur compounds in the diesel fuel are also present. The combination of solid and liquid pollutants is referred to as particulates or total particulate matter (TPM). Note that H_2SO_4 derived from the combustion of sulfur compounds in diesel fuel is included since it is a liquid at the

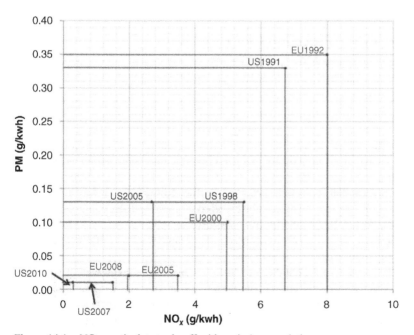

Figure 14.1 NO_x–particulate trade-off with emission regulations.

collection conditions for TPM (52 °C) on a filter. The gaseous pollutants are CO, HC, and NO_x, all of which must be abated to extremely low levels.

> Total particulate matter
> > Dry soot
> > Liquids (oil, fuel) called soluble organic fraction
> > H_2SO_4
> Gases
> > CO, HC, NO_x

The Clean Air Amendment of 1990 in the United States required that by 1994 particulate emissions be reduced to 0.1 g/(bhp h) for trucks and 0.07 g/(bhp h) for buses. The units of g/(bhp h) reflect the fact that trucks do work, while g/mile or g/km are the units used for passenger cars that move people. Since then, standards have become more stringent and thus emission regulations have led to the development of new technologies.

This Federal Test Procedure for trucks simulates the emissions expected during commercial operations. Both truck emission and passenger car requirements are measured using the chassis dynamometer where the exhaust is sampled during various duty cycle modes. Emission standards vary considerably from country to country and from year to year. Earth moving construction equipment is now also required to meet emission standards for off-the-road applications. Soot filters and selective catalytic reduction (SCR) are now being fitted into new vehicles that will be discussed in this chapter.

14.1.2 Analytical Procedures for Particulates

The analysis of the diesel exhaust is much more complicated than that of the IC engine due to the three phases of pollutants present. The Federal Test Procedure defines "particulates" as those that are collected on a filter at 52 °C to condense SOF. The particulates are collected in a dilution tunnel, which cools the exhaust to the required temperature. Because of the hygroscopic nature of the sulfates, the filter must be further conditioned in a controlled atmosphere to equilibrate the water content. The unburned fuel and lubricating oil are extracted from the filter with methylene chloride, making it the soluble organic fraction. It is then injected into a gas chromatograph (GC), where a capillary column separates the fuel from the lube fraction. In this manner, the effectiveness of the catalyst toward converting each component of the SOF can be assessed. Some laboratories volatilize the unburned fuel and lube directly from the filter into the GC. This gives a volatile organic fraction (VOF). The results are essentially equivalent to the SOF method.

Another portion of the conditioned filter is then extracted with water to dissolve the sulfate. The aqueous solution is then injected into an ion chromatograph for sulfate analysis. The dry carbon is obtained by difference, or in some cases can be subjected to thermal analysis for a burn-off after the extraction steps, to complete the material balance.

14.2 CATALYTIC TECHNOLOGY FOR REDUCING EMISSIONS FROM DIESEL ENGINES

14.2.1 Diesel Oxidation Catalyst

A diesel oxidation is positioned downstream from the turbocharger in the exhaust manifold and experiences much cooler temperatures (maximum <650 °C) than gasoline catalysts mainly due to the large amount of excess air present. Because many of the thermally resistant materials developed for the gasoline catalysts have been incorporated into the diesel oxidation catalyst (DOC) and they experience lower temperatures than in gasoline vehicles, deactivation due to thermal stresses is not considered a major problem. However, some diesel exhaust molecules, such as decane ($C_{12}H_{26}$), are difficult to oxidize at the lower inlet temperatures and thus these reactions are much more sensitive to thermal stresses than the molecules such as CO, olefins, and aromatics that are easy to oxidize. Therefore, it is common to evaluate a material for its ability to catalytically oxidize the most stable molecules present, especially after an accelerated aging temperature of 800 °C. Catalysts will also experience peak temperatures close to 800 °C when diesel fuel is injected and oxidized to generate an exotherm sufficiently high to initiate combustion of the soot in the downstream wall flow filter. Thus, thermal stability is still a matter of concern to ensure sufficient life.

In 1994, when the sulfur levels in the fuel were 500 ppm, precious metals similar to those used in gasoline exhaust catalysts (Pt and Pd) would catalyze the oxidation of SO_2 to SO_3 producing H_2SO_4 when mixed with moist atmospheric air. One of the early DOC catalysts was primarily CeO_2–Al_2O_3 with a trace of Pt impregnated on a ceramic monolith found to be active for oxidizing the liquid portion of the particulates with little or no activity for catalyzing SO_2 to SO_3 but with some activity of oxidizing CO and gaseous HCs. Other catalysts were higher levels of Pt poisoned with components such as V_2O_5 that poisoned the SO_2/SO_3 reaction and yet allowed conversion of SOF and some of the gaseous compounds. For European passenger cars, beta zeolite, in combination with CeO_2 and small amounts of Pt, was used to adsorb some of the HC emissions during cold start. The adsorbed HC were retained until a temperature was reached in which the oxidation components in the DOC could convert them to CO_2 and H_2O.

Diesel engines combust larger quantities of oil than their gasoline counterparts, so the catalyst must be more resistant to the oil and its additives. The unburned oils and their inorganic additives deposit within the catalyst washcoat structure under the cooler modes of operation. Unlike the organic portion of the oil, the additives remain after the oil is catalytically oxidized. Oxides of zinc, phosphorous, sulfur, and calcium accumulate on or within the catalyst. The microprobe traces shown in Figure 14.2 were scanned across a 75 μm thick washcoat fillet using a diesel fuel with 500 ppm S. Note the scale is in units of 3 μm.

Modern catalysts (since 2007) still suffer from poisoning by the oil additives, but much less so from sulfur given its decreased levels (20 ppm). This allowed more freedom in the use of precious metals for modern DOCs to indirectly assist in decreasing soot levels in the wall flow filter.

(a)

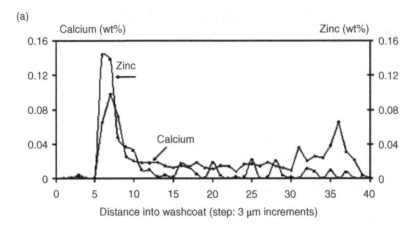

(b)

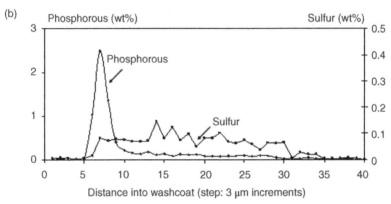

Figure 14.2 Electron microprobe scans of the washcoat of an aged diesel oxidation catalyst. (a) Zn and Ca. (b) P and S. (Reproduced from Chapter 8 of Heck, R.M., Farrauto, R.J., and Gulati, S.T. (2009) *Catalytic Air Pollution Control: Commercial Technology*, 3rd edn, John Wiley & Sons, Inc., New York.)

14.2.2 Diesel Soot Abatement

In 2003, new regulations for dry soot reduction required an additional solution. This led to the introduction of the wall flow or diesel particulate filter (DPF) (Figure 14.3). A DPF is primarily a cordierite honeycomb structure with alternating adjacent channels plugged at opposite ends. Some wall flow filters are made of silicon carbide (SiC) mainly because of its higher temperature resistance (2200 versus ~ 1400 °C for cordierite) and its higher thermal conductivity that allow better heat distribution during soot combustion. Some designs utilize a segmented bed to minimize the possibility of cracking.

Exhaust enters the open channels, but only the gaseous components can pass through the wall exiting via the adjacent channel. Soot that is entrained in the exhaust stream is trapped on the wall, while the gaseous components pass unrestricted. Periodically (e.g. every 1000 km of driving), the filter is heated to a temperature high

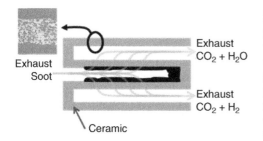

Figure 14.3   Wall flow filter. Soot particulates deposit on the porous wall, while the gaseous components (CO_2, H_2O, NO, and NO_2) and air pass through. The soot is combusted periodically by raising the inlet temperature to >500 °C when a small amount of diesel fuel is injected into the DOC.

enough (~500–550 °C) to initiate combustion of the soot and regenerate the filter. For vehicles containing both a DOC and a DPF, active regeneration is accomplished by periodically injecting diesel fuel upstream of the precious metal-containing DOCs where it is oxidized to generate the exotherm required to initiate soot combustion. Alternatively, for vehicles without a DOC, fuel is injected into the cylinders during the exhaust stroke to promote combustion, thereby raising the exhaust and DPF temperatures. The DOC also functions to oxidize the NO emanating from the exhaust to NO_2. The importance of this will become clearer when we discuss SCR technology. NO_2 is also active toward soot combustion at <300 °C, but is not the major mode for regeneration of the soot collected in the trap:

$$2NO_2 + C \rightarrow 2NO + CO_2 \tag{14.1}$$

Equation 14.1 is referred to as passive regeneration of the soot. Obviously, NO must be present and oxidized to NO_2, but is regenerated and must be removed downstream. Furthermore, the required amount of NO is not always present in the exhaust and thus regeneration cannot depend on this. For this reason, diesel engine exhaust systems are usually equipped with active regeneration with periodic injection of fuel to initiate soot combustion.

The filter will also collect ash (oxides of Zn, Ca, and P) components leading to plugging and high-temperature spikes occurring during soot burn-off can cause mechanical failure. Thus, soot regeneration must be carefully controlled by sensors and frequent regeneration to avoid excessive buildup of soot and the exotherm generated during combustion.

The DPF may also contain a precious metal washcoat (i.e., a catalyzed soot filter (CSF)) to assist with the combustion of soot and to oxidize CO generated during the soot regeneration process and to generate some NO_2 to assist in SCR downstream. In general, DPFs are required to meet the current medium- and heavy-duty diesel regulations. The same is true for the U.S. light-duty vehicle regulations that equipped these devices in 2009.

14.2.3   Controlling NO_x in Diesel Engine Exhaust

The U.S. and European standards for 2010 required significant reduction of all three phases of diesel emissions (i.e., solids, liquids, and gases). In particular, reduction of NO_x will offer considerable challenges due to the oxidizing environment. Two of the most promising technologies for controlling NO_x at the tailpipe are SCR and lean NO_x traps (LNT). Both utilize catalytic processes to eliminate NO_x by selectively reducing it to N_2.

SCR relies on the reduction of NO_x by ammonia (NH_3) over either a vanadia (V_2O_5) or a transition metal-exchanged zeolite-based catalyst. V_2O_5 technology is widely used for stationary applications (power plants) where exhaust temperatures are more or less constant and controlled within the required temperature windows. For diesels, especially heavy-duty trucks, a combination of Cu- and Fe-containing zeolites (chabazite and beta zeta zeolites) is used. The Cu–chabazite gives improved low-temperature performance, but loses selectivity toward N_2 formation at higher temperatures. The NH_3 injected either decomposes to N_2 or is oxidized to more NO/NO_2. An Fe–beta zeolite catalyst becomes active and selective toward NO_x reduction to N_2 at higher temperatures and therefore a combination of both catalysts is needed (Figure 14.4). The major desired reactions are shown in Equations 14.2–14.4:

$$4NH_3 + 4NO + O_2 \rightarrow 4N_2 + 6H_2O \qquad (14.2)$$

$$4NH_3 + 2NO_2 + O_2 \rightarrow 3N_2 + 6H_2O \qquad (14.3)$$

$$2NH_3 + NO + NO_2 \rightarrow 2N_2 + 3H_2O \qquad (14.4)$$

The zeolites possess some unique pore structures and acidic properties that efficiently adsorb NH_3, which dissociates the molecule to H atoms that react with adsorbed NO/NO_2 on the metal sites forming N_2. Any excess NH_3 produced by injection is stored on the acid sites of the zeolites and thus SCR can be continued for a short time while the adsorbed supply of NH_3 is expended.

Deactivation can occur due to excessive temperatures, especially during soot burn-off from the soot filter, where Al in the tetrahedral zeolite structure is extracted from the framework decreasing the zeolite content of the catalyst. This is observed as a shift of the NMR peaks. Sulfur levels have been decreasing over the years, but even at 10 ppm, it can accumulate on the Cu or Fe zeolite SCR catalyst. Fortunately, the high temperatures experienced during soot regeneration desorb the sulfur from the metal sites and catalyst activity is restored.

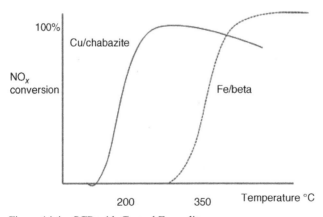

Figure 14.4 SCR with Cu and Fe zeolites.

Although commonly used in stationary source applications, NH_3 is not practical in automobiles or trucks, so an ammonia surrogate such as liquid urea is utilized to generate ammonia *in situ* in the exhaust. Typically, a solution of urea and water (called Adblue) is injected into the exhaust stream before the SCR catalyst, and in the presence of water vapor at $T = 150\,°C$, it is hydrolyzed to ammonia that can participate in the $NH_3/SCR/NO_x$ reduction reactions:

$$CO(NH_2)_2 + H_2O \rightarrow 2NH_3 + CO_2 \qquad (14.5)$$

Although SCR technology has been successfully demonstrated for mobile applications and is in fact used in some heavy-duty diesel trucks, it relies on a newly established supply infrastructure for the distribution of the urea. This infrastructure is now in place and urea distribution is common in service stations.

In addition, an ammonia "cleanup" catalyst (often referred to as an AMOX catalyst) may be required to remove any ammonia "slip" that may pass through the SCR catalyst unconverted. Clearly, integration of the SCR catalyst within the exhaust system may be a challenge. Depending on the specific application, an oxidation catalyst (DOC), a catalyzed soot filter, an SCR catalyst, and an ammonia destruction catalyst (AMOX) may all be required to meet the combined TPM, CO, HC, and NO_x regulations. The relative location of each component within the system is the subject of intense development for vehicle OEMs. In addition, sophisticated engine controls are required to ensure proper operation of all components within the system. Figure 14.5 shows the catalytic unit operations for meeting current diesel emission standards using an SCR system. The diesel oxidation catalyst (a combination of Al_2O_3

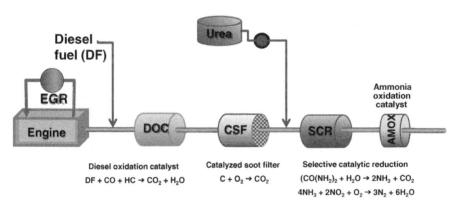

Figure 14.5   Schematic of simplified diesel exhaust aftertreatment system. A diesel oxidation catalyst and wall flow filter (or diesel particulate filter) are contained in one canister, a dosing system for injecting urea to the SCR catalyst. An ammonia decomposition catalyst (Pt/γ-Al_2O_3//ceramic monolith) is installed at the outlet of SCR. The DOC catalyzes the oxidation of CO, HC, and some of the NO to NO_2 and generates sufficient heat (~500 °C) by oxidizing injected diesel fuel to initiate combustion of the soot collected on the wall flow filter. The NO and NO_2 exiting the filter are mixed with inject urea, which hydrolyzes to NH_3 and enters the SCR catalyst. EGR may be used to further reduce the engine out NO. Not shown is a turbocharger than compresses the air as it enters the combustion chamber. This further enhances the power of the engine to move heavy loads.

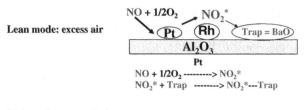

Lean mode: excess air

$$NO + 1/2O_2 \longrightarrow NO_2{}^*$$
$$NO_2{}^* + Trap \longrightarrow NO_2{}^*\text{---}Trap$$

Rich mode: excess fuel

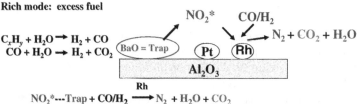

$$NO_2{}^*\text{---}Trap + CO/H_2 \longrightarrow N_2 + H_2O + CO_2$$

Figure 14.6 Chemistry of NO_x reduction in using BaO to capture NO_2 during lean operation.

supported Pt and Pd) serves to oxidize gaseous and liquid pollutants as well as to oxidize injected diesel fuel for soot filter (DPF) regeneration. The DPF (or CSF) traps the liquid and solid particulates, the SCR catalyst reduces the NO_x, and the ammonia decomposition catalyst destroys excess ammonia escaping the SCR.

The second promising technology for controlling NO_x at the tailpipe is lean NO_x trapping. Figure 14.6 illustrates the mechanism of a LNT. The technology utilizes a Pt- and Rh-based TWC catalyst in combination with an NO_2 trapping agent (e.g., an alkaline earth compound such as BaO). During the normal fuel economy mode of lean operation, NO is oxidized to NO_2 over the Pt catalyst and NO_2 is adsorbed by the BaO within the catalyst washcoat. Periodically (e.g., every 60–120 s), the trap is regenerated by introducing a "rich pulse" of reductant (e.g., diesel fuel) into the exhaust stream or by switching the engine operating mode to stoichiometric or slightly rich for 1–2 s. This rich pulse provides the necessary chemical reductant (CO and H_2) to convert the adsorbed nitrate to nitrogen catalyzed by Rh, as shown in Figure 14.6. Although LNT technology has been successfully demonstrated in vehicle applications, the primary disadvantages are that high Pt levels are required to maintain sufficient catalyst durability and a significant fuel penalty ($\sim$3–5%) results from the periodic trap regeneration. In addition, the SO_x derived from the fuel-borne sulfur forms $BaSO_4$ that is much more stable than the corresponding nitrates and are not easily removed during the stoichiometric or rich operation mode (Figure 14.7). Therefore, the trap becomes progressively poisoned by sulfates. Complicated engine control strategies have been developed to desulfate the poisoned trap by operating the engine at a high temperature ($>$550 °C) and rich of the stoichiometric air/fuel ratio for a short period of time. In addition, the air/fuel ratio must be carefully controlled to avoid the formation of H_2S during excessive rich conditions. LNT technology has the capability of removing up to 90% of the NO_x in the exhaust. Having lower sulfur fuels available will favor high NO_x conversion levels and also reduces the requirements for desulfation.

The driving cycle for a vehicle equipped with a lean NO_x trap is shown in Figure 14.8. Under the lean operation where maximum fuel economy is established,

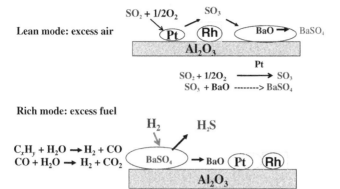

Figure 14.7   Deactivation of the NO_x trap by sulfur oxide poisoning.

NO_2 is adsorbed until a saturation point is reached. This is predictable from the engine map. The engine is then commanded to go slightly rich (or fuel is injected) raising the temperature, which decomposes BaO—NO_2 releasing NO_2 to the Rh catalyst and the *in situ* generated CO and H_2 where it is reduced. The cycle is then repeated:

$$BaO + SO_x \rightarrow BaO\text{—}SO_x \tag{14.6}$$

$$BaO\text{—}SO_x + H_2 \rightarrow \text{No reaction} \, (< 600\,^{\circ}C) \tag{14.7}$$

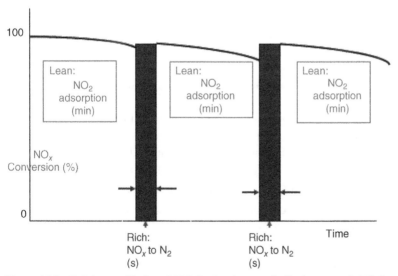

Figure 14.8   Driving profile for a LNT. During lean mode (fuel economy), NO is converted to NO_2 over a Pt catalyst, which is adsorbed in the alkaline trap. Regeneration (solid area) occurs when the engine is commanded to a stoichiometric mode where the NO_x is desorbed from the BaO and the Rh in the TWC reduces it to N_2.

QUESTIONS

1. What are the issues with the emissions from a diesel engine?

2. What are the functions of a diesel oxidation catalyst?

3. Why does a DPF have to be regenerated? List all possible methods to achieve this regeneration.

4. Explain how a LNT works. Design a LNT.

5. What are the system design issues for implementation of the LNT technology?

6. Describe a catalytic system for NO_x and PM control from a diesel exhaust. Draw the system and auxiliary equipment needed for implementation.

7. **a.** Describe how the NO_x trap works in a diesel engine?

   b. What are the anticipated deactivation modes?

   c. What are the pollutants that could be liberated when the engine operates too rich (λ much less than 1).

8. **a.** What happens when the soot build on the filter?

   b. How is the soot regenerated

   c. Why does not the catalyst decrease the ignition temperature?

   d. Why add Pt?

   e. What deactivation modes are expected for filter and catalyst

9. Please refer to the article Farrauto and Voss (1996), Monolith diesel oxidation catalysts, *Applied Catalysis B: Environmental* 10, 29–51. Note that emissions for diesel trucks are measured in units related to work expressed as grams per brake horsepower per hour (g/(bhp h)). Total particulate matter is composed of solids (dry soot and small amounts of sulfates), and liquids (diesel fuel + oil = soluble organic fraction).

   a. What can you conclude from Table 1 regarding the emissions and regulations?

   b. What was the role of the diesel oxidation catalyst?

   c. Describe the laboratory test the authors developed for screening candidate catalysts (see Table 2) in the article.

   d. Explain the data generated and how were they used to evaluate effectiveness of catalyst candidates.

   e. What was the main conclusion for the best catalyst in Table 2.

   f. What were the catalytic results when a cerium solution was impregnated into Al_2O_3 and calcined? This would make the surface rich in CeO_2.

   g. Explain the importance of Figure 5. What does Figure 8 tell us about poisoning?

   h. Why is $Pt/(CeO_2 + Al_2O_3)$ considered a dual function catalyst?

10. How was the truck catalyst modified to meet passenger car standards in Europe? Refer to Figures 9 and 10 in the article.

11. **a.** Describe the NO_x–particulate trade-off for diesel emissions?

   b. What catalysts and operational procedures are to be used to regenerate soot filters?

   c. Why does a TWC cannot be used to abate NO_x from a diesel engine?

   d. What unique performance do V_2O_5 and urea (NH_3) have that make them useful for NO_x reduction in diesel exhausts?

BIBLIOGRAPHY

Farrauto, R. and Voss, K. (1996) Monolithic diesel oxidation catalysts. *Applied Catalysis B: Environmental* 10, 29.

Farrauto, R., Voss, K., and Heck R. (1995) Diesel oxidation catalyst. *U.S. Patent* 5,462,907.

Heck, R. Farrauto, R., and Gulati, S.T. (2009) Chapter 8, in *Catalytic Air Pollution Control: Commercial Technology*, 3rd edn, John Wiley & Sons, Inc., Hoboken, NJ, pp. 238–292.

Hochmuth, J., Wassermann, K., and Farrauto, R. (2013) Car cleaning catalyst, in *Comprehensive Inorganic Chemistry II*, Vol. 7 (eds R. Schlogl and H. Niemantsverdriet), Elsevier, The Netherlands.

Johnson, T. (2008) Diesel engine emissions and their control. *Precious Metal Reviews* 52 (1), 23–37.

Morlang, A., Neuhausen, U., Klementiev, K., Schultze, F-W., Mieke, G., Fuess, H., and Lox, E. (2005) Bimetallic Pt/Pd diesel oxidation catalyst: structural characterization and catalytic behavior. *Applied Catalysis B* 60 (3 and 4), 191.

Robinson, K., Ye, S., Yap, Y., and Kolaczkowski, S. (2013) Application of methodology to assess the performance of a full scale diesel oxidation catalyst during cold and hot start NEDC drive cycles. *Chemical Engineering Research and Design* 91, 1292.

Weibenza, M., Kim, C., Schmieg, S., Oh, Se, Brown, D., Kim, D., lee, J., and Peden, C. (2012) Deactivation mechanism of Pt/Pd-based diesel oxidation catalysts. *Catalysis Today* 184 (1), 197.

NO_x Reduction

Forzatti, P., Lietti, L., Nova, I., and Tronconi, E. (2010) Diesel NO_x aftertreatment catalytic technologies: analogies in LNT and SCR catalytic chemistry. *Catalysis Today* 151, 202–211.

Miller, W., Klein, J., Mueller, R., Doelling, W., and Zuerbig, J. (2000) Urea selective catalytic reduction. *Automotive Engineering International* 125–128.

Morita, T., Suzuki, N., Wada, K., and Ohno, H. (2007) *Study on low NO_x emission control using newly developed lean NO_x catalyst for diesel engines.* SAE 2007-01-0239.

Particulate Filters

Huang, Y., Dang, Z., and Barllan, A. (2006) *Catalyzed diesel particulate matter filter with improved thermal stability.* U.S. Patent 7,138,358.

Konstandopoulos, A., Kostoglou, M., Skaperdas, E., Papaioannou, E., Zarvalis, D., and Kladopoulou, E. (2000) Fundamental studies of diesel particulate filters: transient loading, regeneration and aging. SAE 2000-01-1016.

ALTERNATIVE ENERGY SOURCES USING CATALYSIS: BIOETHANOL BY FERMENTATION, BIODIESEL BY TRANSESTERIFICATION, AND H_2-BASED FUEL CELLS

15.1 INTRODUCTION: SOURCES OF NON-FOSSIL FUEL ENERGY

The combustion of hydrocarbons to CO_2 and H_2O is the subject of our use of catalysts for pollution control. For example, we can combust glucose at low temperatures with the use of the proper catalysts. This reaction is thermodynamically favorable as is evident by the large enthalpy and free energy of the reaction.

$$C_6H_{12}O_6 + 6O_2 \rightarrow 6CO_2 + 6H_2O, \quad \Delta H^\circ = -2870\,\text{kJ/mol}, \quad \Delta G^\circ = -3000\,\text{kJ/mol}$$

(15.1)

Nature reverses this reaction through the photosynthesis process with the energy provided from the sun ($h\nu$).

$$6CO_2 + 6H_2O + h\nu \rightarrow C_6H_{12}O_6 + 6O_2$$ (15.2)

By producing hydrocarbon-containing fuels using photosynthesis, there is no net carbon introduced into the atmosphere upon combustion. The use of fuels in theory becomes carbon neutral, an important goal for energy sustainability.

This is the rationale for converting biomass to fuels. Carbohydrates present in edible products such as corn (starch) and vegetable oils (triglycerides (TRGs)) are convenient and relatively easy to convert to fuels. The conversion of corn, wheat, rice, and so on to ethanol has been practiced for centuries in enzymatic (nature's catalysts)

Introduction to Catalysis and Industrial Catalytic Processes, First Edition. Robert J. Farrauto, Lucas Dorazio, and C.H. Bartholomew.
© 2016 John Wiley & Sons, Inc. Published 2016 by John Wiley & Sons, Inc.

fermentation, while vegetable oils can be converted to biodiesel through the trans-esterification process using homogeneous and occasionally heterogeneous catalysts. These two technologies will be briefly discussed.

Ideally, it is preferred that this be done without disturbing the food chain (nonedible lignocellulose biomass) and thus this will continue to be a goal of scientists and engineers as we attempt to produce a sustainable environment, but for now the edible biomass remains the simplest feedstock to produce energy sources.

Alternative energy implies the use of renewable sources of energy other than those derived from fossil sources. This is necessary for sustainability since fossil fuels are in limited supply. Furthermore, extracting more and more carbon from the earth is increasing the concentration of CO_2 (greenhouse gas) in the atmosphere and having a profound influence on climate change. Thus, it is important to understand what alternative sources of energy are available and the chemistry, physics, and engineering aspects of these technologies.

We are excited about advancements being made in the development and commercialization of solar, wind, geothermal, and hydroelectric sources of energy. Nuclear energy must also be included, although it generates less excitement given its potential safety issues along with the issue of safe storage of nuclear waste. Solar energy will continue to supply renewable energy for photoelectric devices, but let us not forget its role in photosynthesis as a natural source of energy for biomass production. The sun is the source of energy for plant and food growth and this has served the world by feeding its population. Recently, corn, wheat, and rice, all rich in starch, have been exploited via catalytic enzymatic processing for the production of fuel ethanol. Similarly, vegetable oils, extracted from canola, soy, and palm beans, have provided triglycerides that can be catalytically processed to biodiesel fuel. As stated above, there is great hope that nonedible biomass (lignocellulose biomass) will be a source of fuel via biological and/or thermal chemical processing; however, these technologies are still under intensive study with limited but some promising success. Wind, geothermal, and hydroelectric sources of energy are also being utilized, but they will primarily be a supplement to a larger source of renewable energy. One possibility is a large-scale hydrogen economy where natural sources of energy can provide sufficient energy to electrolyze water leading to the clean generation of "green" H_2 and O_2 from water. If the hydrogen is then used as anode fuel for a fuel cell, we can generate heat and power for stationary applications while fueling fuel cell vehicles carbon-free. Furthermore, toxic emissions of CO, HC, NO_x, and soot associated with combustion of fossil fuels can be minimized.

The consequences would be a sustainable energy future with little or no new carbon being extracted from the earth for power generation. The carbon that remains in the earth can be used for producing chemicals necessary for comfortable and healthy lives.

This chapter will discuss the catalytic aspects of edible biomass utilization via fermentation and transesterification. The various types of fuel cells, using electro-catalytic electrodes, will also be discussed.

15.2 SOURCES OF NON-FOSSIL FUELS

15.2.1 Biodiesel

Purified triglycerides are reacted with methanol in the presence of alkaline homogeneous catalysts such as sodium methoxide ($NaOCH_3$) to produce fatty acid methyl esters (FAMEs) or biodiesel.

$$
\begin{array}{c}
CH_2\text{-}O\text{-}CO\text{-}R \\
| \\
CH_2\text{-}O\text{-}CO\text{-}R \quad + \quad 3\ NaOH \quad \xrightarrow[\substack{NaOCH_3 \\ CH_3OH}]{} \quad 3\ CH2\text{-}O\text{-}CO\text{-}R \quad + \quad CH_2OH \\
| \qquad\qquad\qquad\qquad\qquad\qquad\qquad\quad \textit{Biodiesel} \qquad\qquad | \\
CH_2\text{-}O\text{-}CO\text{-}R \qquad\qquad\qquad\qquad\qquad\qquad\qquad\qquad\qquad CH_2OH
\end{array}
$$

Glycerol

The R groups that usually vary in composition in each chain are a mixture of C_{12}–C_{18} straight-chain paraffinic hydrocarbons with varying degrees of unsaturation based on their origin. For example, soybean oils typically have one or two double bonds mainly in the *cis*-position, while palm oil is more or less completely saturated. The more saturated the bonds, the more stable the biodiesel is toward air oxidation and thus the more suitable for biodiesel use. Another important issue is the gelation temperature referred to in the petroleum industry as the pour point. In cold conditions, the straight-chain biodiesel tends to gel with poor flow properties. To counter this, biodiesel companies catalytically isomerize the biodiesel creating branched chains with lower pour points. The catalysts used are acidic zeolites.

15.2.1.1 Production Process TRGs are processed by a series of adsorption steps to remove impurities. Small amounts of free fatty acids (stearic acid) must be removed before the main transesterification reaction since they will neutralize the basic catalyst. They are converted to biodiesel using mineral acid catalysts such as H_2SO_4, which is removed by washing before the main reaction.

The purified oil is mixed vigorously with the methanol to ensure sufficient solubility in the presence of the homogeneous catalyst. The reaction is carried out at room temperature and atmospheric pressure. About 80–85% conversion is achieved in the first vessel due to equilibrium constraints between the biodiesel, glycerol, and TRGs. The glycerol is very dense relative to biodiesel, TRGs, and excess methanol and can be easily removed in a settling tank. This shifts the equilibrium to the product side and the reaction is continued until full conversion is achieved. The glycerol is again separated from the biodiesel and methanol. The product is neutralized with mineral acid and water (to remove the Na salts) and the excess methanol evaporated and recycled to the front end of the process. The glycerol is also purified and used commercially.

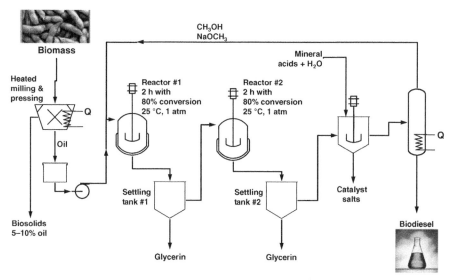

Figure 15.1   Biodiesel production process.

Energy supply to evaporate and recycle the methanol is an expensive step. The oil must be very dry to avoid soap formation.

15.2.2 Bioethanol

Ethanol, from edible sources, is getting enormous visibility as an alternative fuel to fossil-based gasoline for transportation use. The age-old process of natural fermentation by enzymes converts starch and oil in the seeds of corn, wheat, sugarcane, potato, and so on to ethanol used for beverages and/or fuel. Ethanol is renewable, consumes greenhouse gas CO_2 through the photosynthesis process, and can be grown in local environments independent of the geopolitical issues with petroleum. It does, however, disturb the food chain.

Brazil is one of the world's leading growers of sugarcane. After harvesting the milled cane, it is pressed to squeeze out the sucrose that is fermented to glucose. The addition of yeast with its catalytic enzymes converts the glucose to bioethanol. The residue of the cane (called bagasse) is rich in unreactive lignin and is usually burned for heat.

The corn kernel contains a high percentage of an amorphous starch (sugar polymers) that is hydrolyzed and fermented using the enzymes found in common yeast, to catalyze the conversion of sugars to alcohols. Thus, corn, which is a food staple, is being used to produce a fuel rather than edible products. Obviously, nonedible sources of biomass for energy production are needed. Ethanol has only 57% carbon versus 85–88% for gasoline, so the greenhouse gas emissions are reduced during combustion; however, it has a lower enthalpy creating an energy penalty. It does offer the positive advantage of a 13% higher octane rating than regular gasoline (100 versus 87) allowing higher compression ratios and more complete combustion enhancing power, which translates to energy savings. Ethanol (and its blends with gasoline) can be transported through the trucking infrastructure that exists for gasoline. The ethanol component is nontoxic, can be easily stored, is biodegradable, and contains no sulfur.

Gasohol is 10% ethanol and 90% gasoline and is referred to as E10 in the United States and Canada. Flexible fuel vehicles are being produced that allow blends of 85% ethanol with 15% gasoline (E85) to be used. The low percentage of gasoline (BP = 25–200 °C) is necessary due to the lower BP of ethanol (78.5 °C) assisting in cold climate start-up.

15.2.2.1 Process for Bioethanol from Corn The process for producing ethanol from corn involves four major steps: (1) milling, (2) fermentation, (3) distillation, and (4) dehydration. The first step involves milling the corn feed to produce a high surface area substrate ideal for the fermentation reactions. Next, the milled corn is added to water and heated to 50 °C forming a starchy "mash." The mash is cooled and yeast is added to catalyze the transformation of glucose to ethanol. After fermentation, the ethanol must be separated from the water and fermentation solids, which is done primarily via distillation. Ethanol forms an azeotrope with water and to obtain fuel-grade purity a dehydration step is required after distillation.

15.2.3 Lignocellulose Biomass

It is a great hope that technology will be able to utilize the lignocellulose portion of biomass to convert to energy. This would eliminate the competition with the food chain as is the current situation with ethanol and biodiesel. It would also significantly decrease greenhouse gas emissions. This portion of biomass is typically waste such as stems, leaves, corn husks, and wood, and at best is burned for its heating value. The goal is to convert this portion of the biomass to useful energy. One major problem is the high oxygen content that renders the liquids produced from them to be very unstable with a very low heating value. The second major problem is the resistance to attack by enzymes or chemicals to break through the lignin (polymers of aromatic phenols) that is present as fiber-like shells that protect the fermentable hemicellulose and cellulose. Hemicellulose constitutes the cell wall that contains the cellulose. It is composed of reactive C_5 and C_6 polymers. Cellulose has a structure of a polymer of glucose with some reactivity but less than hemicellulose due to hydrogen bonding. The biochemical approach to genetically engineer enzymes that can destroy the lignin without decreasing the fermentable portion of the hemicellulose and cellulose has had limited success. Some success with cellulose-rich plants has been reported but to date no large-scale plants have been built. Alternatively, there is the thermal–chemical approach of gasification with the addition of small amounts of O_2, steam, or CO_2 that produce synthesis gas for future use in Fischer–Tropsch (F-T) technology. The process most favored by the petroleum companies is to generate a liquid product through pyrolysis (absence of air). If the pyrolysis liquid generated could be deoxygenated (~47% oxygen) to improve its stability during transport, the existing petroleum refineries could upgrade it in a manner similar to their current treatment of petroleum feedstocks. Deoxygenation catalysis, using traditional hydrotreating materials such as supported Co/Mo, and supported Ni/Mo catalysts, has not been successful mainly due to coking and poisoning. Researchers are also studying other sources of biomass such as algae oil, but this feed is also very complicated and difficult to process. Hopefully, breakthroughs in technology will be forthcoming. This will remain a major issue for sustainability and hopefully funding for fundamental and applied research will be continued or increased.

15.2.4 New Sources of Natural Gas and Oil Sands

High concentrations of natural gas are contained in the earth's shale and new advanced drilling technology is making it plentiful. The environmental consequences are still being heavily debated regarding the environmental impact, including the use of large volumes of H_2O, toxic chemicals pumped into the ground, and possibly high methane emissions due to leaks. Methane is a powerful greenhouse gas and thus its control using catalysts will be a critical issue for the future. To date, only Pd catalysts appear to have some activity to convert methane to CO_2 and H_2O but further improvements are clearly needed.

Similar arguments hold for oil sands (tar sands) where the environmental impact is of deep concern to many. The easy energy has been extracted from the earth already. The remaining energy is more difficult to extract and therefore represents a greater environmental risk to the world community.

15.3 FUEL CELLS

Why are fuel cells a promising alternative to fossil fuel combustion for transportation and stationary power generation currently used in conventional power generating plants? A hydrogen–oxygen (air) low-temperature fuel cell directly converts its chemical energy to electricity and heat with efficiencies ranging from 40 to 80%. The lower efficiency is for electricity only, while the high efficiency is obtained by utilization of the heat. Figure 15.2 presents a cartoon comparison of the major power generation technologies with the fuel cell.

The top frame shows the combustion of coal (or any other carbon-based fuel) and the consequent generation of heat and CO, HC, NO_x, particulates, and CO_2. The heat generated vaporizes high-pressure steam in a boiler, which is directed to a steam turbine that spins a magnet in the field of a metal coil and induces a current by the electromagnetic induction effect. The efficiency of a steam turbine ranges from 45% for a single-cycle power plant to about 55% for a combined cycle power plant. The middle frame shows the familiar internal combustion engine that combusts a fuel in a piston, which drives the work stroke spinning the flywheel turning an axle producing motion to the wheels of the vehicle. It also generates primary pollutants due to incomplete combustion (CO and HC) and nitrogen fixation (NO_x). For a modern diesel passenger car, the efficiency is no greater than 20–25%. In each of these heat engines, fossil fuel is combusted and pollutants formed. The heat and high-pressure gases are converted to mechanical energy to generate power. The thermodynamics of this heat engine cycle limits efficiency.

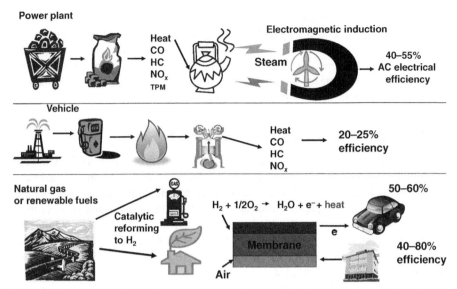

Figure 15.2　A comparison of power generation for a coal-fired power plant, gasoline/diesel internal combustion engine, and a H_2–O_2 low-temperature fuel cell. (Reproduced from Chapter 16 of Heck, R.M., Farrauto, R.J., and Gulati, S.T. (2009) *Catalytic Air Pollution Control: Commercial Technology*, 3rd edn, John Wiley & Sons, Inc., New York.)

Fuel cells require hydrogen for the anode reactions; however, no infrastructure exists for its distribution. Ideally, it should be generated by electrolysis of water using renewable energy from solar, wind, or geothermal sources. These technologies are localized and not always reliable to meet the demands for power and thus another convenient and local source is to convert a readily available infrastructure fuel, such as pipeline natural gas, to hydrogen. Natural gas is currently in great supply in many places in the world and although it is still carbon based it can be a transitional source of hydrogen by catalytic reforming. It must be repeated that the natural gas (mostly CH_4) is only a transitional fuel, while other natural sources of energy are being developed; however, it is rich in H_2 (25% relative to gasoline that has about 10–12% H_2) and is obtained from more politically stable regions of the world. The natural gas is to be converted to hydrogen via a hydrocarbon reforming process described in Chapter 6. One can envision hydrogen available for fuel cell vehicles at service stations where it is generated on sight. Alternatively, H_2 could be produced by reforming biofuel blends such as E85 (85% ethanol and 15% gasoline).

Returning to the bottom frame, it is seen that the fuel cell converts H_2 and O_2 to power without combustion. Since this operation has no mechanical steps, higher efficiencies than the traditional power generation systems are possible. The amount of greenhouse CO_2 produced during the reforming reactions is less than that produced by the heat engines described in top and middle frames by about 40%.

With the current state of cost and power efficiency, some argue against the fuel cell. Some studies suggest, from an energy point of view, that it is more feasible to use natural gas directly as a fuel for IC engines as opposed to converting it to H_2 for a fuel cell powered vehicle. The authors indicate that there is an advantage for H_2 regarding lower CO_2 emissions; however, it is important to recognize that combustion of any hydrocarbon-based fuel generates primary pollutants such as carbon monoxide, unburned hydrocarbons, oxides of nitrogen, and particulates. Clearly, we need a strategy to transition to a hydrogen economy from one that utilizes fossil fuels, which makes economic, political, and environmental sense. The key is to ultimately develop H_2 from renewable sources.

15.3.1 Markets for Fuel Cells

15.3.1.1 Transportation Applications
An ultimate goal for the hydrogen economy is to develop a fuel cell powered vehicle with H_2 stored on-board safely and with a reasonable driving range approaching 350 miles. The only fuel cell that shows promise for this application is the low-temperature Nafion®-based proton-exchange membrane (PEM) fuel cell because of its operation at 70 °C for fast start-up and its high power density relative to the other fuel cells. Simply stated, a 50–75 kW system can potentially fit under the hood of the vehicle. An additional advantage is its low operating temperature of 80 °C that will allow relatively rapid start-up. For this reason, many car companies are aggressively developing and testing PEM fuel cell systems.

A cost factor is the Pt electrocatalyst necessary for the fuel cell electrodes. The cost of such a system is extremely challenging with automobile manufacturers targeting no more than 15–20 g of Pt per vehicle depending on vehicle size. In comparison, a typical gasoline automobile has 2–5 g of precious metal (Pd, Rh) in the three-way converter. Diesel vehicles have a mixture of Pt, Pd, and Rh in different catalytic stages depending on the nature of the emission control system. Some estimate that the current low-temperature vehicle fuel cell contains more than 30 g of Pt, so cost reduction without sacrificing reliability is a key issue. A large number of vehicle manufacturers have announced (2014–2015) that they intend to market a fuel cell vehicle for about $50,000, so it is likely that they have reduced the Pt present in the electrodes. Another cost benefit is the absence of toxic emissions, thereby eliminating the catalytic converter. A secondary automotive application is auxiliary power in which a fuel cell will generate power for the vehicle's lights, heating, and entertainment systems.

High-pressure tanks lined with carbon or glass fibers are one choice for liquid hydrogen storage on the vehicle. Storage of tanked hydrogen is practiced today in fuel cell vehicles as well as hydrogen hybrid engines as practiced by BMW. Most automobile companies envision the local filling station generating high-pressure H_2 from natural gas (or possibly an ethanol-based fuel such as E85), which will charge either the high-pressure tank or hydrogen cartridge in your car. Charging with high-pressure H_2 can be accomplished quickly relative to the slow charging of a battery-only electric vehicle. There is also much activity in chemical storage materials composed of a wide variety of metal hydrides and intermetallic compounds such as LiH. $LiNH_2$, $NaAlH_4$, MnH_2, NiH_2, PdH, MgH_2, and so on are claimed to absorb up to their own weight of H_2. Carbon nanofibers have also received attention for storage;

however, other studies indicate that considerably more research and development is necessary before useful H_2 storage devices can be made of carbon fibers. Molecular organic framework (MOF) materials with high H_2 storage capacity with rapid fill and discharge capability are being evaluated.

The reader is advised to continuously check the websites of the auto companies for updates on progress (check websites for fuel cell vehicles). It is, however, not expected that they will penetrate the mass market to a large extent before 2020.

Construction of H_2 fueling stations is slowly progressing all over the world in preparation for the introduction of commercially viable fuel cell vehicles. The reader is encouraged to search the web for H_2 fueling stations. California has a particularly aggressive program (CaH2Net), but there are many others throughout the world. Clearly, there are major economic and logistical challenges to make such as infrastructure available.

15.3.1.2 Stationary Applications

One additional advantage is that some fuel cells lend themselves to a distributed model whereby power (electricity and hot water) can be generated in the home or commercial building independent of the existing centralized power station. The Japanese program called Enefarm has installed over 50,000 combined heat and power ($1\,kW_e$) low-temperature PEM fuel cell systems. The hydrogen can be produced from natural gas for which extensive infrastructures exist. Centralized power generating plants have large losses in transmission and are subject to brownouts and blackouts due to peak demands. Furthermore, they are subject to weather conditions and are expensive to transmit to homes or offices in remote locations. As new buildings are constructed, power companies must absorb the cost of adding new generating capacity. Clearly, fuel cells have a place in today's energy-hungry world provided they can be implemented with high reliability at competitive prices.

Larger scale stationary power demands can be met with different fuel cells depending on the power demands. For homes requiring less than about $5\,kW_e$, the Nafion-based PEM fuel cell system is the most popular for combined heat and power using mainly reformed natural gas although liquid petroleum gas (LPG) and kerosene are also under consideration. Molten carbonate fuel cells (MCFCs) and solid oxide fuel cells (SOFCs) are also being installed in major company facilities for power generation with natural gas as the source of H_2. Installations such as hospitals, schools, businesses, and centralized housing developments are installing phosphoric acid fuel cell (PAFC) systems. The new World Trade Center in New York City has 6 MW of power from a phosphoric acid fuel cell system. The reader should consult the websites of companies such as Tokyo Gas, Osaka Gas, Toshiba, Bloom Energy, FuelCell Energy, Ballard, Ceres, SAFCell, UltraCell, and Watt Fuel Cell to obtain information on the latest commercial installations.

15.3.1.3 Portable Power Applications

Portable power is likely to be the first major application for fuel cells primarily to replace or recharge batteries. The market is first responder emergency, military, and business communications, none of which are as cost sensitive as products for the mass market. Many conditions exist where grid power is not available and thus fuel cell power offers an alternative to batteries. Many companies are developing fuel cell systems for portable applications based on either direct methanol systems where the methanol is directly fed to the fuel cell anode and then electrochemically oxidized or reformed methanol fuels that deliver H_2 to the anode.

15.4   TYPES OF FUEL CELLS

There are four basic fuel cell systems. The solid polymer electrolyte or the low-temperature PEM fuel cell has an electrical efficiency of close to 40% (when used in a combined heat and power mode the efficiency approaches 80%). The alkaline fuel cell with an electrical efficiency of 50–60% is not suitable for terrestrial applications due to its sensitivity to CO_2 in the atmosphere neutralizing the electrolyte. The electrical efficiency of a phosphoric acid fuel cell is 35–40%. The high-temperature fuel cells such as molten carbonate and solid oxide fuel cells both have electrical efficiencies between 40 and 80% depending on the mode of operation and heat recovery. Each has its own particular application in diverse markets. The basic technologies for all four will be only described in very simple terms later in this chapter.

The first major use of fuel cells came in the 1960s for the space program. The acid-based PEM fuel cell was briefly used in the Gemini space program but was short lived due to the slow kinetics of the reaction relative to the alkaline fuel cell. It is now, however, the primary choice for terrestrial applications (stationary, portable power, and vehicles) due to high current densities and tolerance to CO_2 that neutralizes the electrolyte in the alkaline fuel cell.

Here alkaline electrolyte systems were used operating with liquefied H_2 and O_2 on-board. The use of phosphoric acid fuel cells followed in early 1990 by United Technologies (International Fuel Cells and ONSI) who commercialized the PC25 with some systems of up to 11 MW for large power plants such as that built by Tokyo Electric in 1993. Currently, there about 30 large power plants in Japan, the United States, and Europe utilizing fuel cells. There are approximately 250 units of 50–500 kW throughout the world used for stationary applications such as schools, apartments, and commercial building. So fuel cells have been commercialized but principally for niche markets. The desire now is to broadly produce them for the mass market.

The PEM fuel cell operates at 75 °C and is the number one choice for automobile companies for vehicle applications. The pioneering work of Daimler-Benz Chrysler-Ballard demonstrated NECAR 1 in 1994 at 50 kW with a weight per kW ratio of 21 kg/kW. NECAR 2 was demonstrated in 1996 at 6 kg/kW, NECAR 3 in 1997, and NECAR 4 in 2000. These vehicles were operated with an on-board methanol reformer to generate the H_2. On-board methanol is no longer considered viable for on-board vehicle applications given its toxicity and solubility in water. Many of these companies are no longer in existence specifically for fuel cell technology, but they represent a rich history of the pioneering work that has evolved to the fuel cell vehicles anticipated for the future.

A market expected before fuel cell vehicles are commercialized is that of residential or distributed power for homes. The concept is to reform fuels for which an infrastructure exists, such as natural gas, LPG, and/or kerosene providing H_2 for the anode of the PEM fuel cell for home and business use. Companies such as Plug Power have pioneered in developing this technology for American homes and businesses delivering $5 kW_e$. In Japan, companies such as Tokyo Gas, Nippon Oil, Osaka Gas, Sanyo, Toshiba, Aisin, and Mitsubishi Heavy Industries have targeted $1 kW_e$ systems designed to provide combined hot water and electricity. An additional complication is their model to start and stop once per day adding additional complications to the materials and control strategies for both the fuel cell and the integrated fuel processor.

Molten carbonate fuel cells operate between 600 and 700 °C and are being demonstrated in many locations. FuelCell Energy demonstrated its Direct Fuel Cell® (DFC®) 2 MW power plants in Santa Ana, CA in 1995. The same company, in cooperation with PPL Energy Plus, completed a demonstration in July 2000 of a 250 kW system for over 12,000 h in Danbury, CT generating 1.9 million kWh of power. Additional demonstrations of 250 kW units have been installed in Alabama (Mercedes-Benz), Louisiana, Asia, Germany, and a 1 MW unit in Washington (Kings County) operating on digester gas. Efficiencies of 50–60% are achieved by cogeneration (combined cycle) of electricity and heat. MTU is demonstrating a 250 kW cogenerating system in a hospital in Germany.

Solid oxide fuel cell systems, some of which operate at a low temperature of 500 °C while others closer to 800 °C, are being commercialized for centralized power generation above about 200 kW. In April 2000, Siemens-Westinghouse manufactured a natural gas-fueled 220 kW hybrid SOFC and microturbine for use in Irvine, CA. It provides electricity for 200 homes. The hot exhaust from the fuel cell drives the microturbine such that 55% efficiency is realized. Efficiencies approaching 70% are predicted. The anode fuel is CO and H_2 produced by partial oxidation and reforming of the hydrocarbon fuel conducted within or adjacent to the anode compartment. The anode is relatively thick, while the electrolyte and cathode are very thin. The cathode fuel is air. Their market is 1–25 kW stationary power and 3–5 kW auxiliary power (i.e., air conditioning, music center, etc.) for vehicles. Ceres Power, SAFCell, and Watt Fuel Cell are developing new SOFC systems that operate at 500 °C for residential applications.

The need to replace batteries in cellular phones, portable laptop computers, digital equipment battery chargers, and so on strongly suggests the value of a fuel cell operated on direct or reformed liquid fuel such as methanol, which can easily be changed simply by replacing a small lightweight cartridge. Direct methanol fuel cell portable power devices are available but in limited use. Companies such as UltraCell, DuPont, Casio, MTI, Toshiba, SMART, and Samsung are actively engaged in research, so the reader is encouraged to periodically check their websites for updates on commercialization.

15.4.1 Low-Temperature PEM Fuel Cell

15.4.1.1 Electrochemical Reactions for H_2-Fueled Systems The fuel cell is a galvanic cell in which spontaneous oxidation of a species occurs at the anode and reduction of another species at the cathode. Since the enthalpy of a reaction is a state function (independent of the path), the fuel cell generates the same enthalpy as the combustion of H_2 and O_2 to make water; however, some of the energy is used to generate electricity. The free energy for the reaction provides the driving force for electrons to move through an external circuit where they perform work in the form of electrical power with a portion generated as heat. The closer the system operates toward equilibrium ($\Delta G \sim 0$), the maximum amount of electricity can be generated with the balance of energy ($\Delta H \sim T\Delta S$) in the form of heat, which can be recovered and used productively. The positive voltage (E°_{cell}) output is equated to the total cell free energy by $\Delta G^{\circ}_{cell} = -nFE^{\circ}_{cell}$, a negative value indicating a spontaneous reaction. Here, n is the number of electrons transferred and the Faraday constant (F) is the number of coulombs per mole of electrons. The H_2–O_2 fuel cell (Figure 15.3) operates by the electrocatalytic oxidation of H_2 at the anode and reduction of O_2 at the cathode to form H_2O, electricity, and heat. It will continue to supply power provided H_2 and O_2 are continuously supplied and the electrocatalysts retain their activity.

Both anode and cathode reactions are catalyzed by Pt on carbon; however, their respective compositions of metal and carbon are different. The cathode carbon is more graphitic in nature and thus resistant to oxidation by the O_2 present as a fuel. The Pt levels are two to three times higher as in the anode since the electrochemical rate-limiting step is the reduction of O_2 at the cathode (Equation 15.4).

$$H_2 \rightarrow 2H^+ + 2e^-, \quad E^{\circ} = 0.00 \text{ V} \tag{15.3}$$

$$O_2 + 4H^+ + 4e^- \rightarrow 2H_2O, \quad E^{\circ} = 1.23 \text{ V} \tag{15.4}$$

One undesirable side reaction is the reduction of O_2 forming hydrogen peroxide with a negative voltage that must be subtracted from the single cell voltage of O_2 reduction to H_2O (Equation 15.5).

$$O_2 + H_2O + 2e^- \rightarrow HO_2^- + OH^-, \quad E^{\circ} = -0.07 \text{ V} \tag{15.5}$$

$$\text{Net reaction}: \quad H_2 + \tfrac{1}{2}O_2 \rightarrow H_2O, \quad E^{\circ}_{cell} = 1.16 \text{ V}, \quad \Delta H^{\circ} = -242 \text{ kJ/mol} \tag{15.6}$$

The net voltage of the cell (E_{cell}) is related to standard state cell voltage (E°_{cell}) and the partial pressures of the reactants and products through the Nernst equation.

$$E_{cell} = E^{\circ}_{cell} + (RT/2F_c)\ln(P_{H_2})/(P_{H_2O}) + (RT/2F_c)\ln(P_{O_2})^{1/2} \tag{15.7}$$

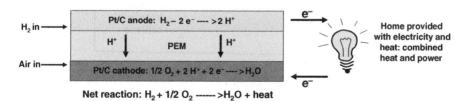

Net reaction: $H_2 + 1/2\ O_2 \longrightarrow H_2O + \text{heat}$

Figure 15.3 A single cell of the PEM fuel cell.

The net reaction (and the electricity and heat) continues provided H_2 and O_2 are continuously fed to the cell. This assumes that the electrocatalysts are not poisoned or deactivated in some manner. The anode and cathode compartments are separated by a solid polymer membrane that allows the H^+ ions generated at the anode to migrate to the cathode where they combine with the O_2 and form H_2O. The membrane prevents the H_2 and O_2 from mixing and provides a path for H^+ ions to migrate from anode where they participate in the cathodic reaction. The membrane will be discussed in more detail later.

15.4.1.2 Mechanistic Principles of the PEM Fuel Cell
The H_2 is first dissociatively chemisorbed onto the Pt electrocatalyst followed by electrocatalytic oxidation to protons and electrons. This is a relatively easy and fast reaction. The reduction of O_2 at the Pt/C cathode is the slower of the two reactions and controls the rate and power output of the fuel cell at low current densities. Thus, the reaction is initially electrochemically controlled, which is favored by higher Pt loadings. A corrosion-resistant carbon is also required given the potentials experienced at the cathode. The rate-limiting steps, however, change with increasing current drawn from the cell as discussed below.

Figure 15.4 shows a voltage–current profile for a typical low-temperature PEM fuel cell. The open-circuit voltage ($E°$) is the theoretical value of about 1.16 V. When current is drawn, there is a drop in voltage and power (power = voltage × current) controlled by the slow O_2 reduction at the cathode. As current is further increased, the voltage–current output more slowly decreases due to the transition of the rate-limiting step from the electrode O_2 reduction to H^+ migration through the membrane. One can envision this caused by increased H^+ traffic though the membrane. Finally, as the current required further increases, the output drops considerably due to polarization or

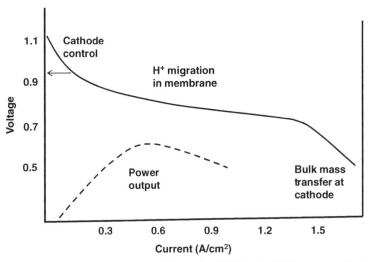

Figure 15.4 Voltage–current profile for the PEM fuel cell. The curve with the maxima represents the power profile.

lack of sufficient supply of gaseous O_2 to the cathode due to buildup of the water layer at the surface of the electrode. The O_2 (air) must undergo bulk mass transfer diffusion through the H_2O layer. Sweeping away the water by the incoming air is a critical issue that must be engineered into the air input flow field. Thus, there are three distinct steps in the fuel cell operation each of which can limit the rate of reaction analogous to those in heterogeneous gas-phase reactions.

A key factor to note is that the fuel cell converts chemical energy directly to electrical energy and heat with no mechanical steps involved and free of thermo-dynamic inefficiencies associated with traditional heat cycles in power generation. For this and its environmental benefits, there continues to be tremendous interest in developing cost-effective fuel cell systems as an alternative technology for conventional fossil fuel powered energy generation.

15.4.1.3 Membrane Electrode Assembly The electrodes consist of highly dispersed Pt (20–40 Å) deposited (about 30 wt%) on nonporous conductive carbon powders (300 Å in diameter). The layer of electrocatalyst about 50 μm thick is admixed with an optimized amount of a Nafion solution to enhance conductivity. It is deposited onto the surface of the membrane (50–175 μm) by spraying, painting, or filtration. The Pt loading is about 0.25 mg/cm^2, although it is desirable to reduce it to less than 0.1 mg/cm^2. The electrodes are hot pressed onto each side (anode and cathode) of the membrane to ensure intimate contact.

Pt on carbon is currently the only viable active electrocatalyst for both anode and cathode for hydrogen–oxygen acid electrolyte fuel cells. The reduction of O_2 on the cathode is a slow step, so about two to three times as much Pt in the anode is usually used. The cathode must use a carbon resistant to corrosion especially under air exposure open-circuit conditions (system off) where the electrode potential exceeds about 1.1 V. Ruthenium is sometimes present in the Pt anode when the CO levels in the H_2 exceed about 10 ppm, but cleaner H_2 from reformers will eliminate the need for it. Direct methanol fuel cells require Ru alloyed with the Pt in the anode to broaden the useful voltage–current range.

The anode and cathode gases are dispersed through gas diffusion layers (GDLs) positioned on top of each electrode. The diffusion layer is also permeable to allow the cathode product water to escape. The GDL is composed of electrically conductive carbon cloth (300–400 μm) woven from carbon fibers that are melt coated with Teflon (40–70%) to render them hydrophobic to prevent flooding by water. The main channels must be kept open for gas permeability. The combination of GDL, electrodes, and solid polymer electrolyte is called the membrane electrode assembly (MEA).

Electrically conductive nonporous graphite bipolar plates have grooved micro-channels on their surface to allow the gases to be delivered uniformly to their respective GDL. Polymer-based bipolar plates are also being developed. Sandwiched between these conductive plates is the single-cell MEA with an open-circuit voltage output of about 1.16 V. These are stacked in series to increase the power or voltage output. For this reason, the opposite side of the bipolar graphite plate is also grooved to permit the other reactant gas to flow as shown in Figure 15.5. The bipolar plates also contain heat transfer lines containing fluid to recover the heat and maintain operation at 70–75 °C.

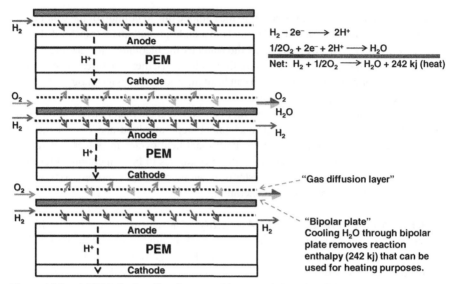

Figure 15.5   A PEM single cell and arranged in a "stack." Each cell is separated by an electrically conductive impermeable bipolar plate that serves as a gas manifold for the cells connected in series.

15.4.2 Solid Polymer Membrane

The most well-known solid polymer electrolyte used is polyperfluorosulfonic acid (PFSA) Nafion (DuPont trade name) developed in 1960 by DuPont. The Pt anodes can tolerate more CO than the low-temperature systems. If successful, this can have far-reaching positive consequences in simplifying the reformer since the water gas shift reactor can be reduced in size and the CO cleanup reactor eliminated.

PFSA

Hydrophobic Teflon® backbone has side chains of strongly acidic $-SO_3^-H^+$ that dissociate forming hydronium ions, H_3O^+. The thickness of the membrane varies between 50 and 175 μm. The membrane is conductive to protons provided sufficient moisture is present to permit ionization of the sulfonic acid group but is impermeable to gases. Inverted spherical micelles are formed due to the interface between the hydrophobic fluorocarbon structure and the hydronium ions forming channels 10–20 Å in size. Through these channels ion transport occurs giving rise to the ion conductivity of the membrane. This model is known as the "cluster network."

The membrane swells with water uptake and has high conductivity at a humidity close to 100% at 80 °C. Such a low temperature is very attractive for rapid electrochemical start-up for transportation and residential fuel cell applications, but low temperatures make the anode more susceptible to poisons such as CO always present in reformate. At low humidity or higher temperatures, the membrane acts as an insulator. Because the membrane has a low permeability of hydrogen and oxygen, it prevents their mixing across its interface. It is used extensively as the membrane in the chloralkali industry where Cl ions are oxidized to Cl_2 and water reduced to hydroxide at the cathode.

Alternative fluorine-containing membranes are also being developed by DuPont, Gore, 3M, and Ballard. Non-fluorine membranes are also being developed. A difference between membranes is related to the number of CF_2 groups (i.e., n) in the backbone of the solid polymer.

Membranes based on polybenzimidazole (PBI) that operate at temperatures up to 200 °C have also been developed. Phosphoric acid is used as the ionically electrically conducting medium as opposed to sulfonic acid groups in Nafion. Unlike F-containing polymer membranes, they require no humidification and operate at temperatures above 180 °C. The commercial availability of these membranes is somewhat questionable since major companies have closed facilities; however, it is believed that there is a supply from Japan (PBI Performance Products, Inc.).

15.4.3 PEM Fuel Cells Based on Direct Methanol

This system also utilizes PEM technology but is differentiated from the hydrogen-fueled system in that it operates directly on methanol with no need to reform the fuel to H_2. The operating temperature is 40–80 °C.

$$\text{Anode}: \ CH_3OH + H_2O \rightarrow CO_2 + 6H^+ + 6e^-, \quad E° = 0.029 \text{ V} \qquad (15.8)$$

$$\text{Cathode}: \ 1\tfrac{1}{2}O_2 + 6H^+ + 6e^- \rightarrow 3H_2O, \quad E° = 1.229 \text{ V} \qquad (15.9)$$

$$\text{Net reaction}: \ CH_3OH + 1\tfrac{1}{2}O_2 \rightarrow CO_2 + 2H_2O, \quad E° = 1.2 \text{ V} \qquad (15.10)$$

The electrolyte is an acidic solid polymer (i.e., Nafion) but it must be impermeable, so the methanol in the anode does not cross over to cathode. Permeability or methanol crossover still remains a problem minimizing the power output. Both electrodes contain large amounts of expensive Pt dispersed on conductive carbons, but the anode is a PtRu alloy on carbon to minimize its deactivation caused by partially oxidized products such as aldehydes and carboxylic acids that results in poisoning of the reaction. The electrochemical oxidation of methanol involves a six-electron transfer. The rate-limiting step for Pt only is reaction of adsorbed CO on Pt reacting with adsorbed OH. A voltage greater than about 0.8 V is needed to overcome this step, but this limits the current to low values. The presence of

Ru, alloyed to the Pt, catalyzes the adsorbed CO and OH reaction extending the voltage to 0.25 V allowing higher current densities.

In addition to the methanol crossover problem, the retention of Ru in the alloy that leaches into the acidic solution remains a problem. Evidence shows that the Ru migrates through the membrane onto the cathode creating a mixed potential that reduces the effectiveness of the cathode. Commercial products from SMART and DuPont are available for specialized low power output applications.

15.4.4 Alkaline Fuel Cell

The Apollo space program (1960–1968) used the alkaline fuel cell system because it had better kinetics and delivered a higher voltage than the acid-based system because there was no peroxide intermediate to lower the power output. The reactions are shown below.

$$\text{Anode}: 2H_2 + 4OH^- \rightarrow 4H_2O + 4e^-, \quad E^\circ_{anode} = -0.828 \text{ V} \tag{15.11}$$

$$\text{Cathode}: O_2 + 2H_2O + 4e^- \rightarrow 4OH^-, \quad E^\circ_{cathode} = 0.401 \text{ V} \tag{15.12}$$

$$\text{Net reaction}: H_2 + \tfrac{1}{2}O_2 \rightarrow 2H_2O, \quad E^\circ_{cathode} - E^\circ_{anode} = 1.23 \text{ V} \tag{15.13}$$

Although its lifetime is only 2000–5000 h maximum, this was sufficient for early space exploration. The water produced was used for drinking by the astronauts. Since liquid H_2 was used, there was no concern for CO_2, which will neutralize the alkaline electrolyte. This limitation renders the alkaline fuel cell impractical for any application where CO_2 is present as in the case of reformate. For these cases, the PEM is preferred.

The electrolyte for the 1981 space shuttle was KOH/asbestos with a mixture of Pt + Pd on carbon bonded by polytetrafluoroethylene (PTFE). The cathode was predominately Au promoted with a small amount of Pt on a nickel grid. Other possible electrode materials are Ni–Ti and Pt–Pd for the anode and Ag and perovskites for the cathode.

15.4.5 Phosphoric Acid Fuel Cell

The first commercialized fuel cells (PC25) were manufactured by United Technologies. Systems were supplied to Tokyo Electric in 1993 for an 11 MW power plant. Currently, there are a number of large power plants, schools, apartments, and commercial buildings all over the world from 250 KW to 10 MW of power using phosphoric acid fuel cells. The new World Trade Center in New York City has 6 MW of electrical power capacity using these fuel cell systems.

The reactions are identical to the PEM fuel cell in that H_2 is oxidized at the anode and O_2 from air reduced at the cathode but at about 200 °C. Because of the higher operating temperatures, the Pt anode can tolerate larger concentrations of CO and consequently CO purification is needed. The electrolyte is 100% phosphoric acid adsorbed on SiC and is sufficiently conductive at 200 °C. The anode is 0.1 mg/cm^2 Pt dispersed on carbon black (i.e., Vulcan XC-72) that is admixed with a Teflon polymer such as PTFE to render it hydrophobic to minimize flooding by water. The electrode is then printed via a doctor blade onto porous graphite paper composed of graphite fibers bonded with phenolic resins. The cathode is Pt on a corrosion-resistant carbon but requires more Pt, that is, $0.5 \, mg/cm^2$, since the kinetics of the electrocatalytic reduction of O_2 are much slower than the anode reaction.

Each electrode assembly, composed of the electrodes, electrolyte, and gas dispersion graphite paper, is stacked similarly to the PEM with cooling plates every four to six stacks. They are stacked using bipolar grooved or channeled conductive plates (necessary for gas flow) bonded on each side by the anode of one assembly and the cathode of the next producing a series stack for increased power output.

The major problem is cost ($4000/KW), relatively low current densities, and longer start-up times compared with the PEM that operates at a lower temperature. For this reason, they are no longer considered the most attractive system for residential or vehicular applications.

The major source of deactivation is corrosion and dissolution of the carbon and sintering of the Pt. Corrosion can be controlled by densifying the carbon black by treating in an inert atmosphere at elevated temperatures, that is, 90 °C.

15.4.6 Molten Carbonate Fuel Cell

The electrolyte is conductive for carbonate ions between 600 and 700 °C. It is typically 50% each of Li and K carbonates stabilized on γ-$LiAlO_2$ with additives of particles and/or fibers of α-Al_2O_3 to give mechanical strength. The selection of Li and K carbonates and the amount of each depends on resistance to solubility of the Ni-containing cathode as well as resistance to corrosion. Increasing the amount of Li carbonate increases basicity, which decreases solubility (Appleby and Nicholson, 1980). The presence of K carbonates decreases corrosion. Other additives such as Ca, Ba, or Sr also decrease solubility. The exact compositions are highly proprietary. Thicknesses are typically 0.5 mm and are prepared by tape casting.

The anode is porous sintered Ni ($1 \, m^2/g$, 50–70% porosity) with about 10% Cr to stabilize against excessive thermal sintering. Typically, the thickness of the catalyst layer is 1.5 mm. Frequently, a small amount of Li is added to further decrease sintering.

The cathode is Ni doped with small amounts of Li to minimize solubility of the Ni in the electrolyte. The surface area is less than $0.5 \, m^2/g$ (65% porosity) and about 0.75 mm thick. An alternative cathode material is $LiCoO_2$ that in limited tests performs essentially comparable to Ni cathodes but is less likely to corrode.

Additional O_2 is brought into the cathode compartment to combine with the CO_2 for the main electrochemical reduction reaction. The carbonate produced carries the current to the anode and participates in the electrochemical oxidation of the H_2. Each cell has a manifold for water, fuel, and air.

$$\text{Anode :} \quad H_2 + CO_3{}^{2-} \rightarrow H_2O + CO_2 + 2e^- \tag{15.14}$$

$$\text{Cathode :} \quad \tfrac{1}{2}O_2 + CO_2 + 2e^- \rightarrow CO_3{}^{2-} \tag{15.15}$$

$$\text{Net reaction :} \quad H_2 + \tfrac{1}{2}O_2 + CO_{2(\text{cathode})} \rightarrow H_2O + CO_{2(\text{anode})} \tag{15.16}$$

$$E = E^\circ + (RT/2F)\ln(P_{H_2})/(P_{H_2O})(P_{CO_2})_{\text{anode}} + (RT/2F)\ln(P_{O_2})^{1/2}(P_{CO_2})_{\text{cathode}} \tag{15.17}$$

The voltage per cell varies between about 0.7 and 1.0 V depending on the current drawn. Individual cells are stacked, making a repeating series generating proportionally higher voltage and power. For a typical megawatt power plant, about 340 cells are stacked. Steam, fuel, and air are added to each cell. Typically, each cell is about 2 ft × 3 ft × 4 ft. Each stack is targeted to have a 40,000 h life, but it is not unusual to replace components periodically due to corrosion, fatigue, and catalyst deactivation at the severe operating conditions.

A unique feature of both the MCFC and SOFC (Section 15.4.7) is the internal reforming process. Due to the high temperatures necessary for these fuel cells, a fossil fuel can be reformed in the anode chamber with the endothermic energy provided by the fuel cell. The hydrocarbon fuel is mixed with steam (about 2–3 moles of H_2O per mole of C) at about 650 °C. The reforming catalyst Ni/Al_2O_3 is positioned adjacent to the anode but usually in a separate compartment, so the fuel is internally reformed with the endothermic heat of reaction provided by the fuel cell. This eliminates the entire unit operations associated with the external reformers for the lower temperature fuel cells. It also puts less demand on the cooling equipment provided heat integration is successfully included in the design.

$$CH_4 + H_2O \rightleftharpoons 3H_2 + CO, \quad \Delta H = 49 \, \text{kcal/mol} \tag{15.18}$$

$$CO + H_2O \rightarrow H_2 + CO_2, \quad \Delta H = -10 \, \text{kcal/mol} \tag{15.19}$$

In most modern molten carbonate and solid oxide fuel cell designs, some precondition of the fuel be either steam reforming or catalytic partial oxidation of the hydrocarbon has been found to improve the performance and life of the SOFC by minimizing coke formation in the anode as well as minimizing the amount of cooling occurring in the anode during internal reforming causing thermal stresses in the fuel cell compartment.

As H_2 is produced, it is oxidized electrochemically shifting the equilibrium for both the reforming and water gas shift reactions, producing more H_2 and CO_2. For this reason, little CO is present in the effluent. The anode exhaust gas composed of mainly H_2, some unreacted hydrocarbon, and CO_2 and H_2O (from reaction at the anode) is mixed with some air and passed through a catalytic oxidizer designed to convert the hydrocarbons and H_2 to H_2O and CO_2. The exhaust is then passed to the cathode compartment to provide the CO_2 necessary for the cathodic reaction. Due to the molten salt vapor pressure, some is carried in the anode exhaust and deposits on the anode oxidizer, leading to its deactivation. Some scrubbing of the alkali carbonates is designed into the process loop to minimize deposition onto the catalysts with subsequent deactivation.

The steam reforming catalyst is a specially designed Ni-based material not unlike what is conventionally used in standard steam reforming plants practiced in the chemical industry (Chapter 6). The higher temperature operation permits cogeneration of electricity and heat, at about 450 °C, which improves the overall system efficiency to almost 60%.

It is subject to deactivation by coke formation, sintering, and poisoning by impurities in the fuel, that is, sulfur compounds, as well as alkali carbonate from the electrolyte. Shields of SiC and other ceramic membranes have been developed that minimize the poisoning effect of the electrolyte.

The current density is much lower than that for the PEM, so its major market is for large-scale power plants for buildings, among others.

15.4.7 Solid Oxide Fuel Cell

The electrolyte is typically a material such as 10% Y_2O_3-stabilized ZrO_2. The Y^{3+} replaces a Zr^{4+} in the lattice, freeing an O^{2-} for conduction from the cathode to the anode. Conduction occurs at a reasonable rate at 1000 °C. The anode is 30% porous and is composed of Ni/ZrO_2 (150 μm thick). The cathode is $LaMnO_3$, doped with about 30% Sr (1 mm thick). Recent developments are available.

The anode reactions are shown in Equations 15.20 and 15.21.

$$H_2 + O^{2-} \rightarrow H_2O + 2e^- \qquad (15.20)$$

$$CO + O^{2-} \rightarrow CO_2 + 2e^- \qquad (15.21)$$

The cathode reaction is shown in Equation 15.22.

$$O_2 + 4e^- \rightarrow 2O^{2-} \qquad (15.22)$$

The net reaction is shown in Equation 15.23.

$$O_2 + H_2 + CO \rightarrow H_2O + CO_2 \qquad (15.23)$$

Typical cell voltage is 0.8 V at 1 A/cm^2 and 1000 °C. The high operating temperatures cause a decrease in the thermodynamic free energy of formation of H_2O (a product of the anode reaction) leading to a 100 mV loss relative to other fuel cells. It has a reported efficiency of over 60–80% for cogeneration systems.

Since the SOFC operates at 800 °C, less active but more stable catalysts can be used. Furthermore, the catalysts (electrocatalysts) are much less sensitive to impurities in the fuel, so it is especially attractive for fuels generated from coal-based gasification plants. It is still desirable to remove most of the sulfur to preserve their life.

As with the MCFC, the Ni-based reforming catalyst must be in close contact with the anode for heat management. No CO_2 recycling is required as it is in the molten carbonate fuel cell. There are no flooding issues or electrolyte migration since the electrolyte is an O^{2-} conductive solid oxide.

A potential problem is the thermal stress due to expansion differences between anode–cathode and the solid electrolyte that leads to delamination. Altair has reported some new electrode/electrolyte designs that minimize this problem.

15.5 THE IDEAL HYDROGEN ECONOMY

We are on the road to a hydrogen economy. Success will free us from substantial dependence on petroleum-based fossil fuels, will render combustion for power generation no longer necessary, and thus will make obsolete the need for primary pollution abatement of CO, HC, NO$_x$, and particulates. The H_2 will be ideally

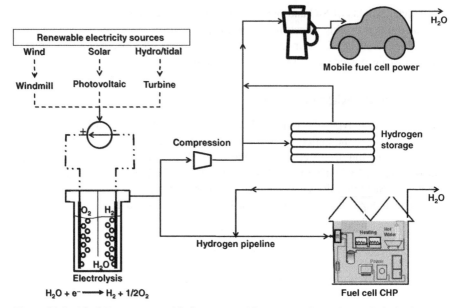

Figure 15.6   Ideal H_2 economy with the sun providing energy for a photovoltaic device generating sufficient voltage to electrolyze water to H_2 and O_2.

generated from renewable energy such as solar photovoltaic or wind, which will provide the power to electrolyze water to H_2. Over time we will slowly see the replacement of heat engines that require fossil fuels with hydrogen or renewable fuels. We are talking about an ideal hydrogen economy that can operate a fuel cell. This idealistic description is represented by the cartoon in Figure 15.6.

Here the sun is the source of energy generating a voltage through a photovoltaic device sufficient to electrolyze water. The released H_2 is fed to the anode of a fuel cell cleanly generating heat and electricity with the only product being water.

QUESTIONS

1. How is biodiesel made and what advantages does it have compared with fossil fuel-derived diesel?

2. What is the biological/agricultural path for preparing ethanol for beverages and fuels in the United States and other countries? What other feedstocks are under consideration to produce fuel ethanol? What are the technical challenges for nonedible plants to be used for fuels and what solutions are being investigated? Briefly discuss the Fischer–Tropsch catalyst process and its role in future fuels.

3. Explain the basic principles of the low-temperature PEM fuel cell.

4. Explain its advantages over conventional power generation technologies.

5. Draw a typical voltage versus current plot for the PEM fuel cell output. On the same plot show how the plot will change if you have improved the activity of the cathode catalyst,

decreased the resistance to H^+ migration through the membrane, and minimized mass transfer resistance of air to the cathode by enhancing turbulence.

6. What are the technical approaches under consideration for converting lignocellulose (nonedible biomass) to fuels? What are the general processes under consideration from biomass to fuels? What are the technological issues that must be addressed?

7. What are the advantages of the low-temperature (80 °C) PEM fuel cell versus other engine processes to generate power?

8. Ideally, how will we generate H_2 from natural sources of energy?

9. How will the voltage be increased for the PEM fuel cell to meet appliance requirements? Sketch the system.

10. Given the voltage–current plots shown in class, describe the possible rate-limiting steps (show reactions when necessary) and how you can increase the performance for each step.

BIBLIOGRAPHY

Fuel Cells

Appleby, A and Nicholson, S. (1980) Reduction of oxygen in lithium-potassium-carbonate melts. *Journal of Electrochemical Chemistry* 112, 71.

Gasteiger, H., Kocha, S., Sompalli, B., and Wagner, F. (2005) Activity benchmarks for Pt, Pt alloys, and non-Pt oxygen reduction catalysts for PEMFCs. *Applied Catalysis B: Environmental* 56 (1–2), 9.

Hoffmann, J., Yuh, Y., and Jopek, A. (2003) Molten carbonate fuel cells and systems: electrolyte and material challenges, in *Handbook of Fuel Cells*, Vol. 4, Part 2 (eds W. Vielstich, A. Lamm, and H. Gasteiger), John Wiley & Sons, Inc., Hoboken, NJ, pp. 921–941.

Kawada, T. and Mizusaki, J. (2003) Solid oxide fuel cells and systems: current electrolytes and catalysts, in *Handbook of Fuel Cells*, Vol. 4, Part 2 (eds W. Vielstich, A. Lamm, and H. Gasteiger), John Wiley & Sons, Inc., Hoboken, NJ, pp. 987–1001.

Membranes for Fuel Cells

Jones, D. and Roziere, J. (2003) Inorganic/organic composite membranes, in *Handbook of Fuel Cells*, Vol. 3, Part 1 (eds W. Vielstich, A. Lamm, and H. Gasteiger), John Wiley & Sons, Inc., Hoboken, NJ, pp. 447–455.

Lin, J., Kunz, R., and Fenton, M. (2003) Membrane/electrode additives for low-humidification operation, in *Handbook of Fuel Cells*, Vol. 3, Part 1 (eds W. Vielstich, A. Lamm, and H. Gasteiger), John Wiley & Sons, Inc., Hoboken, NJ, pp. 456–464.

Mauritz, K. and Moore, R. (2004) State of understanding of Nafion®. *Chemical Reviews* 104, 4535–4585.

Xiao, L., Zhang, H., Jana, E., Scanlon, R., Choe, E., Ramanathan, L., and Benicewicz, B. (2005) Synthesis and characterization of pyridine-based polybenzimidazoles for high temperature polymer electrolyte membranes. *Fuel Cells* 5 (2), 287–295.

Xiao, L., Zhang, H., Scanlon, R., Ramanathan, L., Choe, E., Rogers, D., Apple, T., and Benicewicz, B. (2005) High-temperature polybenzimidazoles for fuel cell membranes via a sol–gel process. *Chemistry of Materials* 17, 5328–5333.

Biomass

Huber, G., Iborra, S., and Corma, A. (2006) Synthesis of transportation fuels from biomass: chemistry, catalysis and engineering. *Chemical Reviews* 106, 4044–4098.

Laval, A. and Farrauto, R.J. (2013) Chapter 8: The convergence of emission control and sources of clean energy, in *New and Future Developments in Catalysis: Remediation and New Fuels* (ed. S. Suib), Elsevier.

Mousdale, D. (2008) *Biofuels: Biotechnology, Chemistry and Sustainable Development*, Taylor & Francis, Boca Raton, FL.

Waldron, K. (ed.) (2010) *Bioalcohol Production: Biochemical Conversion of Lignocellulosic Biomass*, Woodhead Publishing, Oxford, UK.

H$_2$ Generation

Farrauto, R.J. (2014) New catalyst and reactor designs for the hydrogen economy. *Chemical Engineering Journal* 232, 172–177.

Farrauto, R.J., Liu, Y., Ruettinger, W., Ilinich, O., Shore, L., and Giroux, T. (2007) Precious metal catalysts supported on ceramic and metal monolithic structures for the hydrogen economy. *Catalysis Reviews* 49, 141–196.

Nielsen, J. (1993) Production of synthesis gas. *Catalysis Today* 18 (4), 3015.

H$_2$ Storage

Funck, R. (2003) High pressure storage, in *Handbook of Fuel Cells*, Vol. 3, Part 1 (eds W. Vielstich, A. Lamm, and H. Gasteiger), John Wiley & Sons, Inc., Hoboken, NJ, pp. 83–86.

Mueller, U., Schubert, M., Teich, F., Puetter, H., Arndt, K., and Pastre, J. (2006) Metal-organic frameworks—prospective industrial applications. *Journal of Materials Chemistry* 16, 626–636.

Wolf, J. (2003) Liquid hydrogen technology for vehicles, in *Handbook of Fuel Cells*, Vol. 3, Part 1 (eds W. Vielstich, A. Lamm, and H. Gasteiger), John Wiley & Sons, Inc., Hoboken, NJ, pp. 89–100.

INDEX

A

abatement catalysts, 218
abatement system, 222
ABS. *see* acrylonitrile-butadiene-styrene (ABS)
acetic acid, 208
acid-based PEM fuel cell, 283
acrolein, 165
acrylic acid production, 164
    catalysts, 164–165
        deactivation, 166
    process design, 164–165
    reactor design, 165
acrylonitrile, 166
    catalyst, 168
    deactivation, 168
    production, 167–168
acrylonitrile–butadiene–styrene (ABS), 166
activation barrier, 1, 5
activation energies, 1, 2, 3, 4, 5, 6
    apparent, 82–83
    Arrhenius plot for determining, 83
    for non-catalytic thermal reaction of CO
        and O_2, 4
    for Pt-catalyzed reaction, 5
active catalytic
    components, 9
    materials, 31
    species, 7
adhesion, of oxide-based washcoat, 246
adiabatic bed design, 157
adiabatic flame temperature, 236
adipic acid, 184
adsorption, 10
    isotherm for nitrogen for BET surface area
        measurement, 50
adsorption, models, 10–13
AES. *see* Auger electron spectroscopy (AES)

air/fuel ratio, 252
air to fuel ratio, 235–236
Al-containing species, 94
aldehydes, 206
alkaline electrolyte systems, 283
alkylation, ethene & benzene to ethyl
    benzene, 187
all-battery electric vehicles, 230
Al_2O_3 beads, 258
Al_2O_3 carriers, structure and morphology
    of, 58
γ-Al_2O_3 particles
    attrition resistance of, 250
    usage in catalysis, 9
alumina silicates, 31, 35
aluminum hydroxide, 244
ammonia (NH_3), 129, 268
    converter, illustration of, 134
    oxidation (AMOX), 24, 146
    synthesis, 119, 129
        catalyst deactivation, 134
        catalyst design, 130
        process design, 132
        reaction chemistry, 130
        thermodynamics, 129
ammonia oxidation
    catalyst deactivation, 150–151
    catalyst design, 147–148
    reaction chemistry, 146–147
    reactor design, 148–150
ammonia synthesis
    catalyst deactivation, 134
    catalyst design, 130–132
    process design, 132–134
    reaction chemistry, 130
    reactor design, 132–134
    thermodynamics, 129–130

ammonium nitrate, 146
AMOX. *see* ammonia oxidation (AMOX)
Anderson-Shultz-Flory, 141
anode reactions, 293
antioxidants, 213
Apollo space program, 290
apparent activation energy, 82
Arrhenius expression, 3
Arrhenius plot, for determining activation
        energies, 83
Arrhenius profile, 3
L-aspartic acid, 210
ATR. *see* autothermal reforming (ATR)
Auger electron spectroscopy (AES), 63
automobile engine, 237
automotive applications, traditional beaded
        (particulate) catalysts
            failure, 250, 251
automotive catalysts, 7
    converter, 25
    rate control regimes, 247
autothermal reforming (ATR), 111–112

B
base metals, 31, 183
    catalysts, 240, 250
Bayerite, 33
benzene, 187
benzene oxidation to maleic anhydride, 166
BET. *see* Brunauer-Emmett-Teller (BET)
beta zeolite, 36, 37
biochemical approach, 278
biodegradable detergents, 186
biodiesel, 276
    non-fossil fuels, sources of, 276
        production process, 276–277
    production from triglycerides, 276–277
bioethanol
    from edible sources, 277–278
    non-fossil fuels, sources of, 277, 278
    from corn, 278
biofuels from lignocellulose biomass, 278
biomass
    conversion of, 190
    for energy production, 277
    to fuels (alternative energy source),
        274–275
BMT. *see* bulk mass transfer (BMT)
Boehmite, 33
Boltzmann distribution, 4

Brønsted acid sites, 33
Brunauer-Emmett-Teller (BET)
    equation, 49
    surface, 240
        area, 242
bubble/slurry-phase process, 142
bulk mass transfer (BMT), 21, 218, 219
    coefficient, 78
    control, conversion *vs.* temperature
            profile, 21
    rate, 78
        expression, origin of, 79, 80
        as function of temperature, 22
bulk (film) mass transfer rate, 78–80, 220
butane oxidation to maleic anhydride, 166

C
calcinations, 32, 246
carbohydrates, 274
carbon balance, 26–27
    experimental methods for measuring, 27
carbon-free energy, 230
carbon monoxide (CO), 4, 7, 216
    catalyst deactivation, 222–224
    catalytic incineration of, 216–224
    catalyzed monolith/honeycomb
            structures, 219–220
    cleanup reactor, 288
    deactivated catalysts, regeneration of,
            224
    emissions, 215, 235
    monolith/honeycomb structure, 218–219
    reactor sizing, 220–222
    reduction catalyst, 251–255
    removal methods, 116
        methanation, 117
        preferential oxidation, 117
        pressure swing absorption, 116
    simultaneous conversion of, 252
carbon–sulfur polymers, 155
carboxylation
    acetic acid from methanol, 208–209
    acetic acid production, 208–209
carrier materials, 8
    Al$_2$O$_3$, 32–34
    carbons, 37
    SiO$_2$, 34
    TiO$_2$, 34, 35
    zeolites, 34–36
carsuls, 155

catalysis, chemical and physical steps, 19–23
catalyst
  beds, 136
  carrier/support, 8
  chemical and physical properties, 6–10
  chemical & structural properties, 54–65
  in industrial applications, 6
  measurement of particle size, 51–53
  mechanical strength, 53–54
  morphology, 56–58, 61–62
  physical properties, 49–54
  poisoning, 96–99
catalyst converter, in exhaust, 242
  ceramic monoliths, 242–245
  metal monoliths, 245, 246
catalyst deactivation, 134, 180
catalyst design, 130
catalyst forming, 40–45
catalyst manufacturers, 257
catalyst preparation, 37–40
catalyst $PtRe/\gamma-Al_2O_3$, 8
  adding Cl^- to a naphtha reforming, 8
catalyst regeneration, 99, 100, 224
  TGA/DTA in air of coke burn-off from a catalyst, 100
  TGA/DTA profile for desulfation of Pd on Al_2O_3 catalyst, 99, 100
catalyst removal, safety considerations, 116
catalyst selectivity, 147
catalytic abatement, 238
catalytic components, 8, 9
catalytic converters
  for cooking emissions, 225
  precious metal recovery, 248
catalytic cracking, 187
catalytic crystal size, 10
catalytic enzymes converts, 277
catalytic Fe–Ce redox reaction catalyzed by Mn, 3
catalytic materials, 54
  active phases, 31
  carrier or support, 31–35, 37
  chemical and physical morphology structures, 54
    elemental analysis, 54, 55
    scanning electron microscopy, 56, 57
    thermal gravimetric analysis (TGA), 55, 56
    x-ray diffraction (XRD), 57

  structure and morphology of Al_2O_3 carriers, 58
  zeolites, 35–37
catalytic oxidation, 217
catalytic reactions, 1
  during catalytic abatement, 238
  fundamental steps, 19–10, 48
  generic illustration, 248
catalytic reactor, 8
catalytic sites, 5, 19
catalytic species–carrier interactions, 95
catalytic surface area, 9, 10, 19
catalyzed carrier (Al_2O_3), 246
catalyzed coating, 243
catalyzed monolith nomenclature, 248
catalyzed reactions, 3
catalyzed soot filter (CSF), 267
catalyzed *vs.* noncatalyzed reactions, 1–4
Cativa™ process, 208
CATOFIN technology, 185
C_{12}–C_{18} straight-chain paraffinic hydrocarbons, 276
cell density, 244, 245
ceramic monoliths, 219
ceria (CeO$_2$), 37
chain-driven charbroiler, 226
chemical and energy synthesis, 6
chemical equilibrium constant, 77
chemical kinetics, 21, 23
  conversion *vs.* temperature profile, 21
  relative rates as a function of temperature, 22
chemical partial bond, 5
chemical promoters, 7
chemical reactions, 1
chemisorption, 5, 8, 58–61
  isotherm for determining surface area of the catalytic component, 60
Chilean saltpeter (NaNO$_3$), 129
chiral compound, 210
Claus process, 154
  catalyst deactivation, 155
  description, 154
  for production of sulfur, 154–155
Clean Air Act, 239
Clean Air Amendment of 1990, 264
close-coupled catalyst, 256–258, 257
  TWC catalyst, 258
CO_2 abatement, 230–231
coal-fired power plants, 228, 280

cobalt carbonyl catalyst HCo(CO)$_4$, 206
cobalt carbonyls, 206
cobalt-containing catalysts, 240
cobalt homogeneous catalyst,
    hydroformylation process, 207
CO$_2$ emissions, 231, 281
coke formation, 99–100. *see also* catalyst
    regeneration
coke regeneration, 196
combustion, 125, 235
    of coal, 279
commercial ceramic monoliths, 243
complicated fuels, autothermal reforming
    for, 126
constriction, 81
continuous stirred tank reactor (CSTR), 172,
    207
control automobile emissions, 236
conversion of CO *vs.* temperature for a
    noncatalyzed (homogeneous) and
    catalyzed reaction, 5
cooled tube reactor design, illustration
    of, 138
CO oxidation, Pt catalyzed, 4–6
CO$_2$ reduction, 230–231
corn (starch), 274
    bioethanol, 278
    ethanol, 278
    kernel, 277
corn kernel, 277
corrosion, 291
cracking, 31
    hydrocarbons, 197–200
crude oils, 8, 190
    distillation, 191–193
    origin and properties, 190–191
    transformation of, 191
crystalline materials, 58
crystallite size of catalytic species, 58
CSF. *see* catalyzed soot filter (CSF)
CSTR. *see* continuous stirred tank reactor
    (CSTR)
Cu-containing catalyst, 8
Cu/Cr catalyst, 240
CuCr$_2$O$_4$, 241
Cu zeolite SCR catalyst, 268

D

deactivation, 48. *see also* thermally induced
    deactivation

coke formation, 99–100
cooking emissions catalyst, 226
    masking (fouling), 97–99
    metal-carrier interactions, 95
    poisoning, 96–99
    sintering, 89–95
    steam reforming catalysts, 110–111
    thermally induced, 88–89
    VOC catalysts, 222
    water-gas-shift catalyst, 116
dechlorination, 181
dehydroaromatization, 201
dehydrocyclization reactions, 201
dehydrogenation, 185
    alkanes to alkenes, 185–187
    alkyl benzenes, 185–187
    catalytic, 177
    deactivation & regeneration, 186
    endothermic process, 185
    gas-phase, 185
    heterogeneous catalysts, 185
dehydroisomerization, 200
deoxygenation catalysis, 278
depending on the specific application, an
    oxidation catalyst (DOC), 269
desorption, 6
detergent alcohols (C$_{12}$–C$_{18}$), 207
diammonium phosphate, 152
diesel catalyzed soot filter (CSF), 266–267
diesel engines, 265
    abatement, 262–272
    emissions, 262–264
diesel exhaust
    aftertreatment system, 269
    AMOX catalyst for removal of NH$_3$, 269
    analysis, 264
    lean NO$_x$ trap (LNT), 270–271
    reduction of NO$_x$, 267–271
    SCR catalysts for NO$_x$ reduction,
        268–269
    system, DOC, CSF, SCR, & AMOX, 269
diesel-fueled vehicles, 262
diesel oxidation catalyst (DOC), 265–266
diesel particulate filter (DPF), 266, 267
diesel reduction catalyst technology,
    265–271
diesels, fuel economies, 262
differential porosimetry, for a porous
    catalyst, 52
differential thermal analysis (DTA), 55–56

decomposition of barium acetate on
  ceria, 55
mode, 55
dispersion, 58
model of supported catalyst, 17–19, 89
distillation, 191
double-layered washcoated ceramic
  monoliths, optical micrographs
  of, 243
DTA mode. *see* differential thermal analysis
  (DTA) mode

E
edible biomass, 275
effective diffusivity, 81
effectiveness factor, 22
EGO. *see* exhaust gas oxygen (EGO)
electro-catalyst, 281, 285
electrode assembly, 291
electrolyte, acidic solid polymer, 289
electron microprobe, 57
  showing a two-washcoat-layer monolith
    catalyst, 57
electron transfer process, 3
electrostatic precipitator (ESP), 228
elemental analysis, 54–55
Eley-Rideal (E-R) mechanism, 18
  kinetic mechanism, 18
  reaction, 14
emission reducing, from diesel engines
  catalytic technology, 265–271
emission regulations, NO_x–particulate
    trade-off, 263
emissions, 238
  of CO, 215
  from diesel engines, 262–264
  standards, 264
empirical kinetic parameters, experimental
    measurement of, 73–76
  Arrhenius expression, 75
  conversion *vs.* temperature at different
    space velocities, 76
  determining activation energy, 75
  inhibition effects, 76
empirical power rate expressions, 72, 73
empirical rate expression
  accounting for chemical equilibrium
    in, 77
  forward rate, 77
  power rate law, 77

endothermic thermal reaction, 1, 9, 154
Enefarm, 282
energy barrier, 3
energy processes, 7
energy savings, 6
engine dynamometer aging cycles, 255
engine exhaust, 251
enthalpy, 3
entropy, 6
environmental abatement reactor, 220
environmental emission control, 6
Environmental Protection Agency
    (EPA), 236
environmental washcoated monolith catalysts
    physical properties of, 54
      washcoat adhesion, 54
      washcoat thickness, 54
enzymatic catalysis, 209–210
enzymes, 209–210
  bonded to porous glass, 210
  catalyze reactions, 209
equilibrium constant, 3, 104
E-R mechanism. *see* Eley-Rideal (E-R)
    mechanism
ESP. *see* electrostatic precipitator (ESP)
ethanol, 237, 277
ethene oxidation to ethylene oxide, 160–164
ethyl benzene (EB), 186, 187
ethylene, 187
ethylene oxide, 159
  catalyst, 159
  catalyst deactivation, 160
  production process, 160
ethylene oxide production
  catalyst deactivation, 164
  catalyst design, 162, 164
  process design, 160, 162–164
  reactor design, 162, 164
excess slurry, 219
exhaust, 266
  catalyst converter, 242
  piping, 239
exhaust gas oxygen (EGO), 252
exothermic reactions, 1, 5, 9, 249
  temperature profiles, 250

F
fatty acid methyl esters (FAMEs), 276
Faujasite Y zeolite, 197
FCC. *see* fluidized catalytic cracking (FCC)

Fe–Cr–Al high-temperature alloy, 219
Federal Test Procedure (FTP), 237, 238
    particulates, 264
    for trucks, 264
    vehicle emissions, 237
feed temperature, 186
Fe zeolite SCR catalyst, 268
first-order isothermal reaction, 77
    special case for, 77, 78
Fischer-Tropsch synthesis, 140, 230
    bubble slurry reactor for, 142
    catalyst deactivation, 143
    catalyst design, 141–142
    gas for, 120
    loop reactor for, 144
    process design, 142–143
        bubble/slurry-phase process, 142
        packed bed process, 143
        slurry/loop reactor (synthol
            process), 143
    reaction chemistry, 140–141
    reactor design, 142–143
fixed bed reactor, 8
flammability, 217
fluid catalytic cracking (FCC), 197–199, 198
    catalysts, 197–199
        deactivation, 198–199
        regeneration, 199
    fluid bed reactor, 199
    process, 197–199
    schematic of, 199
fluidization, 168
fluidized gas-phase reactor, 213
fluorine-containing membranes, 289
food processing, 225
    catalyst abatement of, 225
    catalyst deactivation, 226
    fumes, catalyst abatement of, 225
    restaurant cooking, 225
formaldehyde, 160
    high-methanol production process, 163
    low-methanol production process, 162
fossil fuel-derived gasoline, 237
fouling, 222
fructose from glucose, via glycose
            isomerase, 210
FTP. see Federal Test Procedure (FTP)
fuel cells, 279–283
    alkaline, 290
    FuelCell Energy, 284

high efficiency energy conversion,
            279–294
high-temperature solid oxide, 293
hydrogen for anode reactions, 280
markets, 281–284
    portable power applications, 282, 283
    stationary applications, 282
    transportation applications, 281, 282
molten carbonate with nickel anode,
            291–293
phosphoric acid, 290–291
types of, 283–293
    alkaline fuel cell, 290
    H_2-fueled systems, electrochemical
            reactions, 284–286
    membrane electrode assembly,
            287, 288
    molten carbonate fuel cell, 291–293
    PEM fuel cells
        based on direct methanol, 289, 290
        low-temperature, 283
        mechanistic principles, 286, 287
    phosphoric acid fuel cell, 290, 291
    solid oxide fuel cell, 293
    solid polymer membrane, 288, 289
    vehicles, 230, 237
fuel economies, 262
fuel-rich operating conditions, 238
furfural, 183

G
galvanic cell, 284
gas diffusion layers (GDLs)
    anode and cathode gases, 287
gaseousammonia, 227
gaseous pollutants, 264
gas hourly space velocity (GHSV), 71, 148
gas–liquid separator, 213
gasohol, 278
gasoline, 235
    close-coupled catalyst, 257
    close-coupled converter, 256–258
    emission control, 240
    engines, 262
    -fueled internal combustion engine, 200
    -relative engine emissions, 236
    spark-ignited engine, 262
gasoline converter
    catalyst deactivation, 255–256
    catalyst performance, 248–250

failure of pellet catalyst, 250–251
precious metal recovery, 242–247
simulated aging, 255–256
three-way catalyst (TWC), 251–255
gasoline engine
converter design, 239–259
monolith converters, 242–247
octane rating (number), 237
gasoline engine cleanup
catalysts, 239–259
catalytic abatement, 238–259
chemistry, kinetics, 238, 247
gasoline engine emissions, regulations,
235–238
gasoline TWC
air/fuel ratio (AFR), λ, 235–236
exhaust gas oxygen sensor, 252–253
oxygen storage component, 253–254
gas-phase oxidation reaction, 4
gas-to-liquid (GTL) technology, 140
GDLs. *see* gas diffusion layers (GDLs)
gelation temperature, 276
geometric surface area (GSA), 218, 243
GHSV. *see* gas hourly space velocity
(GHSV)
glucose from corn starch, via
glycoamylase, 210
glycerol, 276
GTL. *see* gas-to-liquid (GTL) technology

H

Haber–Bosch ammonia synthesis, 113
HC emissions, 235
HCN. *see* hydrogen cyanide (HCN)
HC reduction catalyst, 251–255
HC, simultaneous conversion of, 252
HDS. *see* hydrodesulfurization (HDS)
heat management, 9
heat of reaction, 141
heat transfer limited reaction, 107
heat treatment, 33
heavy-duty trucks, 268
hemicellulose, 278
He–Ne laser beam, 53
Henry's law constant, H_2, 175
heteroatom VOCs, 216
heterogeneous catalysis, 5, 10, 19
chemical and physical steps during,
19–21
sequence of, 20

importance of physical and chemical
properties
fundamental steps involved in, 48, 49
mechanisms, 13, 14
processes limiting the reaction rate
during, 70
heterogeneous catalysts, 6, 31, 48, 205
materials, 19
supported on a high surface area
carrier, 19
physical structure, 6, 7
heterogeneous CO oxidation
physical and chemical steps occurring
during, 19, 20
H_2 fueling stations, 282
high-temperature shift (HTS), 113
homogeneous catalysis, 205–213
homogeneous catalysts, 205–209
HTS. *see* high-temperature shift (HTS)
hydrocarbons, 104, 193
catalytic incineration of, 216–224
combustion of, 274
cracking, 197
fluid catalytic cracking, 197
hydrocracking, 200
oxidations, 5
hydrocracking, 199–200
hydrodemetalization, 193
hydrodemetalization (HDM), 193–197
catalysts, 194–196
deactivation & regeneration, 194–196
porphyrin, 193–194
hydrodesulfurization (HDS), 8, 106, 154,
193–197
catalysts, 194–196
deactivation & regeneration, 194–196
thiophene, 193–194
hydrodynamics, 10
hydroformylation
aldehydes from olefins, 206–208
catalysts, 206, 208
process design, 206–208
hydroformylation, aldehydes from
olefins, 206–208
hydrogenation, 171
of acetophenone, 183
base metal, 183
vs. noble metal catalysts, 183–184
biomass to polymer, 183
catalysts, 177

hydrogenation (*Continued*)
    deactivation, 180
    continuous stirred tank reactor, design
        equation for, 176
    of CO_2 to methane, 18
    design equation, CSTR, 176–177
    of functional groups, 180
    functional groups, catalysts, 180–183
    furfural, 183
    liquid phase, 171, 174–176
        reactors, 171–173
    mass transfer, 175–176
    organic functional groups, 8,
        180–183
    of organic molecules, 171–184
    precious metal catalysts, 183
    reactions, 177
        and catalysts, 177–184
        kinetics, 174–176
    slurry-phase hydrogenation reaction,
        kinetics of, 174
    in stirred tank reactors, 171
    of vegetable oils, 177–180
        catalysts, 177–180
        for edible food products, 177
hydrogen chemisorption, 242
hydrogen cyanide (HCN), 151
    deactivation, 152
    production
        catalyst deactivation, 152
        catalyst design, 151–152
        process design, 151–153
        reaction chemistry, 151–152
        reactor design, 151–152
    production process, 152
hydrogen economy, 293–294
    catalyst and reactor designs, 122
hydrogen, for anode reactions, 280
hydrogen generation
    for fuel cells, 121
        combustion, 125
        complicated fuels, autothermal
            reforming for, 126
        hydrogen economy, catalyst and reactor
            designs, 122
        methanol, steam reforming, 126
        preferential oxidation, 125
        steam reforming, 123
        water gas shift, 124
    industrial process, 105

ammonia synthesis, 119
    catalyst removal, safety
        considerations, 116
    CO removal methods, 116
    Fischer-Tropsch Synthesis, synthesis
        gas for, 120
    hydrodesulfurization, 106
    methanol synthesis, 120
    partial oxidation, 106
    steam reforming, 106
    water gas shift, 112
hydrogen–oxygen (air) low-temperature fuel
    cell, 279
hydrogen production
    ammonia synthesis, 119–120
    CO removal, 116–119
    for fuel cells, 121–122
    new catalyst designs, 122–126
    new reactor designs, 122–126
    via steam reforming, 105–120
hydrophobic Teflon® backbone,
    288
hydrotreating, 193
    petroleum fractions, 193–197
hysteresis, 51

I
IC engines, 280
ideal H_2 economy, 294
ideal hydrogen economy, 293–294
industrial processes, 69, 104
infrared–DRIFTS, 65, 66
infrared spectroscopy (IR), 65–66
inlet air containing pollutants, 222
International Fuel Cells, 283
intraparticle diffusion, 22
irreversible phase transitions, 34
isomerization, 31, 36, 179
isooctane, 235
isotactic polymer, 212–213

K
Kelvin equation, 51
kerosene, 282
kinetically controlled models, 10
kinetic parameters, 18
    determination, 73–77
kinetic *vs.* empirical rate models,
    18
Knudsen diffusion coefficient, 81

L

lambda sensor, 252
Langmuir–Hinshelwood kinetics, 14
  applied to increasing P_{CO} at constant
    P_{O2}, 16
  for CO oxidation on Pt, 14–16
  ideal dispersion of Pt atoms on a high
    surface area Al_2O_3 carrier, 17
  reaction model, 247
Langmuir–Hinshelwood mechanism, 13,
  148
Langmuir isotherm, 11–13
L-aspartic production
  catalyzed by L-aspartase, 210
lean NO_x traps (LNT)
  driving profile, 271
  technology, 270
LHSV. *see* liquid hourly space velocity
  (LHSV)
lignocellulose biomass, 278, 279
linear plot of the BET equation for surface
  area measurement, 50
linear velocity, 83
liquid hourly space velocity (LHSV), 71, 194
liquid petroleum gas (LPG), 282
liquid-phase redox reaction, 2
lubricating oils, 263
  components, 229
  electron microprobe, 255

M

maleic anhydride, 166
  catalyst deactivation, 166
  production, 166
Mars-van Krevelen kinetic mechanism, 14,
  17, 18
masking, 222. *see also* poisoning,
  nonselective
mass transfer, 23
  bulk (film), influence on rate, 19–23
  coefficient, 220
MCFCs. *see* molten carbonate fuel cells
  (MCFCs)
MEA. *see* monoethanolamine (MEA)
measurement
  catalyst chemical properties, 54–65
  catalyst morphology, 56–58, 61–62
  catalyst physical properties, 49–54
  crystallite size, 58–62
  metal dispersion, 58–61

metal oxide bonding, 64–65
  reaction rate, 73–77
  surface composition, 62–64
mechanical strength, 53, 54
mercury intrusion porosimetry, 51–52
mercury penetration as a function of pore size
  of catalyst, 52
metal dispersion and crystallite size, 58–61
metallic catalytic component, 8
metal oxides, 37, 240
metals substrates, 245
methane, 111, 216
methanol, 134, 276
  crossover problem, 290
  quench reactor design, illustration of, 136
  -soluble CoI, 208
  steam reforming, 126
methanol synthesis, 120, 134
  catalyst deactivation, 139–140
  catalyst design, 135
  flow sheet for, 139
  process design, 136–139
    quench reactor, 136
    shell-cooled reactor, 138
    staged cooling reactor, 137
    tube-cooled reactor, 137
  reaction chemistry, 134–136
  reactor design, 136–139
mineral acids, 187
Mn catalyst, 3
mobile applications, 230
modern catalysts, 265
MOF materials. *see* molecular organic
  framework (MOF) materials
mole balance, 177
molecular dimensions, 10
molecular organic framework (MOF)
  materials, 282
molecular sieves or zeolites, 197
molten carbonate fuel cells (MCFCs),
  282, 284
monitoring catalytic activity, in
  monolith, 248–250
monoethanolamine (MEA), 117
monohydrate (boehmite) alumina, 33
monolith catalyst, 239
  monitoring catalytic activity, 248–250
  preparation, 246, 247
  preparation, for gas engine, 246–247
  Pt catalyst, 217, 218

monolith geometries, 220
monolithic catalyst
    cell density, gas engine, 244–245
monolithic catalysts
    preparation, VOCs, 219–220
    reactor sizing, 220–224
monolithic supports, 44–45
monolith properties
    gas engine, 244–245
    VOC incineration, 220–222
monolith reactors, VOC incineration,
        218–220
monolith structures, 9
monolith *vs.* thermal combustion, 217
monsanto acetic acid process, 209
mordenite, 36
morphology, of carrier, 48

N
Nafion®, 281, 288
Nafion-based pem fuel cell system, 282
Nafion solution, 287
nanosized clusters, 19
naphtha reforming, 200–202
    catalyst, 200
    catalyst redispersion, 201–202
    catalyst regeneration, 201–202
    deactivation, 201
    process design, 201
        flow diagram for, 201
natural gas, new sources of, 279
natural oils, 177
nature's catalysts, 209
negative catalysts, 34
net voltage of cell (E_{cell}), 285
Ni/Al$_2$O$_3$ steam-reforming catalyst, 8
nitric acid, 146
    from ammonia oxidation, 146–151
    catalyst deactivation, 150
    catalyst design, 146
    production process, 148
        design, 148–150
    reaction chemistry, 146
nitric oxide (NO), 147
nitrobenzene, 180
nitrogen adsorption/desorption isotherm for
        pore size measurement, 50
nitrogen fixation, 129, 279
nitrogen oxide (NO$_X$), 215
    particulate trade-off, 263

reduction
    catalyst, 251–255
    deactivation, 229–230
    ozone abatement, in aircraft cabin
        air, 229
    selective catalytic reduction (SCR)
        technology, 227–229
    from stationary sources, 226–230
    in using BaO to capture NO$_2$, 270
nitrogen porosimetry, 49–50, 49–51
NMHCs. *see* non-methane hydrocarbons
        (NMHCs)
NO. *see* nitric oxide (NO)
noncatalytic
    free radical reactions, 5
    gas-phase oxidation of CO, 5
    oxidation, 217
    reactions, 3, 5, 6
non-fossil fuels
    sources of, 274–275
non-methane hydrocarbons (NMHCs), 237
nonprecious metals, 31
    failure of, 240, 241
NO$_x$ emissions, 215, 226–227
nuclear magnetic resonance (NMR), 64–65,
        94
nylon, 151

O
octane, 237
octane number, 200
oil sands
    new sources of, 279
olefins, 185, 206
ONSI, 283
open frontal area (OFA), 244
organic functional groups
        hydrogenation, 180–183
O$_2$ storage component (OSC), 253
overall particle porosity, 81
oxidation catalysts, 239
    first-generation, 239, 242
        converters, 239, 240
    nonprecious metals, failure of, 240, 241
    precious metal, deactivation/stabilization
        of, 241, 242
oxidation, of hydrocarbons, 7
oxidation reactor, 167
oxide-based washcoat, adhesion of, 246
oxides, 31

oxygenates, 237
oxygen sensor, 253
oxygen storage component, 253–254
ozone abatement
  in aircrafts, 229–230
  Pd catalyst, deactivation, 229–230
  reactor design, 229–230
ozone-containing makeup air, 229

P
packed bed process, 143
PAFC systems. *see* phosphoric acid fuel cell
  (PAFC) systems
palladium, 217
palm kernel, 178
partial oxidation (PO), 106
  of methane, 111–112
particle size, 23
  distribution, 51–53
  measurement using laser light scattering
    analysis, 53
particulates (PM), 263
  analytical procedures for, 264
  catalysts for fixed bed reactors, 9
PEM fuel cell
  electrochemical reactions, 285–287
  electrode polarization, 286–287
  Faraday equation, 285
  low temperature, 284–290
  membrane electrode assembly, 287–289
  methanol operation, 289–290
  Pt electrode catalysts, 285
  voltage-current profile, 286
petroleum refinery, 191
phase transformation, 35
Phillips loop reactor, 211
phosphine ligands, 208
phosphoric acid fuel cell (PAFC)
  systems, 282
photocatalytic reactions, 35
photosynthesis process, 274
  hydrocarbon-containing fuels, 274
physical properties of catalysts, 48, 49
  nitrogen porosimetry, 49–51
  pore size, 49
    by mercury intrusion, 51
  surface area, 49
plant-derived oils, 177
  canola, 177
  corn, 177

cottonseed, 177
  peanut, 177
  soy, 177
PO. *see* partial oxidation (PO)
poisoning, 96
  catalysts, 96–99
  nonselective, 97–99
  selective, 96–97
polybenzimidazole (PBI), 289
polyethylene
  Phillips process, 210–211
  production, 210–212
  $TiCl_4/MgCl_2$ process, 210–212
  $TiCl_4$ Ziegler-Natta catalyst, 211–212
polymerization of olefins, 210–213
polyolefins
  polyethylene, 210–212
  polypropylene, 212–213
polyperfluorosulfonic acid (PFSA), 288
polypropylene, 210, 212
  production, 212–213
  Ziegler-Natta catalyst, 212
polytetrafluoroethylene (PTFE), 290
pore channel tortuosity, 81
pore diffusion, 21, 22, 23
  bulk, 81
  conversion *vs.* temperature profile, 21
  influence on rate, 19–23
  Knudsen, 81
  rate, 80–82
  relative rates as a function of
    temperature, 22
  theory, 81, 82
pore size, 48
  distribution, 49, 51
  by mercury intrusion, 51
  and volume, measurement, 49–51
porous network, of carrier, 19
precious metal, 183
  -containing DOCs, 267
  deactivation/stabilization of, 241, 242
  oxidation catalyst, 241
  recovery, 248
  recovery, from catalytic converters, 248
  salts, 250
precipitating agents, 219
pre-exponential factor, 3
preferential oxidation (PROX), 117, 125
pre-reforming of light hydrocarbons, 111
pressure, 4

pressure drop, packed bed, 83–84
pressure swing adsorption (PSA), 106
process design, 132
promoters, 7
propene & NH_3 oxidation to
       acrylonitrile, 167–168
propene oxidation to acrolein to acrylic
       acid, 164–65
propylene, 165
propylene glycol, 166
proton-exchange membrane (PEM) fuel
       cell, 281, 283, 285, 290
  single cell, 288
  technology, 289
  voltage-current profile, 286
PROX. *see* preferential oxidation (PROX)
PSA. *see* pressure swing adsorption (PSA)
pseudoboehmite, 32
Pt atoms, 19
Pt catalyst, 5
  electrocatalyst, 281, 286
Pt-Rh catalyst, 24
  alloy catalyst, 147
Pt/Rh ratio, 254
purified oil, 276
pyrometallurgical method, 248

Q
quench reactor, 136
  design, 132
  SO_3 production, 158

R
rate constant, 3
rate equation: approach to equilibrium, 77
rate expression, power law, 73
rate-limiting process (RLP), 82–83
rate-limiting step, 6, 22
reactant compositions, 4
reactant concentration gradients, 22
  within a spherical structured catalyst, 23
reactants, 1, 2, 3, 5, 10, 19
reaction chemistry, 130
reaction kinetic models, 13, 14
  first order, isothermal, 77–78
reaction rate, 3, 6
  definition of, 71, 72
  O_2 concentration, 247
reactive organic gases (ROGs), 215–216
reactor bed pressure drop, 83, 84

reactor engineering, 70
regeneration
  of catalyst species, 1
  coked catalyst, 99–100
  VOC catalysts, 224
regulations, in United States, 236–237
renewable fuel, 230
residence time, 69
Reynolds number, 218, 222
R groups, 276
Rh catalyst, 207, 208, 271
  TWC catalyst, 270
rhodium triphenylphosphine homogeneous
       catalysts, 206
ROGs. *see* reactive organic gases (ROGs)
Ru catalyst, 18
ruthenium, 131, 287

S
safety, removal of reduced Cu and Ni
       catalysts, 116
scanning electron micrograph (SEM), 32
  energy dispersive analyzer, 57
  γ-Al_2O_3 and α-Al_2O_3, 33, 58
  morphology of catalytic materials by,
       56, 57
scanning electron microscopy (SEM), 56–57
Schmidt number, 79, 80
SCR. *see* selective catalytic reduction (SCR)
secondary ion mass spectroscopy
       (SIMS), 63
selective catalytic reduction (SCR), 36
  with Cu and Fe zeolites, 268
  NH_3 reduction reactions, 269
    catalysts, 227–229
    chemistry, 227–228
    coal power plants, 227–229
    reactor design, 228–229
  of NO_x, 227–229
  reactor schematic, 229
  SCRsystems, 228
  technology, 227–229
selectivity, 24–26, 25
  calculations for reactions with multiple
       products, 25, 26
  general equation for, 24, 25
SEM. *see* scanning electron micrograph (SEM)
sequential reactions, 179
shell-cooled reactor, 138
  design, illustration of, 138

Sherwood number, 80
silicon carbide (SiC), 266
simulated aging methods, 255–256
sintering, 33, 186
  of carrier, 92–95
  of catalytic species, 89–91
  metal crystallites, 89–91
  support, 92–95
SiO₂-Al₂O₃
  carrier material, 19
  ratio, 36
slipstream testing, 224
slurry/loop reactor (synthol process), 143
slurry-phase hydrogenation reaction, 174
slurry-phase process, 210
sodium methoxide (NaOCH₃), 276
solar energy, 104
solid catalyst, 4, 5
solid oxide fuel cells (SOFCs), 282,
      284, 293
soluble organic fraction (SOF), 263
space time, 69, 70
space velocity (SV), 69, 70
spark-ignited engine, 236
spectroscopy, in situ and ex situ, 65–66
SR. see steam reforming (SR)
staged cooling design, illustration of, 137
staged cooling reactor, 137
standard cubic feet per minute (SCFM), 245
standard performance tests, 238
starchy mash, 278
stationary source catalyst technology, 227
steam reforming (SR), 104, 106, 123
  catalysts, 108–109, 140, 292
    deactivation, 110–111
  of hydrocarbons, 104–112
  pre-reforming, 111
  process, 106–111
  reactor, 108
stirred tank reactors, 172
stoichiometric point, 235
structure of catalyst, 8
sulfur-containing chemicals, 155, 222
sulfuric acid production, 34, 155
  catalyst deactivation, 158–159
  catalyst design, 156
  process design, 155–158
  reaction chemistry, 155–156
  reactor design, 157–158
sulfur oxide (SOₓ) compounds, 34

Sulfur oxide poisoning
  NOₓ trap, deactivation of, 271
sulfur production via Claus process,
      154–155
supported catalysts, 31
supporting catalytic component, 19
surface area, 48
  measurement, 49–50
  and pore size, 49
surface kinetics
  rate of, 72
  reaction, 69
surface, reaction models, 13–18
syndiotactic isomer, 212
syngas production
  autothermal reforming, 111–112
  Fischer-Tropsch synthesis, 120–121
  methanol synthesis, 120
  partial oxidation, 111–112
  steam reforming, 111–112
synthesis gas, 105, 118, 140
synthetic cordierite, 244

T
tanks, high-pressure, 281
temperature, 4, 23
temperature-programmed oxidation
      (TPO), 56
temperature-programmed reduction
      (TPR), 56
terephthalic acid, 182
textural promoter, 7
thermal gravimetric analysis (TGA), 55–56
  decomposition of barium acetate on
      ceria, 55
thermally induced deactivation, 88, 89
thermal stresses, 265
thermocouples, 9
thermodynamics, 129
  equilibrium, 132
  function, 1
  properties, 3
thin-wall cordierite substrates
  nominal properties, 221, 244, 245
three-way catalysts (TWCs)
  exhaust system, 246
  performance, 256
time–temperature relationships, 33
titania, 34, 35
toluene, 201

tortuosity, 81
total particulate matter (TPM), 263
total wall surface area (TSA)
    of monolith, 218
TPO. *see* temperature-programmed oxidation
    (TPO)
TPR. *see* temperature-programmed reduction
    (TPR)
transition metals, 240
    -exchanged zeolite-based catalyst, 268
transmission electron microscopy
    (TEM), 61–62
    Pt on CeO_2, 62
triglycerides (TRGs), 274
trihydrate (bayerite) alumina, 33
tube-cooled reactor, 137
tubular reactors, 8, 166

U
unburned diesel fuel, 263
unburned hydrocarbons (UHC), 235
United States
    Federal Test Procedure, 238
    regulations, 236–237
urea, 147

V
vacuum oils, 191
vanadia (V_2O_5) catalysts, 24, 25, 35,
    227, 228
vegetable oils, 274
    hydrogenation, 177–180
volatile organic compounds (VOCs),
    215–224
    abatement, 215
        restaurant cooking, 225–226
    applications, 222
    CO, catalytic incineration of, 216–222
    design, 224
    process with heat integration, 223
    technology, 225
volatilization, 159
voltage–current profile, 286
volume of reactor, 69

volumetric flow, 70, 71
Vulcan XC-72, 290

W
wall flow filter, 267
Washburn equation, 51
washcoat, 4, 243
    electron microprobe scans of, 266
    loss, 256
    monolith, physical properties, 54
washing method, in-house, 226
wastewater treatment, 215
water gas shift (WGS), 112, 124
    catalysts, 113–116
        deactivation, 116
    high & low temperature, 114–116
    mechanism, 113
    process, 112–116
water gas shift reaction (WGSR), 134
weight hourly space velocity (WHSV), 71
Weisz-Prater criterion, pore diffusion, 82
WGS. *see* water gas shift (WGS)
WGSR. *see* water gas shift reaction (WGSR)
wide-body aircraft fly, 229

X
x-ray diffraction (XRD), 57–58, 62
    crystal size, 240
    patterns of γ- and α-Al_2O_3, 59
    studies, 242
x-ray photoelectron spectroscopy
    (XPS), 62–64, 99
    spectrum of various oxidation states of
        palladium on Al_2O_3, 64

Z
Zeigler–Natta catalyst, 212
zeolite (HZ), 8, 198
    cage, 35
    lose Si–O–Al bridges, 94
    NMR profile of a thermally aged, 94
    possess, 268
zeolites, 35–37
ZSM-5, 36

Printed and bound by CPI Group (UK) Ltd, Croydon, CR0 4YY

16/04/2025

14658345-0004